ASHRAE Laboratory Design Guide

This publication was prepared under ASHRAE Research Project 969. It was sponsored by TC 9.10.

About the Authors

Ian B.D. McIntosh, Ph.D., is a recent Ph.D. graduate in mechanical engineering from the University of Wisconsin—Madison and is a member of ASHRAE and ASME. While working on his Ph.D. dissertation, he completed an eight-month research internship at Siemens Building Technologies—Landis Division, where he gave business and technical presentations to aid in the marketing of potential product concepts related to fault detection and diagnosis in HVAC subsystems. Currently, he is with Farnsworth Group, Inc., and was the lead engineer in charge of research and development of this publication.

Chad B. Dorgan, P.E., Ph.D., is a licensed professional engineer in Wisconsin, South Carolina, Oregon, and California, and is a recent Ph.D. graduate in indoor air quality from the University of Wisconsin—Madison. Between 1989 and 1993, he coordinated a 60-person operations and maintenance section for the United States Air Force. As an engineering technician, he was involved in the analysis and optimization of over 30 large commercial and industrial facilities, including a large enamel oven complex in Wisconsin and several auto manufacturer's facilities in Michigan. As part of these studies, several publications for EPRI were developed, including *EPRI's Cold Air Distribution Design Guide, Field Evaluation of Cold Air Distribution Systems,* and *Detailed Field Evaluation of a Cold Air Distribution System.* Since 1993, he has led engineering activities at Farnsworth Group, Inc. (formerly Dorgan Associates).

Charles E. Dorgan, P.E., Ph.D., has been involved in refrigeration, air-conditioning, and energy projects since 1960. A professional engineer since 1964, he is registered in Wisconsin, Ohio, Kansas, and South Carolina. He is an ASHRAE fellow. He received his Ph.D. in mechanical engineering from the University of Wisconsin—Madison in 1979. He has authored or contributed to more than 25 books and more than 100 articles and other publications, plus various educational presentations and engineering reports. Since 1980, he has had partial appointments at the University of Wisconsin, while working for three different consulting firms, including Dorgan Associates.

ASHRAE Laboratory Design Guide

Ian B.D. McIntosh
Chad B. Dorgan
Charles E. Dorgan

American Society of Heating, Refrigerating and Air-Conditioning Engineers, Inc.

ISBN 1-883413-97-4

©2001 American Society of Heating, Refrigerating
and Air-Conditioning Engineers, Inc.
1791 Tullie Circle, N.E.
Atlanta, GA 30329
www.ashrae.org

All rights reserved.

Printed in the United States of America

Cover design by Tracy Becker.

ASHRAE has compiled this publication with care, but ASHRAE has not investigated, and ASHRAE expressly disclaims any duty to investigate, any product, service, process, procedure, design, or the like that may be described herein. The appearance of any technical data or editorial material in this publication does not constitute endorsement, warranty, or guaranty by ASHRAE of any product, service, process, procedure, design, or the like. ASHRAE does not warrant that the information in the publication is free of errors, and ASHRAE does not necessarily agree with any statement or opinion in this publication. The entire risk of the use of any information in this publication is assumed by the user.

No part of this book may be reproduced without permission in writing from ASHRAE, except by a reviewer who may quote brief passages or reproduce illustrations in a review with appropriate credit; nor may any part of this book be reproduced, stored in a retrieval system, or transmitted in any way or by any means—electronic, photocopying, recording, or other—without permission in writing from ASHRAE.

ASHRAE STAFF

Special Publications

Mildred Geshwiler
Editor

Erin S. Howard
Assistant Editor

Christina Johnson
Editorial Assistant

Michshell Phillips
Secretary

Publishing Services

Barry Kurian
Manager

Jayne Jackson
Production Assistant

Publisher

W. Stephen Comstock

Table of Contents

Acknowledgments ix

Chapter 1 Introduction

Overview 1
Organization 1

Chapter 2 Background

Overview 7
Laboratory Types 7
Laboratory Equipment 10
References 12

Chapter 3 Laboratory Planning

Overview 13
Risk Assessment 13
Identification and Understanding of Hazards 14
Hazard Analysis Methods 18
Risk Assessment Guidelines 24
Responsibilities 25
Documentation 25
Environmental Requirements 25
Appliances and Occupancy 29
Pressure Relationships 30
Ventilation and Indoor Air Quality Considerations 32
Laboratory Codes, Standards and References 34
Integration of Architecture and Engineered Systems 37
Development of Planning Documents 43
References 43
Bibliography 45

Chapter 4 Design Process

    Overview 47

    Design Process 48

    Special Space Considerations 53

    References 65

Chapter 5 Exhaust Hoods

    Overview 67

    Chemical Fume Hoods 67

    Perchloric Acid Hood 73

    Biological Safety Cabinet 74

    Other Hood Types 77

    Hood Safety Certification and Continuous Monitoring 80

    Selection of Exhaust Hoods 80

    References 81

Chapter 6 Primary Air Systems

    Overview 83

    Zone Air Distribution 83

    Zone Heating 86

    Exhaust Air System 88

    Supply Air System 96

    Duct Construction 102

    Energy Efficiencies 109

    References 109

Chapter 7 Process Cooling

    Overview 111

    Types of Water Cooled Loads 111

    Water Treatment and Quality Requirements 112

    Temperature and Pressure Requirements 113

    Pumping System Configurations 113

    References 116

Chapter 8 Air Treatment

    Overview 117

    Requirements 117

    Fan Powered Dilution 119

    Filtration 120

    Scrubbing 122

    Condensing 123

    Oxidization (Incineration) 123

    References 123

Chapter 9 Exhaust Stack Design

Overview 125
Elements of Stack Design.. 125
Dispersion Modeling Approach 133
References 135

Chapter 10 Energy Recovery

Overview 137
Air-to-Air Heat Recovery... 137
Water-to-Air Heat Recovery... 141
Selection Parameters 142
References 143
Bibliography.. 143

Chapter 11 Controls

Overview 145
Equipment Control 145
Room Control 151
Central System Emergency Situations 155
References 156

Chapter 12 Airflow Patterns and Air Balance

Overview 157
Airflow Patterns 157
Testing, Adjusting, and Balancing 161
Laboratory Testing Requirements 164
References 168

Chapter 13 Operation and Maintenance

Overview 169
Decontamination of Existing Laboratories... 169
Maintenance of Equipment and Systems 169
Cost Information 173
Training 174
Reference 175

Chapter 14 Laboratory Commissioning Process

Overview 177
The Commissioning Process 177
Planning Phase 178
Design Phase 178
Construction Phase 179

    Acceptance Phase 179
    Operation Phase 180
    Commissioning of Existing Buildings 180
    References 180

Chapter 15 HVAC System Economics

    Overview 181
    Initial Price of System 181
    Life Cycle Cost Analysis 182

Chapter 16 Microbiological and Biomedical Laboratories

    Overview 185
    Biological Containment 185
    Animal Overview 186
    Design of Laboratory Animal Areas 190
    References 194
    Bibliography 194

Appendix 195

References 197

Annotated Bibliography 201

Index 209

Acknowledgments

This guide was developed through a cooperative effort of the Farnsworth Group, Inc., formerly known as Dorgan Associates, and laboratory industry experts. The input of the ASHRAE Project Monitoring Subcommittee, chaired by John Mentzer (Giffels Associates, Inc.) and including members Pete Gardner, John Varley, Bob Weidner, and Randall Lacey, was instrumental in achieving a successful guide. Special thanks to Jack Wunder for his detailed review of the draft guide and access to the laboratories at the University of Wisconsin–Madison.

The input provided by several key people, including designers, manufacturers, and others, was invaluable in making the guide whole. These include Todd Hardwick, Otto Van Geet, Victor Neuman, Carl Lawson, Greg De Luga, Geoffrey Bell, Luis Solarte, Daniel Ghidoni, Patrick Chudecke, Edward Fiance, Robert Haugen, Tom Begley, Bob Parsons, and Gary Butler.

Dorgan Associates' staff members who deserve recognition for their contributions include Svein Morner, Ph.D., and Zachary Obert for their detailed input on writing various sections of the guide and endless hours of reviewing background material for ensuring technical accuracy of the guide. Chad Grindle's efforts in creating and editing the hundreds of graphics required to make the guide understandable and useful are appreciated. Sincere thanks go to Joan Dorgan for proofreading. Finally, we would have never finished the guide without the tireless efforts of Suzanne Bowen in word-processing and proofreading to ensure consistency in formatting, and general legibility was maintained.

Chapter 1
Introduction

OVERVIEW

The American Society of Heating, Refrigerating and Air-Conditioning Engineers, Inc. (ASHRAE), commissioned this design guide in response to the need for a comprehensive reference manual for the planning, layout, and design of laboratories. It is intended that the information in this guide assist engineers, owners, and system operators in determining the needs of their laboratory facility and how to best match these needs to mechanical system options.

The Guide is written in a manner that progresses from general to specific to reach a wide target audience of designers, architects, engineers, owners, operations and maintenance personnel, and others in the heating, ventilating, air-conditioning, and refrigerating (HVAC&R) industry. The beginning chapters of the Guide present general background information and prescribed design frameworks, whereas the later chapters are more specific, providing detailed design and application information.

A primary benefit of this structure is that many types of readers can use the Guide. Whereas an owner may only read the first two chapters, an engineer may focus on the design chapters.

To improve the comprehension of the material for those unfamiliar with laboratory design, the guide is organized around a typical project, progressing through the basic steps of planning, design, construction, and operation and maintenance (O&M). This Guide consists of sixteen chapters.

ORGANIZATION

- Chapter 2, "Background" provides basic background information on laboratories including their various types and typical equipment found in them. The intent of this chapter is to provide a basic understanding of laboratories, their importance, as well as their different functions and needs.
- Chapter 3, "Laboratory Planning" describes the many important issues that are addressed during the planning phase of a laboratory project. At the very core of this phase is the need to ensure the utmost safety for the laboratory occupants via risk assessment and hazard analyses while achieving experimental integrity, good comfort and indoor air quality, the capacity to meet operating loads, and successful integration of architecture and engineering systems. These various important considerations are then carefully documented in the owner's program of requirement and the owner's design intent.
- Chapter 4, "Design Process" outlines the design process with guidance provided to designers and engineers on the key steps in meeting or exceeding the owner's design intent. Also included in this chapter is a review of special space design considerations for the different system types encountered in a laboratory.
- Chapter 5, "Exhaust Hoods" details the operability, types, and applications of exhaust hoods.

- Chapter 6, "Primary Air Systems" describes the many HVAC system options that are available for conditioning and contaminant removal, including supply systems, room or space air diffusing, exhaust systems, duct construction, and energy efficiencies.
- Chapter 7, "Process Cooling" examines the different system options available to meet supplementary process cooling needs. This includes the types of water-cooled loads found in laboratories, water treatment and quality, temperature and pressure requirements, and pumping system configurations.
- Chapter 8, "Air Treatment" presents several methods of treating the air as it is either expelled to the outside environment or drawn into the laboratory space from the outside.
- Chapter 9, "Exhaust Stack Design" deals with the final conduit through which air is expelled and diluted into the environment – exhaust stacks. Required elements for good exhaust stack design in addition to modeling techniques available for design verification are described.
- Chapter 10, "Energy Recovery" addresses the recovery of the temperature/humidity content of the exhaust air or other sources of energy that would otherwise be wasted, to realize economic savings. The two main energy recovery technologies discussed are air-to-air and water-to-air heat recovery. In addition, key parameters are presented that are used to properly select heat recovery options.
- Chapter 11, "Controls" discusses control for constant volume fume hoods, variable volume fume hoods, and other exhaust equipment. Room control discussion includes the theory of room control, outside air needs, minimum ventilation air changes per hour, control stability, variable and constant volume strategies, control of critical spaces, and building pressurization.
- Chapter 12, "Airflow Patterns and Air Balance" provides an understanding of airflow patterns throughout the laboratory environment and how they need to be maintained in a specific direction and velocity to protect against malicious pollutants or contamination. This is followed by descriptions of the proper air and hydronic balancing and testing procedures.
- Chapter 13, "Operation and Maintenance" details how to properly operate and maintain laboratory systems to ensure the continual safety of occupants, quality of laboratory experiments, and efficiency of HVAC and laboratory equipment. To this end, this chapter discusses the training needed for O&M personnel and laboratory users. In addition, operating costs and the importance of continual fume hood and biological safety cabinet testing and certification is also addressed.
- Chapter 14, "Laboratory Commissioning Process" introduces the quality method known as the commissioning process, which commences during the planning phase and follows through to the design, construction, acceptance, turnover, and operation phases of the laboratory building project. In so doing, the commissioning process ensures that the diverse requirements for the laboratory and the owner's design intent are met.
- Chapter 15, "HVAC System Economics" covers both the initial costs (costs associated with planning, design, and construction) and life-cycle costs (costs associated with operation and maintenance) of the laboratory facility. Although more emphasis is often given to the initial cost due to budgetary constraints, this chapter also considers the life-cycle cost in designing the laboratory. This is important since a substantial investment is required to effectively operate and maintain the laboratory systems and equipment over a life span of 15 to 30 years.
- Chapter 16, "Microbiological and Biomedical Laboratories" provides specific information about laboratories that specialize in biological containment and that house animal areas. Due to the importance of containing and controlling biohazards and product protection within these special laboratories, issues such as system reliability, redundancy, proper space pressurization, envelope design, and sanitation are addressed.

Reference Sources

American Society of Heating, Refrigerating and Air-Conditioning Engineers, Inc.

ASHRAE is an international professional society with more than 50,000 members worldwide. The Society was organized for the sole purpose of advancing the arts and sciences of heating, ventilation, air-conditioning, and refrigeration for the benefit of the public. The Society achieves these goals through sponsorship of research, standards development, continuing education, and publications. For additional information, contact the Society at:

American Society of Heating, Refrigerating
and Air-Conditioning Engineers, Inc.
1791 Tullie Circle, NE
Atlanta, GA 30329-2305
Tel: (404) 636-8400
Fax: (404) 321-5478
Internet: www.ashrae.org

*Building Officials and Code
Administrators International, Inc.*

Building Officials and Code Administrators International, Inc. (BOCA), is a nonprofit association that has been establishing and enforcing building codes since 1915. The association currently has over 16,000 members fully dedicated to the development and enforcement of codes that benefit the public safety and health. For additional information contact BOCA at:

Building Officials and Code
Administrators International, Inc.
4051 W. Flossmoor Rd.
Country Club Hills, IL 60478
Tel: (708) 799-2300
Fax: (708) 799-4981
Internet: www.bocai.org

Centers for Disease Control and Prevention

The Centers for Disease Control and Prevention (CDC) is a federal institution whose primary goal and pledge is "to promote health and quality of life by preventing and controlling disease, injury, and disability." To achieve this pledge, the CDC uses federal and private funding for research and development in several fields and sciences at its 11 different laboratories and locations. For additional information the CDC may be contacted at:

Centers for Disease Control and Prevention
1600 Clifton Road, NE
Atlanta, GA 30333
Tel: (800) 311-3435
Internet: www.cdc.gov

*Institute of Environmental Sciences
and Technology*

The Institute of Environmental Sciences and Technology (IEST), founded in 1953 as a nonprofit organization, is an international society in the area of contamination control and publishes recommended practices for testing cleanrooms, HEPA/ULPA filters, and clean air filtration systems. For additional information, contact IEST at:

Institute of Environmental Sciences and Technology
940 E. Northwest Highway
Mount Prospect, IL 60056
Tel: (847) 255-1561
Fax: (847) 255-1699
Internet: www.iest.org

Institute of Laboratory Animal Resources

ILAR, or the Institute of Laboratory Animal Resources, has been a national leader in the research, publication, and collection of information pertinent to animal care and use in the laboratory setting since 1952. For additional information contact ILAR at:

Institute of Laboratory Animal Resources
2101 Constitution Avenue, NW
Washington, D.C. 20418
Tel: (202) 334-2590
Fax: (202) 334-1687
E-mail: ILAR@nas.edu
Internet: www4.nas.edu/cls/ilarhome.nsf

National Institutes of Health

The National Institutes of Health (NIH) is a federally funded laboratory dedicated to the research and development of medicine to aid the health of people around the world. Founded in 1887, the NIH funds national and international research in numerous medical fields to accomplish this dedication to health. For additional information contact the NIH at:

National Institutes of Health
Bethesda, MD 20892
Tel: (301) 496-4000
Research funding E-mail: grantsinfo@nih.gov
Internet: www.nih.gov

National Fire Protection Association

The National Fire Protection Association (NFPA) is an international leader in the development of fire, electrical, and life safety for the public. This nonprofit, member association was founded in 1896 to develop consensus codes and standards, as well as training, research, and education in fire and other hazard protections in various buildings. NFPA has over 67,000 members and is associated with 80 national and professional organizations from across the globe. For more information contact the NFPA at:

National Fire Protection Association
1 Batterymarch Park
P.O. Box 9101
Quincy, MA 02269-9101
Tel: (617) 770-3000
Fax: (617) 770-0700
Internet: www.nfpa.org

National Research Council

The National Research Council (NRC) was formed by the National Academies of Science in 1916 to bridge the gap between the science and technology communities and the federal government. Today, the NRC continues this initiative by supplying information and advice to the federal and public sectors on the current issues and advances in science and technology. For additional information contact the National Academies of Science at:

National Research Council
2101 Constitution Avenue, NW
Washington, D.C. 20418
Tel: (202) 334-2000
Internet: www.nas.edu

National Sanitation Foundation International

The National Sanitation Foundation International (NSF), founded in 1944, has been committed to public health safety and protection of the environment by developing standards, product testing, and conformity assessments. NSF International developed a Biological Safety Cabinet Standard in 1976 for class II, Type A and B classifications. The most current version is known as NSF Standard 49-1992. For additional information contact NSF International at:

National Sanitation Foundation International
789 Dixboro Road
Ann Arbor, MI 48113-0140
Tel: (734) 769-8010
Fax: (734) 769-0109
Email: info@nsf.org
Internet: www.nsf.org

Occupational Safety and Health Administration

The Occupational Safety and Health Administration (OSHA) is a federal body whose primary purpose is to protect and save the lives of American workers through the development and implementation of various standards and regulations related to the work environment. Currently OSHA employs over 2,100 inspectors at 200 nationwide locations to cover the ever-increasing workforce of over 100 million in the United States. For additional information related to laboratory standards and codes, contact them at:

U.S. Department of Labor
Occupational Safety and Health Administration
Office of Public Affairs – Room N3647
200 Constitution Avenue
Washington, D.C. 20210
Tel: (202) 693-1999
Internet: www.osha.gov

Scientific Equipment and Furniture Association

The Scientific Equipment and Furniture Association (SEFA) was organized in 1988 by laboratory equipment manufacturers to enhance individual member company performance and improve the quality of laboratory facilities. SEFA accomplishes these goals by establishing industry standards for laboratory equipment, installation, and testing. For additional information contact SEFA at:

Scientific Equipment and Furniture Association
7 Wildbird Lane
Hilton Head Island, SC 29926-2766
Tel: (843) 689-6878
Fax: (843) 689-9958
E-mail: mjrp@hargray.com
Internet: www.sefalabfurn.com

United States Department of Health and Human Services

The U.S. Department of Health and Human Services (DHHS) is the federal administration primarily in charge of funding and developing programs, research, and education that benefits and protects the health of Americans. Some of the agencies that fall under the supervision or funds of the DHHS include the CDC, Food and Drug Administration (FDA), and Health Resources & Services Administration (HRSA). For additional information about the DHHS pertaining to laboratory design contact the department at:

The U.S. Department of Health and Human Services
200 Independence Avenue, SW
Washington, D.C. 20201
Tel: (202) 619-0257
Toll Free: (877) 696-6775
Internet: www.hhs.gov

Trade Magazines

Trade magazines often carry articles covering numerous technical and design-related issues that pertain to laboratory design. Additional references may be found in the most recent issues of such publications as:

ASHRAE Journal
Heating/Piping/Air-Conditioning (HPAC)
Consulting-Specifying Engineer
Engineered Systems
Air Conditioning and Refrigeration News

University Research Theses and Papers

Universities are a primary source of information on specific details of laboratory design. Individual university libraries and organizations can be accessed to assist in locating information on specific topics.

Others

ADA
Americans with Disabilities Act
U.S. Department of Justice
Disability Rights Section
P.O. Box 66738
Washington, D.C. 20035-6738
Toll Free: 1-800-514-0301 or 1-800-514-0383 (TTY)
Internet: www.usdoj.gov/crt/ada/taprog.htm

ANSI
American National Standards Institute (ANSI)
11 West 42nd Street, 13th Fl.
New York, NY 10036
Tel: (212) 642-4900
Fax: (212) 398-0023
Internet: www.ansi.org

AIHA
American Industrial Hygiene Association
2700 Prosperity Avenue, Suite 250
Fairfax, VA 22031
Tel: (703) 849-8888
Fax: (703) 207-3561
Internet: www.aiha.org

ACGIH
American Conference of Governmental Industrial Hygienists, Inc.
1330 Kemper Meadow Dr., Suite 600
Cincinnati, OH 45240
Tel: (513) 742-2020
Fax: (513) 742-3355
Internet: www.acgih.org

CETA
Controlled Environmental Test Association
4110 Lake Boone Trail
Raleigh, NC 27607
Tel: (919) 787-5181
Fax: (919) 787-4916
Internet: www.cetainternational.org

EPA
U.S. Environmental Protection Agency
Headquarters Information Resources Center
401 M. Street, SW
Washington, D.C. 20460
Tel: (202) 260-5922
Fax: (202) 260-5153
Internet: www.epa.gov

NRC
U.S. Nuclear Regulatory Commission
One White Flint North
11555 Rockville Pike
Rockville, MD 20852-2738
Tel: (301) 415-7000
Internet: www.nrc.gov

Chapter 2
Background

OVERVIEW

The intent of the information presented in this chapter is to provide a brief summary of the laboratory types and systems that are typically encountered in laboratory facilities. This helps form a common framework for the terminology and systems in this guide.

LABORATORY TYPES

There are numerous potential applications for capturing pollutants to minimize adverse impacts on occupant health. Therefore, there are numerous types of laboratories. The more common laboratory types introduced in this section are:

- General Chemistry Laboratory
- Radio-Chemistry Laboratory
- Teaching Laboratory
- Research Laboratory
- Hospital or Clinical Laboratory
- Biological Containment Laboratory
- Animal Laboratory
- Isolation/Cleanrooms
- Materials Testing Laboratory
- Electronics/Instrumentation Laboratory
- Support Spaces

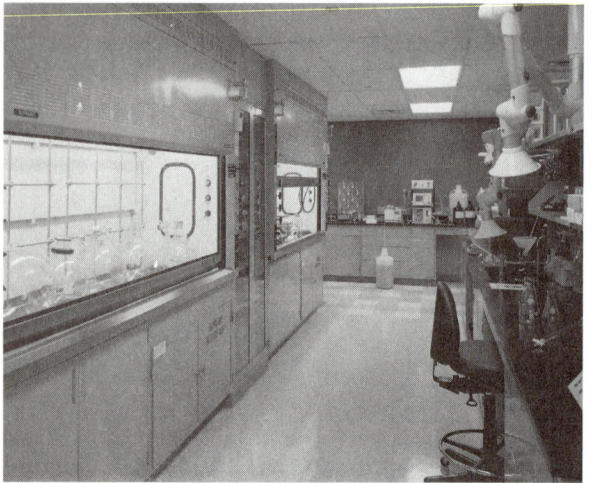

Figure 2-1 *General chemistry laboratory.*

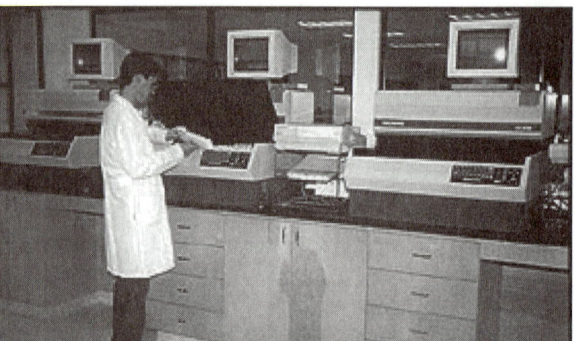

Figure 2-2 *Radio-chemistry laboratory.*

Figure 2-3 *Teaching laboratory.*

General Chemistry Laboratory

A general chemistry laboratory is used for analysis and general experimentation with a wide variety of chemicals. General chemistry laboratories are not used for extremely hazardous materials, such as carcinogens, or large quantities of explosive materials. There are typically few specialized equipment needs for general chemistry laboratories beyond chemical fume hoods and limited direct equipment exhaust.

Radio-Chemistry Laboratory

A radio-chemistry laboratory is similar to a general chemistry laboratory, with additional design requirements to contain direct radiation (x-rays, gamma rays, etc.) and prevent the release of radioactive chemicals and materials. Laboratories that use radioactive materials have strict requirements that are determined by the Nuclear Regulatory Commission (NRC), OSHA, and the EPA.

Teaching Laboratory

Teaching laboratories are designed to provide a safe learning environment for large groups (usually 30, sometimes more) of high school and undergraduate students. Teaching laboratories typically use the same types of materials as a general chemistry laboratory, although in much smaller quantities. Graduate student work is typically associated with research laboratories, which are not considered teaching laboratories. Primary needs of teaching laboratories are for the instructor to maintain eye contact with all students and to have quick access to the controls of the laboratory equipment to prevent hazards. Sound control should be practiced to maximize communication.

Research Laboratory

Research laboratories perform general and analytical experimentation with a wider variety of materials in larger quantities than a general chemistry laboratory. These laboratories often use high-end analytical instruments and equipment and require more accurate control of temperature, humidity, air distribution, and other laboratory conditions. The HVAC design in these areas must be tolerant of change.

Figure 2-4 *Research laboratory.*

Hospital or Clinical Laboratory

Hospital laboratories provide chemical and biological testing of specimens associated with patient care. These laboratories do not use large quantities of dangerous materials, and typically do not perform general research. Although it depends on the specific type of hospital, these laboratories usually contain chemical fume hoods, local exhaust, and vented Class II biological safety cabinets.

Biological Containment Laboratory

Biological laboratories are used to contain various toxic and infectious biological materials. These laboratories are defined by biosafety levels 1 through 4, with level 1 containing known biological substances of minimal risk and level 4 containing exotic materials, often unknown, which pose a known or potential high risk of life-threatening diseases to laboratory personnel and community. These laboratories may require decontamination procedures.

Animal Laboratory

Animal laboratories need to maintain clean and humane conditions for animals and provide safe separation of the animals from laboratory personnel. Design of animal laboratories should be performed with close interaction with veterinarians and researchers who will be working in the laboratory. There are numerous standards for animal laboratories that must be followed, such as those of the American Association for Accreditation of Laboratory Animal Care (AAALAC), which follows National Institutes of Health (NIH) standards.

Isolation/Cleanrooms

Cleanrooms are designed to provide a working environment with minimal airborne particulate materials and constant temperature, humidity, air pressure, and airflow patterns. Their environments require controlled

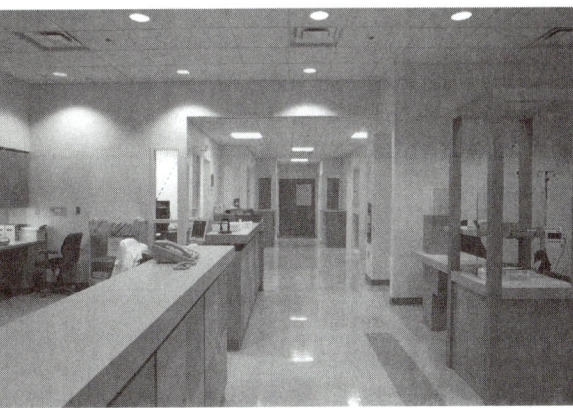

Figure 2-5 *Hospital laboratory.*

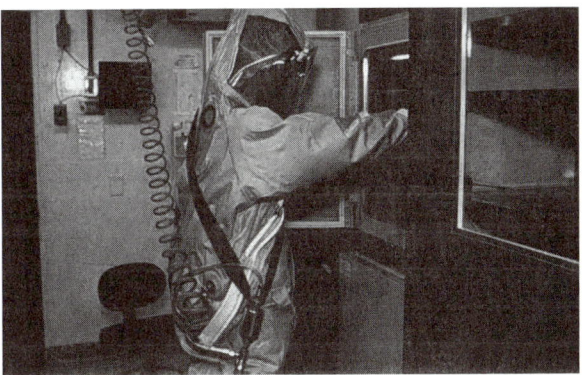

(Courtesy of Central Virginia Governor's School.)

Figure 2-6 *Biological containment laboratory.*

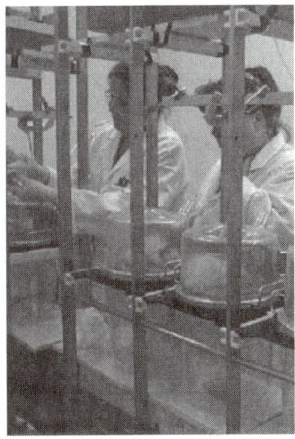

(Courtesy of ABC Laboratories.)

Figure 2-7 *Animal laboratory.*

atmospheres to regulate gases, negative and positive ions, and the use of robotics. In recent years, there has been development of new standards of air cleanliness as a consequence of the greater need for ultra clean environments. The most current standards are ISO 14644-1, *Cleanrooms and Associated Controlled Environments, Part 1: Classification of Airborne Particulates* and ISO 14644-2, *Cleanrooms and Associated Controlled Envi-*

(Courtesy of Component Systems, Inc.)

Figure 2-8 *Isolation room.*

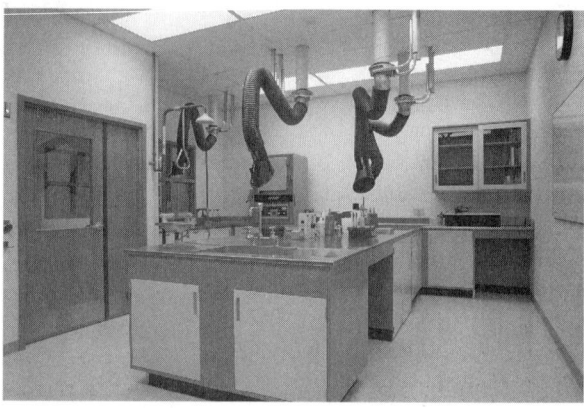

(Courtesy of Torcon, Inc.)

Figure 2-9 *Materials testing laboratory.*

ronments, Part 2: Testing and Monitoring to Prove Continued Compliance with ISO 14644-1*. These two standards, developed by the International Standards Organization (ISO), are components of a ten-part series of standards that will cover more cleanroom environmental parameters and practices than the earlier Federal Standard 209E, *Airborne Particulate Cleanliness Classes in Clean Rooms and Clean Zones*.

Materials Testing Laboratory

Materials testing laboratories are used for physical experimentation with materials, such as wear and strength testing. The types of equipment needed for a materials laboratory depends largely on the types of testing performed. Materials laboratories usually have some process cooling requirements and may have limited chemical exhaust requirements similar to a general chemistry laboratory.

Electronics/Instrumentation Laboratory

An electronics laboratory is similar to a cleanroom, in that low levels of airborne particulates and precise control of temperature, humidity, and airflow are needed. Electronics laboratories, such as those that produce electronic chips and semiconductor wafers, often need to prevent the release of hazardous chemicals by using local exhaust while still maintaining a positive room pressure.

Support Spaces

Laboratory support spaces contribute to research and experimentation activities in a laboratory by providing general laboratory maintenance, material and chemical storage, and equipment and material preparation. Support spaces may require laboratory exhaust equipment, such as fume hoods and biological safety cabinets, depending on the type of support provided.

(Courtesy of National Institute of Standards and Technology.)

Figure 2-10 *Electronics laboratory.*

LABORATORY EQUIPMENT

The design of a laboratory depends largely on the type of equipment that is needed to safely protect the laboratory personnel. The most common types of equipment include fume hoods, biological safety cabinets, and storage cabinets. These are discussed in further detail in Chapter 5, "Exhaust Hoods."

Fume Hoods

Fume hoods are used in a variety of laboratories to protect personnel performing general chemistry research or testing. They provide protection by maintaining inward airflow to the opening of a hood, which contains and exhausts the airborne materials generated during experimentation, if used properly. Fume hoods must be tested annually for face velocity and directional airflow.

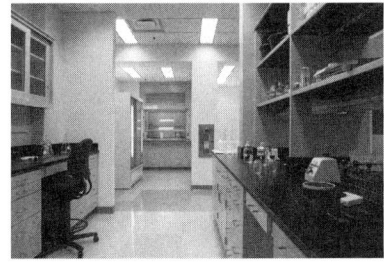

Figure 2-11 *Support space.*

Figure 2-12 *Support space.*

Figure 2-13 *Support space.*

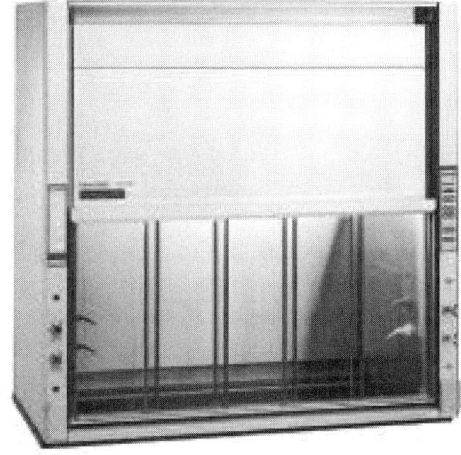

(Courtesy of Baker Company.)

Figure 2-14 *Chemical fume hood.*

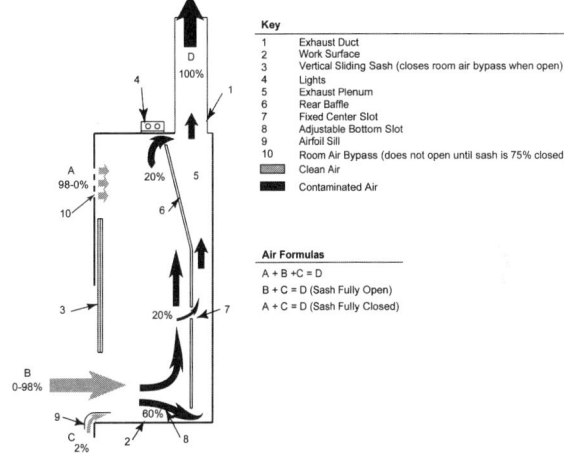

Figure 2-15 *Chemical fume hood schematic (diagram modified with permission from DiBerardinis 1993).*

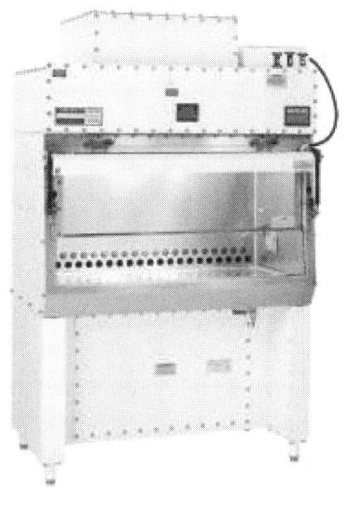

(Courtesy of Baker Company.)

Figure 2-16 *Biological safety cabinet.*

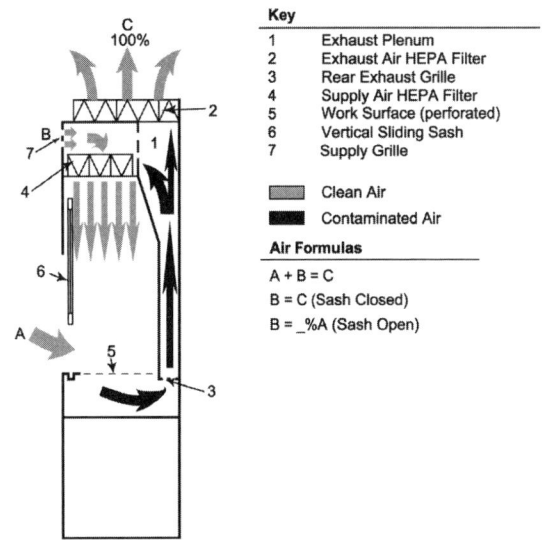

Figure 2-17 *Biological safety cabinet schematic (diagram modified with permission from DiBerardinis 1993).*

(Courtesy of Labconco Corp.)

Figure 2-18 *Flammable and solvent storage cabinet.*

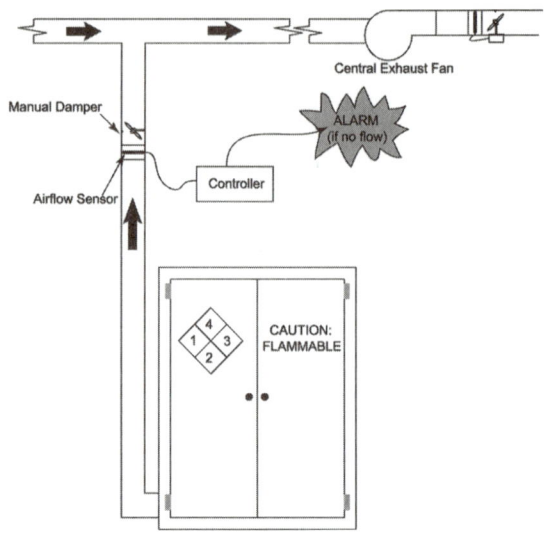

Figure 2-19 *Flammable and solvent storage cabinet schematic.*

Biological Safety Cabinets

Biological safety cabinets are used to control the release of toxic particulates and infectious biological aerosols. They operate similarly to fume hoods in that they provide protection by maintaining an inward flow of air, but they also make use of various levels of filtered supply and exhaust air to protect the materials in the cabinet and to prevent the release of the materials to the atmosphere. Biological safety cabinets require annual certification to maintain function.

Flammable and Solvent Storage Cabinets

Storage cabinets are used to contain fumes from large quantities of flammable materials, prevent excessive internal temperatures in the presence of fire, and contain spilled flammable liquids to prevent the spread of fire. Proper construction, correct venting (if required), conspicuous labeling, and storage of materials at or below the maximum permissible limits will ensure a cabinet's safe use and function. *NFPA 30 Standard, Section 4-3.4* (NFPA 1996), stipulates that "if the storage cabinet is to be vented for any reason, the cabinet shall be vented directly to outdoors."

REFERENCES

DiBerardinis, L.J., J.S. Baum, M.W. First, G.T. Gatwood, E. Groden, A.K. Seth. 1993. *Guidelines for laboratory design: Health and safety considerations*, 2nd ed. New York: John Wiley & Sons, Inc.

National Fire Protection Association (NFPA). 1996. *NFPA 30 – Flammable and combustible liquids code*. Quincy, Mass: NFPA Publications.

Chapter 3
Laboratory Planning

OVERVIEW

Laboratory spaces and buildings are unique compared to other types of spaces and buildings due to the distinctive requirements to maintain the health, comfort, safety, and productivity of the various entities within. Therefore, the successful design, construction, and O&M of a laboratory facility is achieved when sufficient planning is accomplished before undertaking the project. This chapter provides the critical information and guidance needed during the planning phase of a laboratory building project, including:

- Risk assessment
- Environmental requirements
- Appliances and occupancy
- Pressure relationships
- Ventilation and indoor air quality considerations
- Laboratory codes and standards
- Integration of architecture and engineered systems
- Development of planning documents

The goal of this chapter is to present and discuss the pertinent issues that must be considered and documented when planning a new or renovated laboratory facility. The fundamental concepts and principles presented in this chapter are the foundation for the remaining guide. A good understanding of this information is critical for the success of a laboratory design and a delivered laboratory that is functional.

RISK ASSESSMENT

The first step in the planning process for a laboratory is the completion of a *risk assessment* document to minimize the safety issues related to the acquisition, handling, usage, storage, transportation, and waste disposal of hazardous materials. In this document a *hazard* is defined as a danger intrinsic to a substance or operation, and a *risk* is the probability of injury associated with working with a substance or carrying out a particular laboratory operation (NRC 1995). Therefore, a risk can be reduced but a hazard cannot.

The completion of a risk assessment involves the identification and understanding of the various hazards present in each laboratory and the application of appropriate hazard analysis methods to assess the risk associated with hazardous materials and suspected products of laboratory experiments and procedures. In addition, the responsibilities of key individuals must be clearly defined and adequate documentation must be available to handle all foreseeable situations. It is highly recommended to have a one- to two-day workshop that uses a nominal group technique (NGT) or equivalent method with a qualified facilitator to identify hazards and develop consensus range and best assessment of risks and consequences.

IDENTIFICATION AND UNDERSTANDING OF HAZARDS

Before the risk assessment can be completed, the materials that will be present in each space and their subsequent hazards must be thoroughly documented. Therefore, an understanding of the typical hazards present for the primary categories of materials (chemical, biological, radiological, and physical) is paramount. While it is important to have sufficient knowledge about these materials, their inherent hazards, and the risks associated with their use, it is impossible to provide all of this knowledge within the scope of this guide. Instead, the information in this section provides an overview of the hazards with references to publications for the details needed to properly determine the hazards of a material or to perform a risk/hazard/consequence workshop.

Chemical

Hazards associated with chemicals used in a laboratory range widely based on the chemical composition, the procedures used, and, the physiological and toxicological effects. The severity of the toxicological and related physiological effects is also dependent on the exposure, duration, level, and immuno-compromise state (i.e., health condition) of the exposed personnel. Depending on their particular effects after exposure, chemicals can be classified into the following subcategories:

- Flammables
- Toxins
- Carcinogens
- Compressed Gases

Flammable chemicals are materials that easily ignite in the presence of a flame, spark, or heat source. The flammability of a chemical is based on a rating called the NFPA Hazard Rating Index. This rating scale, as defined in Table 3-1 through Table 3-5 for some example substances, provides an indication of the severity of the health, flammability, and reactivity hazard of a substance, with a higher rating equating to a higher risk. While it is simple to identify the risk associated with one chemical, the mixing and subsequent reaction of chemicals during an experiment is often more difficult. Therefore, during the identification of the hazards and their risks, it is critical to understand the experiments that are to be completed. Good reference sources for the hazards of chemicals and their associated risk are Bretherick 1981, NRC 1995, and OSHA 1990a.

Table 3-1: Health Hazard Definitions

4	Materials that with very limited exposure can be fatal or cause major residual injuries even with prompt medical treatment, including those too dangerous to be approached without specialized protective equipment. This degree includes: • Materials that can penetrate ordinary rubber protective clothing. • Materials that under normal or fire conditions give off gases extremely hazardous (e.g., toxic or corrosive) through inhalation or any form of contact.
3	Materials that with short-term exposure can cause serious temporary or residual injury even with prompt medical attention, including those materials that require full body protection. This degree includes: • Materials that give off highly toxic combustion products. • Materials corrosive or toxic to living tissue.
2	Materials that with intense or continued exposure could cause temporary incapacitation or residual injury without prompt medical care, including those requiring respiratory protection with independent air supplies. This degree includes: • Materials that give off toxic combustion products. • Materials that give off highly irritating combustion products. • Materials that under normal or fire conditions give off toxic vapors that do not have warning properties.
1	Materials that on exposure cause irritation and minor residual injury without medical attention, including those that require the use of an approved canister type gas mask. This degree includes: • Materials that under fire conditions emit irritating combustion products. • Materials that cause skin irritation without destruction of tissue.
0	Materials that under fire conditions pose no hazard beyond that of combustible material.

(Source: National Fire Protection Association)

Table 3-2: Flammability Hazard Definitions

4	Materials that rapidly or completely vaporize at atmospheric pressure and ambient temperature or are readily dispersed in air and burn readily. This degree includes: • Gases. • Cryogenic materials. • Any liquid or gaseous material that is liquid while under pressure and has a flash point below 73°F (22.8°C) and has a boiling point below 100°F (37.8°C) (Class IA flammable liquids). • Materials that due to physical form or environmental conditions can form explosive mixtures with air and are readily dispersed in air, such as dust of combustible solids and mists of flammable or combustible liquid droplets.
3	Liquids and solids that can be ignited under almost all ambient temperature conditions. Materials in this degree form hazardous atmospheres with air under almost all ambient temperatures, though unaffected by the temperatures, and are readily ignited in almost all conditions. This degree includes: • Liquids having a flash point below 73°F (22.8°C) and have a boiling point below 100°F (37.8°C) and those liquids having a flash point above 73°F (22.8°C) and below 100°F (37.8°C) (Class IB and IC flammable liquids). • Solid materials in the form of coarse dust, which burn rapidly but generally do not form explosive atmospheres with air. • Solid materials in a fibrous or shredded form that burn rapidly and create flash fire hazards (e.g., cotton, sisal, hemp, etc.). • Materials that burn with extreme rapidity, usually due to self-contained oxygen (e.g., dry nitrocellulose and many organic peroxides). • Materials that ignite spontaneously when exposed to air.
2	Materials that require moderate heat or exposure to relatively high ambient temperatures for ignition. Materials in this degree under normal conditions do not form hazardous atmospheres with air, but under high ambient temperatures or with moderate heating may release vapors that will form hazardous atmospheres with air. This degree includes: • Liquids having a flash point above 100°F (37.8°C), but not exceeding 200°F (93.4°C). • Solids and semisolids that readily give off flammable vapors.
1	Materials that require preheating before ignition can occur. Materials of this degree require considerable preheating, under all ambient temperature conditions, before ignition and combustion can take place. This degree includes: • Materials that burn in air when exposed to a temperature of 1500°F (815.5°C) for less than 5 minutes. • Liquids, solids, and semisolids having a flash point above 200°F (93.4°C). • Most ordinary combustible materials.
0	Materials that do not burn, including any materials that will not burn when exposed to a temperature of 1500°F (815.5°C) for more than five minutes.

(Source: National Fire Protection Association)

Table 3-3: Reactivity Hazard Definitions

4	Materials are readily capable of detonation, explosive decomposition, or explosive reaction at normal temperatures and pressures by themselves. This includes materials that are sensitive to mechanical or localized thermal shock at normal temperatures and pressures.
3	Materials capable of detonation or explosive reaction by themselves but require a strong initiating source or must be heated under confinement before initiation. This includes materials that are sensitive to thermal or mechanical shock at elevated temperatures and pressures or those that react explosively with water without confinement or heat.
2	Materials that are normally unstable and readily undergo violent chemical change but do not detonate. This degree includes materials that can undergo chemical change with rapid release of energy at normal temperatures and pressures or can undergo violent chemical change at elevated temperatures and pressures. It also includes materials that react violently with water or may form potentially explosive mixtures with water.
1	Materials that are normally stable but can become unstable at elevated temperatures and pressures or those that release nonviolent energy upon reaction with water.
0	Materials that are normally stable, even under fire conditions, and do not react with water.

(Source: National Fire Protection Association)

Table 3-4: Special Notice Hazard Definitions

OX	Materials that are oxidizing agents and easily give up oxygen, remove hydrogen from other compounds, or attract negative electrons.
W	Materials that are water reactive and undergo rapid energy release on contact with water.

(Source: National Fire Protection Association)

Table 3-5: Example Chemical Hazard Ratings

Compound	Health	Flammability	Reactivity	Special Notice
Acetaldehyde	3	4	2	
Acetyl Chloride	3	3	2	W
Ammonia, Anhydrous	3	1	0	
Amyl Nitrate	2	2	0	OX
Sodium	3	3	2	W
Toluene	2	3	0	
Zinc (powder or dust)	0	1	1	

(Source: National Fire Protection Association)

Toxins are materials that physiologically affect the occupants or animals within a laboratory. The severity of a toxin is dependent upon the mode of entry (skin, inhaled, ingested, or injected), the exposure level and duration, and the physiological site of action. Exposure to toxins, such as carbon monoxide, may result in headaches, nausea, mental confusion, or fainting at relatively high doses (500 to 1000 ppm). At lower doses (25 ppm), toxins such as formaldehyde may result in serious injury and even death through pulmonary edema, a condition characterized by fluid accumulation in the lungs. Therefore, to properly identify a toxin, its mode of entry, exposure level, and physiological effect, all must be known and documented. Toxic ratings for humans are provided in Table 3-6.

Carcinogens are materials that, as a result of exposure, cause the growth of cancerous cells within an animal. The carcinogenicity of a material is determined typically through experimentation and case histories (e.g., asbestos workers). Typically, the carcinogenicity of a material is dependent on the exposure level and duration, the path of exposure (inhalation, skin, or intravenously), the human or animal species, and the gender and health of these species. Also, the specific genetic makeup of the individual can complicate the impact a material has on an individual. For example, lung cancer is significantly higher in asbestos workers who smoked relative to those who did not smoke. Men are much more susceptible to colon cancer than are women, whereas women are much more susceptible to breast cancer than are men. Therefore, the exposure, path, and species all must be considered to determine the risk of carcinogens.

Compressed gases have unique hazards, as there is the physical danger of explosion but also the additional hazard of flammability, toxicity, and carcinogenicity of the material. During an accident, the minimum acceptable exposure level can be passed quickly due to the sudden release of a gas. Also, the hazard level typically increases, especially the flammability, as the gas is released due to the increased oxygen level and the increased likelihood of an ignition source. Therefore, when identifying the hazards of compressed gas, the location, pressure, and composition of the gas are critical. The Compressed Gas Association (CGA) has published standards and other literature, which offer technical information and recommendations for safe practices in the use of compressed gases.

One final note on the classification of chemicals is that they typically have multiple ratings. For example, gasoline is highly flammable, moderately toxic, and is carcinogenic. Water is not flammable, but it is considered toxic at high enough doses. Therefore, for each chemical, it is important to thoroughly document the hazards and associated risks.

**Table 3-6:
Probable Lethal Dose for Humans**

Toxicity Rating	Animal LD_{50} (per kg)	Lethal Dose When Ingested by 150 lb (70 kg) Human
Extremely Toxic	Less than 1.76×10^{-3} oz (5 mg)	A taste (less than 7 drops)
Highly Toxic	1.76×10^{-3} to 1.76×10^{-2} oz (5 to 50 mg)	7 drops to 1 teaspoon
Moderately Toxic	1.76×10^{-2} to 0.176 oz (50 to 500 mg)	1 teaspoon to 1 ounce
Slightly Toxic	0.176 to 1.76 oz (500 mg to 5 g)	1 ounce to 1 pint
Practically Nontoxic	Above 1.76 oz (5 g)	Above 1 pint

LD_{50} is defined as the mean lethal dose for animals.
(Source: NRC 1995)

Biological

Biological hazards (biohazards) are those associated with exposure to blood and infectious materials that may contain bloodborne pathogens or microorganisms that cause bloodborne diseases. Exposure can occur via the respiratory tract, eye, mucous membrane, skin contact, or skin puncture. These hazards are usually present in laboratories that are involved in clinical and infectious disease research but may also be present in those that handle human and animal bodily fluids and tissues. Testing and quality control laboratories associated with sewage and water treatment may also deal with biohazards.

The Centers for Disease Control and Prevention (CDC) and the National Institutes of Health (NIH) developed the publication *Biosafety in Microbiological and Biomedical Laboratories* (U.S. DHHS 1999), which offers guidance on identifying and controlling various biohazards.

Radiological

Radiological hazards can be subdivided in to ionizing and non-ionizing radiation. The ionizing radiation can cause cell damage in living tissue. The primary forms of ionizing radiation are x-rays, α-rays, β-rays, γ-rays, Bremsstrahlung, neutrons, and positrons. Non-ionizing radiation is actually electromagnetic radiation, which is the region of the spectrum commonly defined as the radio frequency region. The effect of radiation exposure is dependent on the dose (level and time) and source with the effects divided into stochastic effects, which are random in occurrence, and into nonstochastic effects, which occur only when the amount of radiation exceeds a particular upper dose limit.

Physical

Although many chemical hazards have the obvious and evident physical effects based on their explosivity or flammability, there are additional physical hazards such as:

- *Laser light* – The primary hazard is loss of sight; however, at high intensity, laser light can burn through skin, muscle, bone, and steel.
- *Magnetic fields* – While the primary hazard with magnetic fields is with electronic equipment, the fields can affect animals, especially those with electronic or metallic implants. Even humans with pacemakers could be affected.
- *High voltage* – The primary hazard is electrocution, with burns and associated effects.
- *Ultraviolet light* – Similar to laser light, ultraviolet light can result in damage to the eyes. In addition, sunburn and skin cancer can result with improper exposure.
- *High noise* – Loss of hearing is a serious problem in the industrialized world.
- *Extreme heat or cold* – Burns and frostbite are the result from inadequate protection. Hot and cold surfaces and vessels can result in handling accidents without protective gloves.

All of these physical hazards are dose dependent, with the hazard increasing with intensity and duration. Therefore, it is critical that this information be accurately identified for the equipment and processes within each laboratory.

HAZARD ANALYSIS METHODS

A hazard analysis is the study of a facility, process, building, service, or operation carried out to ensure that hazards are identified, understood, and properly controlled. Although there are several regulatory requirements for laboratories using hazardous chemicals, federal mandates for clinical and pharmaceutical laboratories, and industry guidelines for laboratory HVAC design, none of them specifies deliberate methods to analyze hazards (Varley 1998). Therefore, various hazard analysis methods that have been used in the chemical and oil processing industries can be used to evaluate laboratory design concepts and to verify that the HVAC design conforms to the various plans dealing with chemical hygiene, radiation, biological, and physical safety.

The hazard analysis methods that can be applied to laboratory situations include:

- Hazard and operability (HAZOP) studies
- Failure mode, effects, and critical analysis (FMECA)
- Fault tree analysis
- Event tree analysis
- Cause-consequence analysis
- Human error analysis

The discussion in the following sections is intended to detail the procedural steps involved in the various hazard analysis techniques. This discussion is not exhaustive in presenting all the tools and information needed to fully master risk analysis. Therefore, additional details can be found in the literature sources, such as Varley 1998, AIChE 1985, Lawley 1974, and Vesley et al. 1981.

Hazard and Operability Studies

Deviations from the design intent of a given laboratory system or subsystem often cause unwanted hazards or obstruction to system operability. A hazard and operability study is a method to identify these hazards and operability barriers that occur. To complete the HAZOP study, the following fundamental information should be collected:

- Preliminary process and instrumentation diagrams (P&IDs)
- Relevant material safety data sheets (MSDS)
- Preliminary operating procedures (control sequence and modes of operation)
- Preliminary equipment list

Next, the project manager (owner's representative) is responsible for organizing and piloting a HAZOP study team, which should include the following individuals:

- HAZOP study team leader
- Local safety officer
- Biological safety officer
- Affected laboratory users
- Commissioning authority
- Facilities representative
- Project manager
- Engineer
- Recorder

It is essential that key members of the team have an operational knowledge of the system or systems to be analyzed.

The study commences by dividing the laboratory systems or equipment into discrete nodes. Figure 3-1 depicts a P&ID for the supply and exhaust system for a sample laboratory module. Figure 3-2 illustrates the

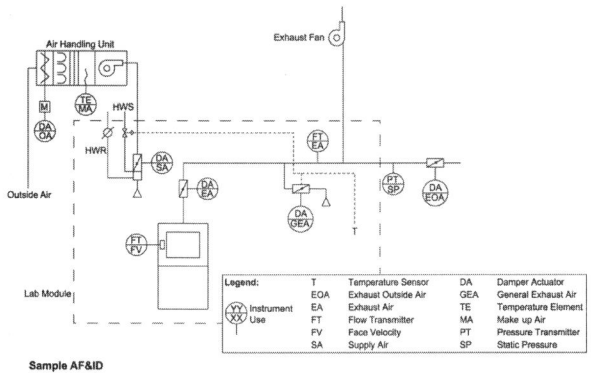

Figure 3-1 *Example P&ID for a ventilation system (Varley 1998).*

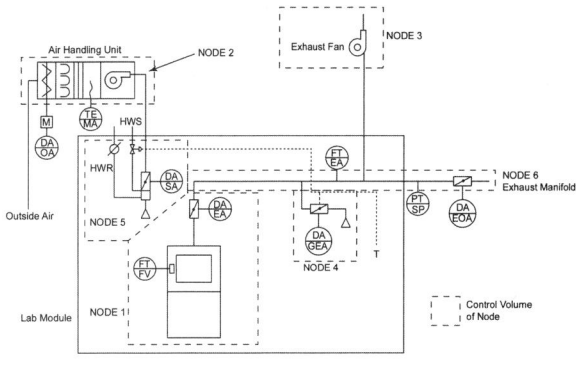

Figure 3-2 *Identification of P&ID nodes (Varley 1998).*

P&ID with the identification of nodes, which are essentially control volumes drawn around discrete subsystems. This is to simplify team discussions in dealing with each discrete subsystem separately. For optimal results, the maximum time spent on a node should be approximately 30 minutes, with no more than four hours per session. It has been found that shorter, more frequent meetings enable a more focused discussion and improved quality.

The HAZOP study is accomplished on each discrete node by having the individual team members first identify all items that could break or go wrong in the nodes. Then the team evaluates each of these scenarios to determine the result and whether or not the design intent has been met. To simplify the process, predefined guide words have been developed to aid in developing the scenarios. The definitions for guide words used in a HAZOP study are summarized in Table 3-7. For each scenario where the design intent has not been met, the team recommends solutions and carefully records the result.

Table 3-8 presents the results of a HAZOP example for the design of a manifolded VAV exhaust system (Node 1 in Figure 3-2). This node includes a fume hood, fume hood sensor, controller, and exhaust valve. The design intent of this node was "the provision of a hood to enclose mixing operations and solvent cleaning of laboratory apparatus."

Failure Mode, Effects, and Criticality Analysis

Failure mode, effects, and criticality analysis (FMECA) is used to systematically identify and tabulate the effects of various component failures in a system, product, or process as well as to determine the seriousness or criticality of the actual failures and the likelihood of their occurrence. FMECA is not usually conducted for all equipment but reserved for those containing hazardous materials that jeopardize human safety when in a failure mode. The ventilation, exhaust,

Table 3-7: Definition of Guide Words

Guide Words	Meaning
No	Design intent is not achieved.
Less	Result is less than the design intent.
More	Result is more than the design intent.
Part of	Result is qualitatively less than the design intent.
As well as	Result is in addition to the design intent.
Reverse	Result is the opposite of the design intent.
Other than	Complete substitution for design intent.

(Source: Varley 1998)

and fire protection systems are examples of systems that are very critical to the safety of the individuals within the laboratory environment.

A FMECA is completed as detailed in the following procedure:

1. Obtain individual equipment design and any relevant information about operation and interactions between equipment.
2. Form the FMECA team, to include the following participants:
   - FMECA technique team leader
   - Equipment designers
   - Operation and maintenance personnel
   - Local safety officer
   - Affected laboratory users
   - Commissioning authority
   - Project manager
   - Engineer
   - Recorder
3. Develop list of all system components to be analyzed, typically those with human safety concerns.
4. Develop the following categorical descriptions in a table format:

Table 3-8: HAZOP Example

Guide Word	Deviation	Causes	Consequences	Recommendations
No	No Flow	VAV valve failure, blockage, power failure	Chemical exposure	1. Evaluate failure modes of VAV valve.
Less	Less Flow	Same as no. Hoods all open at the same time	Chemical exposure	1. Establish procedure for low flow situation.
Reverse	Reverse Flow	Fan failure	Chemical exposure	1. Establish procedure for this failure. 2. Evaluate failure mode of damper. 3. Minimize leakage areas in duct. 4. Explore possibility of installing LEL detectors. 5. Investigate whether explosion venting for the duct should be installed.
Late	Late valve action	Slow controller response	Same as no	1. Establish procedure to verify hood performance prior to use. 2. Establish appropriate controller response time.
As well as	Spill in hood as well as normal hazards	Spill	Chemical exposure, fire and explosion hazard	1. Establish procedure for clean up. 2. Investigate installing emergency devices in hood and room. 3. Establish emergency response procedure for incident in hood and outside of hood.

- Operating modes of each listed component (from Step 3)
- Failure modes of each listed component (from Step 3)
- Effects and consequences of failure mode relating to equipment operation and personnel safety
- Probability of failure
- Criticality of failure
- Possible corrective actions to mitigate the effects of the failure

5. Complete FMECA table as each component is studied and discussed. Recommendations and results must be carefully recorded.

Table 3-9 illustrates the FMECA procedure for a laboratory fume hood. For this example, the exhaust ductwork was one of the more critical components of a fume hood. Further, corrosion was identified as a possible failure mode, resulting in malicious fumes spilling into the laboratory environment and adversely affecting laboratory occupants. The severity of the impact is dependent on the level of exposure and the toxicity of the spill. The likelihood of this failure occurring is dependent on the duct material and vapor corrosivity.

The corrective action for such an equipment failure could commence with an orderly shutdown of the laboratory operations and end with a more permanent replacement of the affected ductwork. Avoiding the problem would include detailed periodic inspection of all exhaust ductwork.

Successful completion of the FMECA influences design intent modifications that will avoid potential hazardous situations. For instance, in the fume hood example, a more robust ductwork material can be applied that would withstand the corrosive nature of the vapors to which it is exposed.

Fault Tree Analysis

Fault Tree Analysis (FTA) was first developed by H.A. Watson and associates at the Bell Telephone Laboratories in 1961 during the safety evaluation of the Minutemen launch control system and was later modified by the Boeing Company to check aircraft safety. In laboratory planning, FTA is used as a hazard analysis method that proceeds a HAZOP study or FMECA. Whereas the latter two methods are qualitative, FTA is quantitative. It is quantitative because probability values can be

Table 3-9: FMECA Example

Component	Operating Mode	Failure Mode	Effects	Probability	Criticality	Corrective Action
Exhaust Duct	On	Corrosion	Leakage	Depends on duct material and vapor corrosivity	Minor or significant	Possible shutdown
Exhaust Duct	On	Obstructed	Flow stoppage or reversal	Medium or high	Significant or immediate	Possible shutdown
Front Face	Setup	Stuck open	Incorrect ventilation	Low	None or minor	Remove hazardous material
Front Face	On	Slides down	No airflow	Low	None or minor	Possible shutdown

(Source: Ruys 1990)

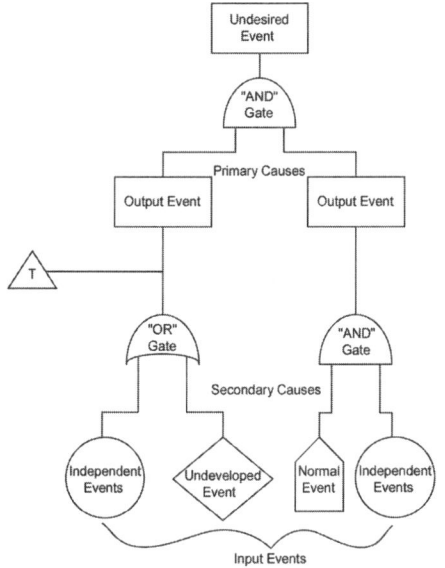

Figure 3-3 *Generic fault tree.*

assigned to each event at the root cause level and, therefore, the total likelihood of the undesired event occurring may be computed.

A fault tree is a logic diagram that is used to systematically investigate an undesired event. The primary causes that make this event possible are broken down to secondary and other levels with the hope of identifying the root cause. All of the events and combination of events that may lead to the undesired event are recorded and evaluated during the FTA. The end result is the identification and subsequent control of various system hazards.

Definitions of the three main types of symbols used in FTA are summarized in Figure 3-10. The key symbols are logic gate events (output, independent, normal, and undeveloped) and transfers. The logic gates demonstrate how events at one level of the fault tree can merge

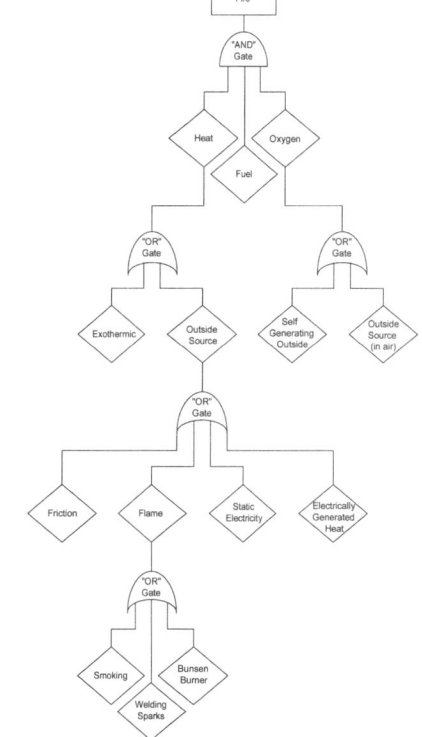

Figure 3-4 *Fire accident fault tree.*

to cause an event on the next higher level. The two common logic gates used are the "AND" and the "OR". With the AND gate, all incoming items must be true to go to the next level, whereas with the OR gate, only one has to be true. The transfer symbol represents the transfer of information between two points in the fault tree. This is used to avoid the repetition of a sequence of events that may be common to different regions of a fault tree. A layout depicting how the various symbols relate to each other in a fault tree is illustrated in Figure

Table 3-10: Fault Tree Analysis Symbols

Symbol	Name	Type	Brief Description
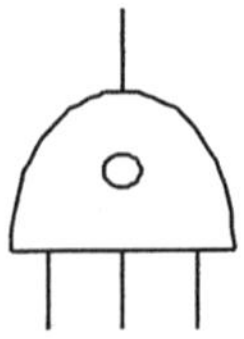 AND gate symbol	AND	Logic Gate	The "AND" gate represents a condition in which *all* of the input events (shown below the gate) are present for the output event (shown above the gate) to occur.
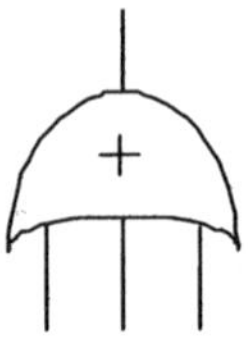 OR gate symbol	OR	Logic Gate	The "OR" gate represents a condition in which *any* of the input events (show below the gate) are present for the output event (shown above the gate) to occur.
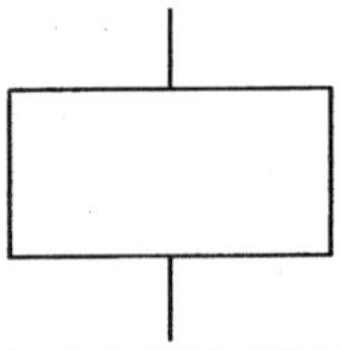 Rectangle symbol	Rectangle	Output Event	Output events are those that can be developed or analyzed further to determine how they may occur. The rectangle is the only symbol that can have logic gates and input events below it.
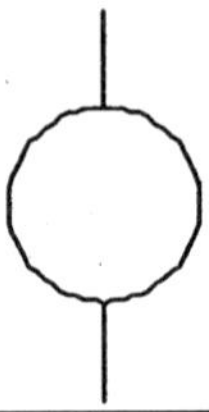 Circle symbol	Circle	Independent Event	Independent events do not rely on other components within the analyzed system for their occurrence.
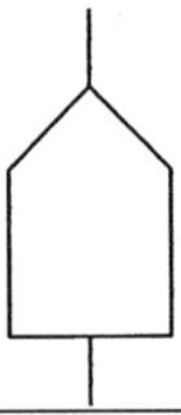 House symbol	House	Normal Event	Normal events are those that are expected to occur during system operation. Houses are typically shown at the lower fault tree levels.
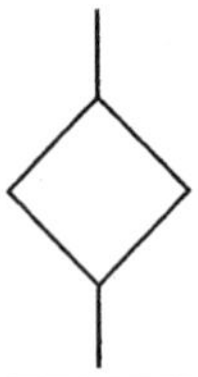 Diamond symbol	Diamond	Undeveloped Events	An undeveloped event is one that is not fully developed due to insufficient information or being considered too insignificant to warrant further analysis.
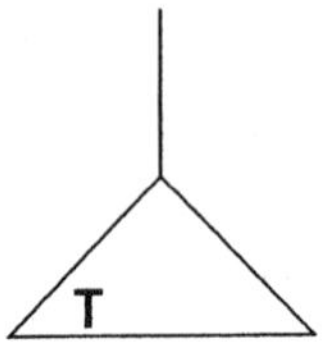 Triangle symbol	Triangle	Transfer Symbol	Transfer symbols are used to signify a transfer of a fault tree branch to another location within the tree. This serves to eliminate repetition of a sequence of events that is the same for various areas of a fault tree.

3-3. An example of a fault tree used for analyzing a fire accident in a laboratory is depicted in Figure 3-4.

An effective FTA is accomplished by following five fundamental steps:

Step 1 - Define the undesired event to be analyzed.

Step 2 - Obtain a working knowledge of the systems, subsystems/components of interest.

Step 3 - Construct the fault tree.

Step 4 - Evaluate the fault tree and identify possible hazards.

Step 5 - Control the hazards identified by elimination or mitigation.

Step 1 – Defining the Undesired Event. There is only one undesired event in a single fault tree, and as a result, care must be taken in defining this event as one that involves a serious accident posing a threat to personal safety. In laboratories, explosions, fires, or escape of toxic or carcinogenic fumes are examples of an undesired event. Therefore, the first step is to identify all of the events that are considered undesirable in a given system. Then, these events should be grouped according to similar causes or accident types. Finally, the one event that can be chosen to represent all the events of a particular group becomes the undesired event.

Step 2 – Understanding the Systems. After identifying and defining the undesired event, the next step is to gain an understanding of the systems, subsystems, and components using the FTA method. This includes all relevant information about the system's operations and component interactions and its environment. As a minimum, as applicable, the following should be accomplished:

- Observations of the actual system in operation
- Discussions with system experts
- Reading operations and maintenance manuals
- Obtaining system drawings, layouts, specifications, schematics, and pictures

Step 3 – Constructing the Fault Tree. This step involves constructing the fault tree. It is a logic procedure that generates a diagram highlighting all the possible causes of an undesired event. The procedure begins with the undesired event located at the top of the tree. Then, systematically working backwards, the primary causes of this event are identified and are located in the next lower level of the fault tree. The appropriate logic gate is used to indicate the logic relationship between these causes in generating the undesired event. This basic procedure is repeated down the tree, to lower levels, until events are reached that cannot be broken down any further.

Step 4 – Evaluating the Fault Tree. Once the fault tree has been constructed, it is evaluated. The key determinations usually made during this evaluation include:

- The most likely sequence of events that lead to the undesired event (at the top of the tree)
- The overall likelihood of the undesired event
- The combination of events most likely to lead to the undesired event
- The events that contribute the most to this combination of events

To compute the probability of the undesired event and to determine if it exceeds acceptable levels, probabilities are first assigned to the events on the lowest level. By continuing the evaluation process up the tree, through the various logic gates, the likelihood is eventually obtained for the undesired event.

Step 5 – Controlling the Hazards. If the likelihood of the undesired event is considered unacceptable, controlling the hazards within the system will reduce the accident potential. By using the most likely sequence of events leading up to the undesired event as identified in Step 4, the prevention or control of any of the events in this sequence can be evaluated to diminish the overall likelihood of the undesired event to occur. Hazard control will typically be achieved through a combination of engineering, education, and administrative solutions. Where these control methods are not adequate, protective apparel is frequently used.

Event Tree Analysis

An event tree analysis (ETA) studies the sequence of events after an undesired event, whereas the FTA studies the sequence of events prior to an undesired event. The event tree graphically represents this chronological sequence of events after an accident occurred. It considers the probable success or failure of personnel response as well as the safety system response to the unwanted accident. An example event tree for the laboratory fire FTA is shown in Figure 3-5.

Cause-Consequence Analysis

The cause-consequence analysis (CCA) method is a systematic tool that is commonly used to describe and analyze the accidents that may occur in complex process plants (Nielsen et al. 1975). Since CCA envelopes the entire event, it is based on the combination of the fault tree and the event tree methods. CCA charts may be developed where the cause chart is essentially a fault tree used to describe all the causes of an undesired event. The consequence chart is like an event tree that sequentially illustrates various sequences of events that an undesired event can promote due to the failing of key accident prevention arrangements. Therefore, the guidance for constructing fault trees and event trees directly applies to the construction of cause-consequence charts.

Human Error Analysis

The field of human engineering deals with the design of man-made objects (equipment) so that people

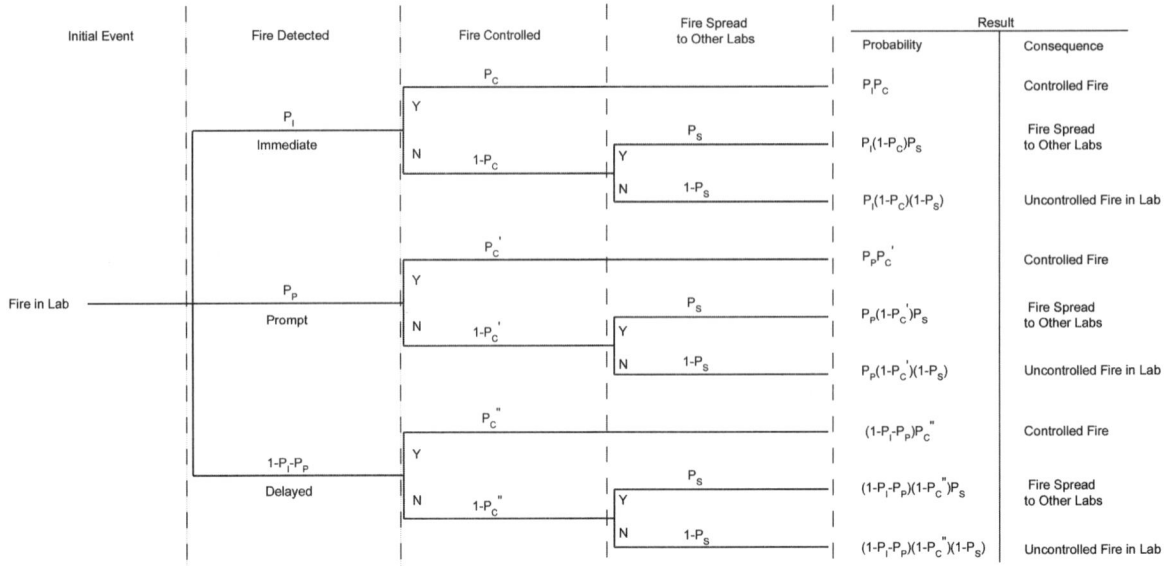

Figure 3-5 *Laboratory fire ETA example (diagram modified with permission from Ruys 1990).*

can use them effectively and safely to create environments suitable for human living and work (Huchingson 1981). It supports the idea that if equipment is poorly designed or requires capabilities that the human does not have, then the human either cannot perform the required tasks or will make errors while trying to do so. Errors may deteriorate the operation of the system as a whole or jeopardize the safety of the human operator. Therefore, a human error analysis (HEA) of laboratory systems can be invaluable as it attempts to identify, quantify, and prevent, or limit, human errors that can occur at the man-machine interface.

The following procedure is completed to perform the HEA for a laboratory:

1. Consider the various man-machine interfaces within the laboratory and identify potential sources of human errors. Historical knowledge of common errors as well as predictions of errors that may be unique to a particular application should be considered.
2. Quantify the likelihood (probability) of the errors identified. This aids in performing a cost benefit analysis, which demonstrates that preventing errors with severe consequences will be worth the cost of prevention and in some cases, the saving of lives.
3. Classify the identified errors. The errors are classified into the following human error categories based on the presumed cause of the error (Huchingson 1981):
   - Failure to perceive stimuli
   - Inability to discriminate among differing stimuli
   - Misinterpretation of the meaning of the perceived stimuli
   - Not knowing what response to make once a stimuli is perceived
   - Responding out of sequence to a stimuli
   - Being physically unable to make the required response to the perceived stimuli
4. Develop mechanisms to prevent or limit the occurrence of these errors for each classification. Examples of mechanisms include:
   - Proper and effective use of instructive and warning signs and symbols
   - Proper and adequate equipment supervision
   - Sufficient training, emergency drills, and reinforcement of strict rules
   - Implementation of life safety and emergency systems

This outlined HEA procedure may be able to provide some human error recovery. However, where no error recovery measures are apparent, immediate steps can be made toward laboratory design modifications, improvements to training programs, restructuring of experimental protocols, and changes in human job functions and tasks.

RISK ASSESSMENT GUIDELINES

Having an understanding of the various types of hazards present in laboratories as well as knowledge of some of the techniques and methods used to analyze them, it is beneficial to consider a few analysis guidelines that are typically applied during the laboratory-planning phase. These are:

1. Develop a chemical hygiene plan (CHP). Federal law (i.e., OSHA) requires that a CHP be written for all laboratories that use hazardous chemicals. The CHP typically includes protective guidance against health hazards associated with laboratory chemicals.
2. Material Safety Data Sheets (MSDS) must be compiled for the inventory of all hazardous materials to be used within the laboratory. MSDSs are required by federal law and have become the primary mode through which information about the potential hazards of materials, obtained from commercial sources, are transmitted to the users of laboratories. The information provided on MSDSs includes:
   - Name, address, and phone number of supplier and date MSDS was prepared and revised
   - Name of the material
   - Physical and chemical properties
   - Physical hazards
   - Toxicity data
   - Health hazards
   - Storage and handling procedures
   - Emergency and first aid procedures
   - Disposal considerations
   - Transportation information
3. The project manager (owner's representative) should assemble a hazard analysis team to study potential hazardous scenarios. The diverse expertise that is made available in a team is invaluable to the success of hazard and risk mitigation. The members of the team should be as indicated in the previous sections. However, at the very least, each team should include a safety officer, a laboratory user, and the risk analysis team leader.

RESPONSIBILITIES

Due to the high need for safety in the laboratory, various responsibilities are expected of all individuals that use the facility regardless of their role or their rank. Although the responsibility for overall safety lies with the project manager, the safety officer, or the principal investigator of experiments, students, technicians, and other staff employees are the ones that actually determine the level of safety practiced. Therefore, all laboratory users must foster a sense of responsibility that gives safety the highest priority at all times.

DOCUMENTATION

To assist in providing vital information to laboratory personnel about the hazards they may work with or become exposed to, adequate and appropriate documentation and training is essential. Documentation should cover the correct step-wise protocols for planning experiments or conducting other laboratory procedures, as well as the relevant federal, state, and local laws and regulations and the prevailing codes and standards. In addition, sufficient documentation must be available for quick and decisive action in case of emergency. Prior to use of any laboratory or new equipment/sequence, training should be provided and documented.

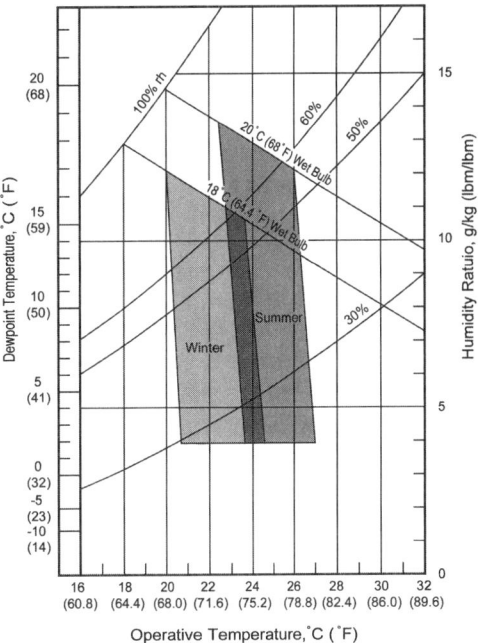

Figure 3-6 *ASHRAE summer and winter comfort zones.*

ENVIRONMENTAL REQUIREMENTS

Laboratories have special environmental requirements that must be identified in the planning phase to ensure the laboratory can be designed and constructed to meet them. Also, the potential for changes in laboratory use should be identified, so that the heating, cooling, ventilating, and exhaust systems can be designed to meet the current and future environmental requirements without substantial modifications.

Unfortunately, laboratory environmental conditions are often inhomogeneous because of the variation in the needs of the different types of occupants and functions. For example, persons that work in an animal laboratory will have very different thermal and ventilation comfort needs than say, rodents, being bred for a particular experiment in the same room.

Environmental requirements include personnel comfort and experimental quality, both of which must be addressed for a laboratory to be successful.

Personnel Comfort

In order to ensure comfortable conditions for laboratory occupants, the environment must be controlled within acceptable tolerances. Therefore, the required setpoints and ranges for temperature, humidity, indoor air quality, lighting, odors, allergens, and noise must be discussed and recorded by the planning and design team during planning for each functional area within the building.

Temperature. C.G. Simmons and R. Davoodpour have found that a comfortable temperature for typical laboratory work areas is typically 70°F to 72°F (21°F to 22°C) for all seasons. In the case of administrative offices, conference rooms, classrooms, corridors, and public spaces, 72°F to 75°F (22°C to 24°C) in the summer and 70°F to 72°F (21°C to 22°C) in the winter has been found to be comfortable (Simmons 1994). In order for the laboratory HVAC systems to be able to maintain these temperatures, discussions must include the setpoints, types, and placement of laboratory thermostats and automatic control system components. Figure 3-6 provides typical summer and winter comfort zones (ASHRAE 1997). This chart is for office space and includes extremes where only 90% of occupants with adequate clothing will be comfortable. For indoor air quality and comfort, the outer limits of this chart should not be used for design. Further, relative humidity should be controlled between 40% and 50% in laboratories.

It is important to note that Figure 3-6 uses operative temperature and not dry-bulb temperature. Since operative temperature accounts for the radiative temperature in addition to the dry-bulb temperature, any radiative sources, such as burners or cold windows, affect comfort.

Humidity. For personnel comfort, laboratory work areas should be kept at a maximum relative humidity of 50% in the summer and a minimum relative humidity of 35% in the winter, unless outside temperatures are less than 30°F for long periods. Comfortable summer and winter humidity are the same for administrative offices, conference rooms, classrooms, corridors, and public spaces.

A maximum relative humidity of 50% in the summer will ensure that occupants remain comfortable and minimize the potential for the growth of mold, mildew, and other organisms that thrive in moist environments and can negatively impact indoor air quality. Maintaining a minimum relative humidity of 35% in winter will minimize skin dryness, allow occupants to feel warm at a slightly cooler temperature, and minimize static electricity. These recommendations are based on comfort and productivity studies accomplished through recent research efforts (Berglund 1986). Berglund further documented that lowering the humidity from 60% to 50% improves comfort satisfaction and maintains equivalent comfort from 60% to 50%, and the space temperature can be raised 1°F (0.5°C).

Indoor Air Quality. Recirculated air may be used in offices, conference rooms, classrooms, corridors, and public spaces and may be required by design for cleanrooms. On the contrary, laboratory work areas should be designed to use 100% outside air in order to prevent the transmission and recirculation of hazardous particulates, aerosols, fumes, and vapors throughout the building. For this reason, laboratory and nonlaboratory areas are usually served by separate air distribution systems. The outside air is typically tempered and filtered or scrubbed based on the regional or local environmental conditions. In some cases, transfer air may be used as part of the laboratory exhaust needs.

During planning, the requirements for cleanliness of supply air, as well as the potential contaminant sources in the laboratory, should be determined by the planning staff so that the equipment necessary to provide these requirements can be selected. The contain-

Table 3-11: Recommended Illuminance Levels

Type of Activity	Illuminance Range, Footcandles	Description
Public spaces with dark surroundings	2-5	General lighting throughout spaces
Simple orientation for short temporary visits	5-10	
Working spaces where visual tasks are only occasionally performed	20-50	
Performance of visual tasks of high contrast or large size	50-100	Illuminance on task
Performance of visual tasks of medium contrast or small size	100-200	
Performance of visual tasks of low contrast or very small size	200-500	
Performance of visual tasks of low contrast or very small size over a prolonged period	500-1,000	Illuminance on task, obtained by a combination of general and local (supplementary lighting)
Performance of very special visual tasks of extremely low contrast and small size	1,000-2,000	

(Source: IESNA 1989)

ment and exhaust of pollutant is critical to maintaining the indoor air quality. Issues relative to this are detailed in the "Pressure Relationships" section later in this chapter.

Lighting Levels and Quality. Lighting levels and quality should be adequate for the occupants to perform their desired tasks. Where natural light is provided, there should be considerations of the effects on experiments and storage areas for chemicals. Therefore, solar loads on sensitive equipment and reflections in cathode ray tubes (CRTs) that make reading instrument dials difficult should be minimized. The type of lights (fluorescent, incandescent, etc.) affects the EMF radiation on sensitive equipment, and the impact on the HVAC system size should be considered during laboratory planning.

Lighting levels vary throughout a laboratory building depending on the functional use of the space. Work areas, both in laboratories and offices, often make use of general lighting and task lighting details with recommended levels of illuminance (see Table 3-11) for various tasks. Full spectrum lights should be used for general lighting and cool white lights for task areas such as fume hoods.

The ideal lighting arrangement from the perspective of capital cost, HVAC, productivity, and operating cost is to have general indirect lighting in the space with task lighting as required. This provides greater flexibility than only having direct general lighting.

Odors. Properly planning the types of materials to be used in a laboratory, properly designing exhaust systems, and ensuring personnel correctly operate exhaust hoods, all minimize the offensive odors generated within the laboratory areas. In addition, the transmission of odors can be avoided if the correct pressure relationships between laboratory areas and their adjacent spaces are maintained. Ventilated storage cabinets can also be used to minimize offensive odors. If the storage cabinets are provided with flame arrestors, they must be used and cleaned periodically to maintain airflow.

While odors may indicate a release of a chemical in a laboratory, which may be hazardous, the reverse is not necessarily true. The absence of odors does not ensure

Table 3-12:
Odor Thresholds, ACGIH TLVs, and TLV:Threshold Ratios of Selected Gaseous Air Pollutants

Compound	Odor Threshold,[a] ppmv	TLV, ppmv	Ratio
Acetaldehyde	0.067	40	597
Acetone	62	500	8.1
Acetonitrile	1600	40	0.025
Acrolein	1.8	0.1	0.06
Ammonia	17	25	1.5
Benzene	61	0.5	0.01
Benzyl chloride	0.041	1	24
Carbon tetrachloride	250	5	0.02
Chlorine	0.08	0.5	6
Chloroform	192	10	0.05
Dioxane	12	25	2
Ethylene dichloride	26	10	0.4
Hydrogen sulfide	0.0094	10	1064
Methanol	160	200	1.25
Methylene chloride	160	50	0.3
Methyl ethyl ketone	16	200	12.5
Phenol	0.06	5	83
Sulfur dioxide	2.7	2	0.74
Tetrachloroethane	7.3	1	0.14
Tetrachloroethylene	47	25	0.5
Toluene	1.6	50	31
1,1,1-Trichloroethane	390	350	0.9
Trichloroethylene	82	50	0.6
Xylene (isomers)	20	100	5

(Source: *2001 ASHRAE Handbook—Fundamentals*, ch. 13)

[a] All thresholds are detection thresholds (ED_{50}).

that there are no unsafe levels of chemicals in a laboratory, as some chemicals have little or no odor or are toxic in small quantities before their odor can be detected. Table 3-12 contains a list of odor thresholds, threshold limit values (TLV), and safety ratios. The TLV is the concentration of a compound that should have no adverse health consequences if, for and 8-hour period, a worker is exposed regularly. The safety ratio is the TLV divided by the odor threshold. OSHA sets the TLV as the safe level of exposure. A safety value greater than 1 indicates that the chemical is smelled before it is dangerous. However, a value less than 1 indicates that the problem occurs before it is detected. In addition, animal odors in research laboratories may not be hazardous but will have a very undesirable impact on researchers and occupants.

Allergens. Allergens are chemicals and substances that evoke an adverse reaction by the immune system, which can be either immediate or delayed. Allergic reactions in laboratories occur when an individual is exposed to small quantities of a chemical, particulate, or other substance after having been previously sensitized to the allergen. As other allergens, such as pollen and dust, affect individuals to widely varying degrees, individuals are also affected differently by chemical and other laboratory allergens. Common chemical allergens that can cause reactions include diazomethane, chromium, nickel, bichromates, dicyclohexylcarbodiimide, formaldehyde, various isocyanates, benzylic and allylic halides, and certain phenol derivatives.

Anaphylactic shock is an example of a severe immediate reaction that can result in death in the absence of prompt and correct treatment. Reddening, swelling, or itching of the skin are common symptoms in the case of delayed allergic reactions occurring hours or even days after exposure. Sufficient safety equipment (gloves, eye protection, fume hoods, etc.) and emergency procedures and preparations must be made if allergens are likely to be a hazard associated with chemicals and experiments conducted in a laboratory. In animal laboratories, the allergens are typically animal-based, resulting from the care and handling of the animals.

Noise. During planning, various building systems, equipment and associated components, materials, and furnishings must be reviewed and analyzed since they are all potential sources of noise or may help transmit noise. This is due to their sound levels being functions of the fabrication quality and the manner in which they are installed (e.g., flexible couplings, spring mounted, vibration pads, etc.). Exhaust fans, for example, have been known to have noise problems due to the high outlet velocities required for safety. The Occupational Health and Safety Administration (OSHA) has specific policies that govern workplace noise levels. Noise criterion (NC) and room criterion (RC) data are available in the *1997 ASHRAE Handbook—Fundamentals*.

Laboratory activities can also generate significant noise, depending on the type of laboratory. During planning, if laboratory operations are anticipated to generate noise over 85 decibels on a regular basis, plans should be made to protect personnel from excessive noise exposure both inside the laboratory and in the surrounding spaces. An acoustic engineer can provide advice on how to minimize the exposure to laboratory personnel or to minimize noise for instructional spaces.

Experimental Quality

To maintain the integrity of experiments, the temperature, humidity, and air quality of the environment must be considered a priority. There are two environments that should be considered: the macro-environment, which is the actual interior space of the laboratory that surrounds the researcher, and the micro-environment, which is a localized experimental space that has its own unique requirements. For both the macro-environment and micro-environment, vibration should also be considered.

Temperature. For all laboratories, considerations must be made to avoid major swings in the room and laboratory hood temperatures. Highly sensitive experiments also typically need to avoid relatively minor temperature fluctuations, which can disrupt the controlled environment of the experiment. Exposure to direct heat or sunlight can also affect experiments conducted in the macro-environment. For example, experiments using temperature-sensitive equipment should not be exposed to direct solar radiation. The automatic control system, laboratory layout (including the location of heat-producing equipment, windows, etc.), and air distribution methods should be properly planned to ensure that the temperature in a laboratory is maintained to appropriate tolerances.

For experiments conducted in micro-environments, which are often subjected to extremely low temperatures, extremely high temperatures, or precisely controlled temperatures, considerations need to be made to ensure that the required temperatures can be maintained. For example, appropriate temperature control and interlocks should be available to avoid heating or cooling experiments beyond desired limits.

Humidity. Like temperature, anticipated conditions that affect the desired humidity must be considered and counteracted to maintain experimental quality. Equipment that produces large amounts of moisture (usually due to heating/boiling of liquids) should be properly vented. Reliable automatic humidity control in addition to vapor barriers in partition walls, floors, and ceiling that minimize loss of water vapor will help ensure a stable environment.

During planning, the necessary limits on humidity should be determined and the HVAC system planned accordingly. This may require the addition of dehumidification equipment for summer operation and the addition of a humidification system for winter operation. Cleanrooms for electronics research or manufacturing and laboratories with sensitive electronic instruments may have special minimum humidity requirements in order to prevent static electricity from damaging experiments and equipment.

Air Quality. To maintain good air quality in laboratory spaces, planning should focus on potential sources of pollutants and contaminants and develop design intent items that minimize them. These include:

- Effective exhaust systems to expel and contain malicious fumes
- Proper pressure relationships between adjacent spaces
- Low velocity diffusers that do not interfere with fume hood and biological safety cabinet containment
- Properly sealed ductwork and envelope
- Specific pollutant sensors to maintain regulated levels

Air quality concerns related to experimental quality are often present in animal laboratories. Proper ventilation and air quality is needed to ensure that the results of experiments are accurate and not due to poor indoor air quality. Air quality can also be a concern in adjacent laboratories of different functional uses, as the contaminants from one experiment may affect the work in the adjacent laboratory.

Vibration. Vibration can originate from within or from outside a laboratory. Interior vibration from mechanical and electrical equipment (e.g, fans, blowers, pumps, chillers, and transformers), by laboratory equipment (e.g., centrifuges and fume hoods), and pedestrians can cause interference with sensitive equipment in the laboratory. Correct balancing of mechanical equipment, using vibration isolators for both mechanical equipment and sensitive instruments, and making the building as stiff as possible can be used to minimize these vibrations. Exterior vibrations can be caused by traffic, construction work in nearby buildings, wind, and seismic disturbances. This type of vibration is harder to control, and, therefore, the location selected for a laboratory during the planning stage is important in order to minimize the possible external sources of vibration.

APPLIANCES AND OCCUPANCY

While the overwhelming cooling and heating load for any laboratory building is typically the conditioning of the supply air that is exhausted through the hoods, a significant amount of heat and moisture originates from appliances, lighting, and the occupants that occupy the internal laboratory environment. The heat gain varies with use, exterior influences (weather), the building envelope, and the heat storage capacity of spaces.

Appliance Loads

The equipment used in laboratories creates sensible and latent heat gains that must be removed by the HVAC system to maintain comfortable conditions. These heat gains are released into the room and influence the thermal environment in the room. While exhaust hoods located above primary heat generating equipment can remove most of the heat generated by that piece of equipment without influencing the temperature of that room, exhaust hoods are not always feasible.

As with any load calculation, the equipment load and use must be estimated. Typical range of heat generation of equipment used in laboratory areas and adjacent offices is provided in the appendix. This information is based on recent ASHRAE research involving the measurement of heat gain in space and the radiant/convective split from equipment in buildings (Hosni et al. 1999). In addition to the equipment shown in Appendix A, there may be large self-contained units in laboratories that must be coordinated with the building supply and exhaust systems. This includes low-temperature freezers and animal cages.

Lighting

In addition to the cooling load resulting from various laboratory equipment, lighting generates heat that adds to the cooling load and needs to be met by the HVAC system. The lighting levels required vary depending on the tasks performed in the area, with the lighting power depending upon what type of ballast and fixtures are used and other strategies such as daylighting. Task lights are often used for specific requirements such as experimental operations that have small detailed work.

Ambient Lighting. Ambient lighting ensures that the overall level of lighting satisfies the need of the most common tasks performed in that area. The lighting levels recommended for areas that are typically found in buildings containing laboratories are shown in Table 3-13. Lighting should be provided without causing excessive glare and brightness contrast.

Task Lighting. Task lighting describes the light provided for specific experimental purposes in need of higher lighting levels than the surrounding tasks. Ideally, to avoid contrast problems, the ratios listed in Table 3-14 should not be exceeded. The benefit of task lights compared to illuminating the whole area is reduced energy consumption for lighting and thereby reduced HVAC system capacity.

Occupants

The occupants (people and animals) contribute to the cooling load of the space. Occupant densities, diversities, activity level, and scheduling are the key criteria in determining the occupant load requirements.

Occupant Densities. The occupant densities for laboratories vary significantly with the type and purpose of a particular laboratory. The space requirement for most laboratories is between 50 ft^2 and 250 ft^2 per per-

Table 3-13: Typical Lighting Requirements

Room Type	Lighting Level, Foot-Candle (Lux)	Maximum Allowed Lighting Power, Btu/h ft² (W/m²)	Typical Lighting Fixtures and Lamps
Laboratory	75-100 (807-1076)	11.3 (35.5)	Recessed fluorescent with parabolic lenses or prismatic lens
Operating rooms		23.9 (75.3)	
Animal laboratories:		—	Gasketed, prismatic lens. Fluorescent fixtures
Night	0 (0)		
Day	30-50 (323-538)		
Cleaning	75-100 (807-1076)		
Fume hoods	100 (1076)	—	By fume hood manufacturer (fluorescent lens)
Walk-in environmental rooms	50-100 (538-1076)	—	Surface, fluorescent or incandescent lens
Office	50 (538)	5.5 (17.2)	Recessed fluorescent with parabolic lenses or prismatic lens
Secretarial	75 (807)	5.5 (17.2)	Recessed fluorescent with parabolic lenses or prismatic lens
Library	75 (807)	6.5 (20.5)	
Corridors	20-30 (215-323)	2.7 (8.6)	Similar to that of surrounding areas
Storage	10-200 (108-2152)	2.1 (6.5)	Strip fluorescent
Mechanical	20-30 (215-323)	2.4 (7.5)	Industrial fluorescent
Computer rooms	30-50 (323-538)	7.2 (22.6)	Recessed parabolic or indirect fluorescent

(Source: Ruys 1990, Wisconsin Commercial building code 1998)

Table 3-14: Recommended Maximum Illumination Ratios for Offices

Areas	Recommended Maximum Illumination Ratios
Task to adjacent surroundings	3:1
Task to remote darker surfaces	5:1
Task to remote lighter surfaces	1:5

(Source: IES 1993)

son (15.24 to 76.2 m²/person). The laboratory units or departments must accurately document and request the need for space and the associated occupancy density.

Diversities. The diversity in occupancy of laboratories and support spaces should be determined by the specified usage of the laboratory. While the system for each space must be sized for 100% load, central systems can be sized with a diversity to account for uneven loading throughout the building. However, with 100% outdoor air and high exhaust rates, the diversity rarely falls below 90%. For instructional laboratories, turn down ratios should be considered during non-occupied periods because of limited use educational programs.

Activity Level. The sensible and latent heat gain of humans is dependent on their activity level. Some of the activities that pertain to laboratories are shown in Table 3-15. Heat generated by normally active animals is presented in Table 3-16.

Scheduling. Occupants spend varied amounts of time within a given laboratory depending on their needs and the scheduled hours of operation. The users and planners of the laboratory should clearly document the expected schedules. This information is critical to avoid design errors, flaws in the control system, inefficient energy management, and a system that is incapable of meeting the user's requirements.

PRESSURE RELATIONSHIPS

Maintaining proper pressure relationships between adjacent spaces in a laboratory building is critical to ensure airflow is in the proper direction, from clean areas to dirty areas. Therefore, pressure relationships need to be determined during the planning phase. Planning includes the identification of individual pressure zones and the development of a preliminary pressure map. The process of creating a pressure map is described in Step 1 of Chapter 4, "Design Process." A primary problem with the design, construction, and operation of laboratory facilities is that the air moves from space to space based on simple pressure relationships—it goes from higher pressure

Laboratory Planning

Table 3-15: Occupant Heat Gain

Activity	Total Heat Generation, Btu/h (W)	Sensible Heat Generation, Btu/h (W)	Latent Heat Generation, Btu/h (W)	Radiant Sensible Heat Btu/h (W)
Seated, very light work	390 (115)	220 (65)	135 (40)	92-205 (27-60)
Moderately active office work Standing, light work, walking	445 (130)	255 (75)	190 (55)	130-198 (38-58)
Walking, standing	495 (145)	255 (75)	240 (70)	130-198 (38-58)
Light bench work at factory	750 (220)	275 (80)	480 (140)	119-167 (35-49)

Note: Values are based on 24°C room dry-bulb temperature. For 27°C room temperature, the total value remains the same, but the sensible heat is reduced by 20% and the latent heat increased accordingly. Total heat generation is based on a normal percentage of men, women and men. The sensible heat that is radiant will vary with air velocity.

Table 3-16: Animal Heat Gain

Species	Weight, lbm (kg)	Total Heat Generation, Btu/h (W)	Sensible Heat Generation, Btu/h (W)	Latent Heat Generation, Btu/h (W)
Mouse	0.046 (0.021)	1.65 (0.49)	1.11 (0.33)	0.54 (0.16)
Hamster	0.260 (0.118)	6.00 (1.76)	4.02 (1.18)	1.98 (0.58)
Rat	0.62 (0.281)	11.6 (3.40)	7.77 (2.28)	3.83 (1.12)
Guinea pig	0.90 (0.41)	15.2 (4.46)	10.2 (2.99)	5.03 (1.47)
Rabbit	5.41 (2.46)	58.5 (17.15)	39.2 (11.49)	19.3 (5.66)
Cat	6.61 (3.00)	68.1 (19.93)	45.6 (13.35)	22.5 (6.58)
Nonhuman primate	12.0 (5.45)	106 (31.12)	71.3 (20.9)	35.1 (10.3)
Dog	22.7 (10.31)	161 (47.2)	105 (30.7)	56.4 (16.5)
Dog	50.0 (22.7)	355 (104)	231 (67.6)	124 (36.4)

(Source: ASHRAE Fundamentals, Ch. 10, 1997)

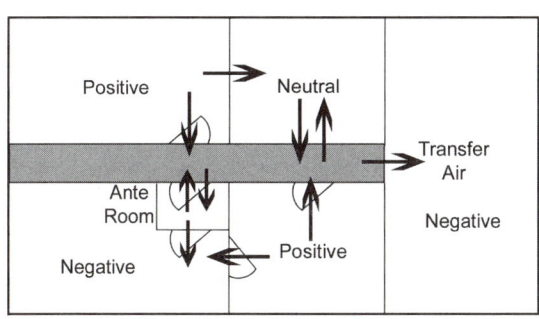

Figure 3-7 *Pressure relationship terminology.*

areas to lower pressure areas. Since there is no such thing as "smart air" that follows the arrows on drawings, special attention must be paid to the layout and design of the HVAC systems and to wall construction.

There are three primary problem areas that are often overlooked when developing pressure maps. First, the outdoor wind direction places a positive pressure on the upstream wall and a negative pressure on the roof and downstream wall. Second, due to stack effect, as one moves upward in a building, the relative pressure from indoors to outdoors goes from negative to positive, resulting in air entering the lower floors and exiting the top floors. Finally, all buildings are composed of compartments (rooms) and chases (elevators and stairs), which allow air to quickly move from one location to another location in a building. Figure 3-7 demonstrates these principles.

Therefore, proper sealing of spaces and accounting for external forces is required in the proper layout of a system in laboratory facilities. To simplify the planning process, the need for negative, positive, and neutral pressure rooms, anterooms (air locks), and transfer air should be discussed. Figure 3-7 details each of these graphically.

Negative Pressure Room

A negative pressure room is one that is at a lower pressure compared to adjacent spaces and, therefore, the net flow of air is into the room. For this to happen, the volume flow rate of the supply air must be less than that of the combined exhaust flow rates from all general and equipment exhausts. Laboratories that use hazardous chemicals and materials need to be maintained at a negative pressure so that fumes, particulates, odors, and other by-products of experiments do not migrate from the experiment to adjacent occupied spaces.

Positive Pressure Room

A positive pressure room is one that is at a higher pressure compared to adjacent spaces and, therefore, the net airflow is out of the room. The volume flow rate of the supply air must be greater than that of the combined exhaust flow rates from all general and equipment exhausts for a room to maintain a positive pressure. Cleanrooms and sterile facilities are examples of laboratories that need positive pressurization to prevent infiltration that could compromise experimental requirements for cleanliness. If hazardous materials must be used in a positively pressurized laboratory, such as an electronics manufacturing laboratory, exhaust equipment needs to be designed and used carefully to ensure that the exhaust equipment removes all hazardous fumes. More commonly, an airlock or other buffer zone can provide a negative containment area around a positive pressure room that contains hazardous material.

Neutral Pressure Room

A neutral pressure room has no specific pressurization requirements. Typically, nonlaboratory spaces such as office areas and corridors are maintained at a neutral pressure. Some rooms that have no pressurization requirements of their own must be maintained at either a positive or negative pressure so that the pressure requirements for adjacent rooms are satisfied and airflow from clean to dirty is maintained. These rooms are identified when creating a pressure map.

Anteroom

An anteroom is a transition room between areas of substantially different pressures or that is used to gain access to a room that must maintain its pressure even during disturbances such as opening a door. The use of anterooms provides assurance that pressure relationships are constantly maintained and air remains flowing from clean to dirty, and they reduce the need for the HVAC control system to respond to large disturbances.

Transfer Air

Transfer air moves from one space to an adjacent space through a transfer grille or other air distribution outlet device. Transferring air should only be done between spaces that have a similar level of cleanliness or from clean areas to those of lesser cleanliness. Typically, transfer air is used between subdivided rooms in a common area to prevent large pressure differences as the supply and exhaust airflows on either side change.

VENTILATION AND INDOOR AIR QUALITY CONSIDERATIONS

During planning, careful consideration of ventilation and indoor air quality is needed due to the hazardous nature of the materials used in laboratories and the need to provide a safe working environment for laboratory personnel. Various codes and standards discussed in the next section specify different types of minimum air change requirements for laboratories. The differences in ventilation requirements from occupied to unoccupied mode should also be considered during planning, along with determining how to satisfy both the cooling load and the air change requirements for a laboratory. Finally, potential pollutants and sources should be identified and air treatment requirements determined during the planning stage.

Minimum Supply Air Changes

Minimum ventilation rates for laboratories are typically defined in air changes per hour, which is the amount of supply air (if calculating supply air changes) or exhaust air (if calculating exhaust air changes) provided in one hour divided by the total room volume. In areas where no air recirculation is used (100% outdoor air), the supply air change rate will equal the outdoor air change rate.

According to the *NFPA 45 Standard, Appendix A: A-6-3.4* (NFPA 45 2000), the air changes of room air currents in the vicinity of fume hoods should be as low as possible, ideally less than 30% of the face velocity of the fume hood. Air supply diffusion devices should be as far away from fume hoods as possible and have low exit velocities.

Minimum Exhaust Air Changes

Exhaust air changes are typically determined by the flow rates needed to satisfy the supply/outdoor air change requirements or the flow requirements of the exhaust equipment. In addition, the airflow rate needed to prevent undesirable conditions from developing in the exhaust ductwork should be considered. For example, *ANSI/AIHA Z9.5 1992* recommends that where practical, the airflow changes in each exhaust duct should be sufficient to prevent condensation of liquids of condensable solids on the walls of the duct. Also, exhaust airflow changes for heat-producing equipment should be sufficient to keep the temperature in the duct below 400°F.

Minimum Outdoor Air Changes

For areas that do not use recirculated air, the outdoor air changes and the supply air changes are the same. Several standards specify outdoor/supply air change rates, including *OSHA 29 CFR Part 1910*, p. 3332, 4. (f) (OSHA 1990b), which recommends 4 to 12 air changes per hour as normally adequate ventilation if local exhausts hoods are used as the primary method of control.

Minimum Total Air Changes

Total air change is the ventilation rate of a space due to both infiltration and mechanically supplied outdoor air.

Typically, the total ventilation rate is determined by using a tracer gas test, where tracer gas is evenly distributed throughout the laboratory and the rate of decay in the tracer gas concentration is used to calculate the air changes per hour.

Recirculation Considerations

ANSI/AIHA Z9.5 Standard and *NFPA 45 Standard* stipulate that laboratory ventilation systems shall be designed to ensure that chemicals originating from laboratory operations shall not be recirculated. However, air from nonlaboratory areas, such as general office spaces, may be recirculated, but the risks associated with the accidental release of chemicals from a laboratory space should be considered.

Occupied vs. Unoccupied Mode

In order to conserve energy, setback control strategies may be used in laboratories to reduce the air changes per hour during unoccupied periods. The *NFPA 45 Standard* recommends a minimum ventilation rate for unoccupied laboratories (e.g., nights and weekends) of 4 air changes per hour. Occupied laboratories typically operate at rates in excess of 8 air changes per hour, as appropriate for the intended uses of the laboratory.

Cooling Load vs. Air Change Requirements

The total airflow rate for a laboratory is specified by either the cooling load generated by internal heat gains (people, light, equipment, etc.), the maximum connected exhaust loads, or by the minimum air change requirements. Designers and engineers must obtain the optimum HVAC system design that can meet both of these criteria in order to provide a comfortable laboratory with acceptable indoor air quality.

When the ACH rate determines the supply air volume, then some form of tempering the air prior to entering the space is required. The primary reason for this is to avoid overcooling the space.

Potential Pollutants and Sources

During the planning phase, an appreciation for the various hazardous substances in the laboratory and the risks associated with their use are obtained. These substances are the root source of potential pollutants and therefore should be carefully managed before they end up in the supply or exhaust airstream. The following is a chronological list of management tasks that should be completed prudently throughout the life of hazardous substances that enter the laboratory:

- Source reduction
- Acquisition
- Inventory and tracking
- Storage
- Recycling

Source Reduction. Source reduction is minimizing the quantity of materials that are ordered, stored, and used in a laboratory. Practicing source reduction exposes laboratory workers to fewer risks and reduces the life-cycle cost of using a material. Purchasing smaller quantities of chemicals in smaller containers eliminates the risks to laboratory workers that are associated with transferring chemicals from the large container in which they were purchased to a smaller container for use in the laboratory. Also, the risks to laboratory staff associated with breakage of a storage container are less when smaller containers are used. Storing smaller amounts of chemicals reduces the amount of chemical storage space needed, which can result in more usable floor space. Ordering large containers of chemicals often requires additional equipment (pumps, funnels, smaller containers, etc.) to transfer the chemicals into smaller containers for use by the laboratory staff while conducting experiments and increases the risk of a spill or accident. Minimizing the use of chemicals is also an effective strategy in reducing the risks to laboratory workers. For example, sometimes an experiment can be carried out on a smaller scale, or sometimes a less hazardous chemical may be substituted. Also, hazardous waste can be reduced to a less hazardous material in some cases by treating it with another chemical, sometimes just water.

Acquisition. The acquisition process of chemicals should be considered during the planning phase so that the needed equipment and facilities can be developed to safely receive incoming materials. It is important that the delivery of chemicals only take place in an area that is equipped to handle them, such as a loading dock or receiving area, so that indoor air quality is maintained at acceptable levels. Chemicals should not be delivered to office areas within a laboratory, as those areas are not properly equipped. Therefore, such an area should be set aside during planning, which includes adequate equipment to handle the delivery of chemicals and is clearly marked and accessible to the delivery personnel. If possible, freight-only elevators should be designated for transporting materials from floor to floor in a laboratory to keep hazardous materials away from occupants. Finally, some materials, specifically flammables and combustibles, are regulated by various codes to a maximum amount that can be stored in a given area. Therefore, the need for such materials should be coordinated with the layout of the laboratory floor plan to ensure that codes and standards can be satisfied.

Inventory and Tracking. An inventory and tracking system can be implemented in a variety of ways, depending on the size of the laboratory and number of staff, from a simple index card system to an electronic

**Table 3-17:
Suggested Compatible Storage Groups**

Inorganic Family	Organic Family
Metals, hydrides	Alcohols, glycols, amines, amides, imines, imides
Halides, sulfates, sulfites, thiosulfates, phosphates, halogens	Acids, anhydrides, peracids
Amides, nitrates (except ammonium nitrate), nitrites, azides	Hydrocarbons, esters, aldehydes
Hydroxides, oxides, silicates, carbonates, carbon	Ethers, ketones, ketenes, halogenated hydrocarbons, ethylene oxide
Sulfides, selenides, phosphides, carbides, nitrides	Epoxy compounds, isocyanates
Chlorates, perchlorates, perchloric acid, chlorites, hypochlorites, peroxides, hydrogen peroxide	Peroxides, hydroperoxides, azides
Arsenates, cyanides, cyanates	Sulfides, polysulfides, sulfoxides
Borates, chromates, manganates, permanganates	Phenols, cresols
Nitric acid, other inorganic acids	
Sulfur, phosphorus, pentoxide	

Note:
- Store flammables in a storage cabinet for flammable liquids or in safety cans.
- Separate chemicals into their organic and inorganic families and then related and compatible groups, as shown.
- Separation of chemical groups *can be by different shelves within the same cabinet.*
- Do **NOT** store chemicals alphabetically as a general group. This may result in incompatibles appearing together on a shelf. Rather, store alphabetically within compatible groups.
- This listing is only a suggested method of arranging chemical materials for storage and is not intended to be complete.

(Source: NRC 1995)

database. Along with an inventory system, detailed, consistent, and accurate labeling of chemical containers is imperative. A successful inventory and tracking system for chemicals in a laboratory can provide valuable information, which can benefit the ordering process and emergency procedures.

As with any inventory system, an inventory of laboratory chemicals is only as good as the data that it contains. Therefore, a verification of the information about the chemical inventory should be conducted, which can be performed at the same time as regular physical inspections of the stored chemicals. Regular inspections of stored materials can improve indoor air quality by identifying materials that have become unstable or are in deteriorating containers and are at risk for accidental release.

Storage. In order to provide acceptable indoor air quality, chemicals in a storage room need to be stored in compatible groups with similar storage requirements so that they will not react with each other. Flammable and combustible liquids, gas cylinders, highly reactive substances, and highly toxic substances may have special storage needs, such as ventilated storage cabinets. Table 3-17 contains suggested compatible storage groups.

Recycling. Recycling of laboratory materials includes materials such as packaging and containers, solvents, mercury, and heavy metals. The potential of chemical recycling should be considered during planning, so that necessary equipment is installed to protect workers during the recycling process. Many forms of recycling require a waste treatment permit under the Resource Conservation and Recovery Act (42 USC 6901) or are governed by state or local regulations.

Supply Air/Exhaust Air Treatment Requirements

Because most laboratory environments are inherently hazardous, pollutants and their sources cannot be completely eradicated. Therefore, air treatment may be needed to provide the desired air quality by either preventing pollutants from leaving the laboratory and entering the environment (e.g., biological laboratories) or preventing pollutants in the environment from entering the laboratory (e.g., cleanrooms). During planning, the work to be performed in the laboratory should be evaluated for supply or exhaust air treatment needs, including the use of filtration, scrubbing, condensing, and oxidation technologies. Details on the various air treatment applications and technologies available can be found in Chapter 8, "Air Treatment."

LABORATORY CODES, STANDARDS, AND REFERENCES

The prevailing codes and standards, along with informative references relevant to laboratory design, are presented in this section. These codes, standards, and references offer guidance on restrictions and minimum requirements developed and published by nationally recognized organizations, and they are summarized in Table 3-18. These include the National Fire Protection Association (NFPA), Occupational Safety and Health Administration (OSHA), National Institutes of Health (NIH), Building Officials and Code Administrators International, Inc. (BOCA), Scientific Equipment and Furniture Association (SEFA), American National Standards Institute (ANSI), and American Society of Heating, Refrigerating and Air-Conditioning Engineers (ASHRAE).

Table 3-18: Summary of Codes, Standards, and References Pertaining to Laboratories

Name	Reference	Summary of Purpose
ADA, 28 CFR 36	(ADA 1994)	Furnishes special considerations that must be given to accommodate laboratory workers with physical impairments. This includes wheel chair accessibility, work bench height, and access to controls.
ANSI Z88.2-1980, American National Standard Practices for Respiratory Protection	(ANSI 1982)	Helps respirator users to establish, implement, and administer an effective respiratory protection program.
ANSI/AIHA Z9.5, Laboratory Ventilation	(AIHA 1993)	Establishes minimum requirements and procedures for the design and operation of laboratory ventilation systems used to protect personnel from overexposure to harmful or potentially harmful contaminants.
ANSI/ASHRAE Standard 110-1995, Method of Testing Performance of Laboratory Fume Hoods	(ASHRAE 1995)	Provides a method to quantify fume hood performance. It tests the competence of a fume hood at a given point in time to establish a baseline for quantifying a fume hood's performance. Repeat testing is done to track and evaluate future performance compared to this baseline.
ANSI/ASHRAE Standard 55-1992, Thermal Environmental Conditions for Human Occupancy	(ASHRAE 1981)	Forms the basis for the indoor design temperature and humidity for most spaces.
ANSI/ASHRAE Standard 62-1999, Ventilation for Acceptable Indoor Air Quality	(ASHRAE 1989)	Forms the basis for the minimum outside air requirements for most spaces and stipulates when treatment of outside air and exhaust air is necessary.
ASHRAE 90A-1980, Standard for Energy Conservation in New Building Design	(ASHRAE 1980)	Provides guidelines for designing energy efficient HVAC systems.
ASHRAE Handbook, HVAC Applications, Chapter 13, Laboratories	(ASHRAE 1999)	Provides a condensed version of the information provided in the chapters of this *ASHRAE Laboratory Design Guide*.
BOCA National Building Code, 10th Edition	(BOCA 1987a)	States regulations in terms of measured performance, thus making it possible to accept new materials and methods of construction, which can be evaluated by accepted standards, without the necessity of adopting cumbersome amendments for each variable condition.
BOCA National Mechanical Code, 6th Edition	(BOCA 1987b)	Sets forth regulations for the safe installation and maintenance of mechanical facilities where great reliance was previously placed on accepted practice and engineering standards.
Biosafety in Microbiological and Biomedical Laboratories	(U.S. DHHS 1993)	Provides Biosafety Level 1 to 4 procedures and guidelines for the manipulations of etiologic agents in laboratory settings and animal facilities. The four levels of control that are defined range from safely dealing with microorganisms that pose no risk of disease for normal healthy individuals to dealing with the high risk of life-threatening diseases.
SHEMP (Safety, Health and Environmental Management Program) Operations Manual for Laboratories	(EPA 1998)	This document provides guidance on management and administration, hazard identification and evaluation, laboratory Safety, Health and Environmental Division programs, engineering controls, protective clothing and equipment, work practice controls and laboratory emergency situations.

Table 3-18: Summary of Codes, Standards, and References Pertaining to Laboratories

Guide for the Care and Use of Laboratory Animals, No. 86-23	(NRC 1996)	Assists institutions in caring for and using animals in ways judged to be scientifically, technically, and humanely appropriate.
Guidelines for Laboratory Design	(DiBerardinis et al. 1993)	Provides reliable design information related to specific health and safety issues for new and renovated laboratories. Factors such as efficiency, economics, energy conservation, and design flexibility are considered.
Handbook of Facilities Planning, Volume 1, *Laboratory Facilities*	(Ruys 1990)	Gives hands-on guidance to a wide range of topics including differences and similarities between various laboratories, laboratory planning, communication enhancing design, and the definition of building systems.
Industrial Ventilation, A Manual of Recommended Practices, 23rd Edition	(ACGIH 1998)	Recommends best practices, including research data and information on the design, maintenance, and evaluation of industrial exhaust ventilation systems. Basic ventilation principles and sample calculations are also presented.
NFPA 101, Life Safety Code	(NFPA 2000a)	Addresses all the construction, protection, and occupancy features needed to minimize danger to life from fire, smoke, and panic. Forms the basis for law in many national jurisdictions.
NFPA 30, Flammable and Combustible Liquids Code	(NFPA 2000b)	Provides the most up-to-date requirements for dealing with flammable and combustible liquids and is therefore useful to design engineers, enforcing officials, insurers, and laboratory workers.
NFPA 45, Fire Protection for Laboratories using Chemicals	(NFPA 2000c)	Provides the minimum fire protection requirements for fire safe design and operation in educational and industrial laboratories using chemicals.
NFPA 801, Facilities Handling Radioactive Materials	(NFPA 1998)	Identifies guidelines for decreasing the risk of explosion or fire and the severity of contamination from a fire or explosion at facilities (except nuclear reactors) that handle materials that are radioactive.
NSF 49, Class II (Laminar Flow) Biohazard Cabinetry	(NSF 1992)	Provides comprehensive information and guidance on the principles and applications of air filtration, which supplies the level of particulate cleanliness required by HVAC systems.
OSHA 29 CFR 1910.1330, Occupational Exposure to Bloodborne Pathogens	(OSHA 1990c)	Provides worker protection from exposure to bloodborne pathogens.
OSHA 29 CFR 1910.1450, Occupational Exposure to Hazardous Chemicals in Laboratories	(OSHA 1990b)	Provides protection for all laboratory workers engaged in the use of hazardous chemicals.
OSHA 29 CFR 1990, Identification, Classification, and Regulation of Carcinogens	(OSHA 1991b)	Determines various criteria and procedures for the identification, classification, and regulation of potential occupational carcinogens that exist in each workplace in the United States and that are regulated by the Occupational Safety and Health Act of 1970 (the Act).
Prudent Practices in the Laboratory, Handling and Disposal of Chemicals	(NRC 1995)	Recommends several prudent practices that stimulate a culture of safety for chemical laboratory operations. Provides information and cross-references on how to handle compounds that pose special hazardous risks.

Table 3-18: Summary of Codes, Standards, and References Pertaining to Laboratories

SEFA 1.2, *Laboratory Fume Hoods, Recommended Practices*	(SEFA 1996)	Provides information on design, materials of construction, use, and testing of laboratory fume hoods. These tests establish the average face velocity and adequacy of the airflow throughout the overall open face area of fume hoods.
SEFA 2.3, *Installation of Scientific Furniture and Equipment, Recommended Practices*	(SEFA 1997)	Provides information for architects, specifying engineers, contractors, and other purchasers about the installation practices recommended by manufacturers of scientific laboratory furniture and equipment.
SEFA 8, *Laboratory Furniture, Recommended Practices*	(SEFA 1999)	Provides manufacturers, specifiers, and users with tools for evaluating the safety, durability, and structural integrity of laboratory casework and complementary items.

INTEGRATION OF ARCHITECTURE AND ENGINEERED SYSTEMS

For the organization of the internal laboratory spaces, the architect and engineer must generate and coordinate various concepts to efficiently integrate architecture and engineered systems. This coordination effort has the objective of developing systems that complement each other while providing a safe and healthy environment at a reasonable cost. The main items that require integration between architects and engineers are building concepts, utility distribution, laboratory layout approaches, specific layout issues such as fume hood location, and budgeting. The locations of positively and negatively pressurized areas and levels of cleanliness need to be planned to minimize controls and ducting.

Building Concept

Building concept issues include basic choices about the laboratory building layout that will affect later design and system opportunities. These issues include choosing the air intake locations to avoid contamination, selecting individual or manifolded supply and exhaust air systems, and considering operation and maintenance (O&M) requirements.

Air Intake Location. The location of fresh air intakes needs to be coordinated between the architects and engineers in order to minimize the possibility of re-entrainment of contaminates from the laboratory exhaust discharge, loading docks, cooling tower discharge, vehicular traffic, and adjacent property pollution into the fresh air intakes. Selection of air intake locations should consider issues such as the prevailing wind direction and seasonal events such as trees losing their leaves or snow, which may clog the air intake. The choice of air intake locations will affect later planning decisions, such as the location of mechanical rooms, location and layout of utility corridors, and the exterior aesthetics of the laboratory building.

Adequate exit velocities, height, placement, and types of exhaust stacks and fans must also be considered along with the selection of outside air intake location. While a minimum of 30 feet from air discharge openings to intakes is recommended to reduce potential reentrainment problems, the maximum possible separation is good practice (DiBerardinis et al. 1993). Additionally, state and local codes may specify a minimum separation distance between exhaust discharge and supply air intakes. For additional details on air intake locations relative to exhaust systems, see Chapter 9, "Exhaust Stack Design."

Individual and Manifolded Air Systems. The supply air to the laboratories, adjoining office spaces, storage areas, and corridors can be provided by a central system that serves all these areas or via separate (distributed) systems dedicated exclusively for either laboratory or nonlaboratory use. Similarly, exhaust air from laboratory and other spaces can be expelled through either individual or manifolded exhaust systems, although separate exhaust systems for laboratory and nonlaboratory systems are almost always used. The choice of either individual or manifolded air systems will determine later planning and design choices, such as the space needed for utility corridors and shafts and their location.

There are advantages and disadvantages to using either individual or manifolded systems, which must be weighed carefully before a selection is made. Manifolded exhaust systems have a lower initial cost and maintenance requirements, due to the reduced number of fans and stacks that are needed, and have better atmospheric dispersion due to the momentum of the large air mass. They also are better for energy recovery applications and easier for installing backup fans. However, manifolded exhaust systems may have incompatible exhaust streams, such as perchloric acid fumes, or have additional control and balancing needs to ensure design airflows are maintained, which are not needed for individual exhaust systems.

O&M Issues. Operation and maintenance issues, such as required clearances and access locations for HVAC and laboratory equipment needs to be thoroughly planned. Laboratories may need to sustain design per-

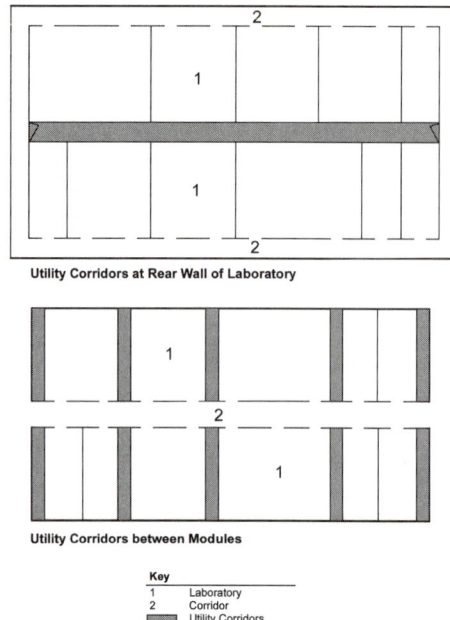

Figure 3-8 *Utility corridor configurations (diagrams modified with permission from DiBerardinis 1993).*

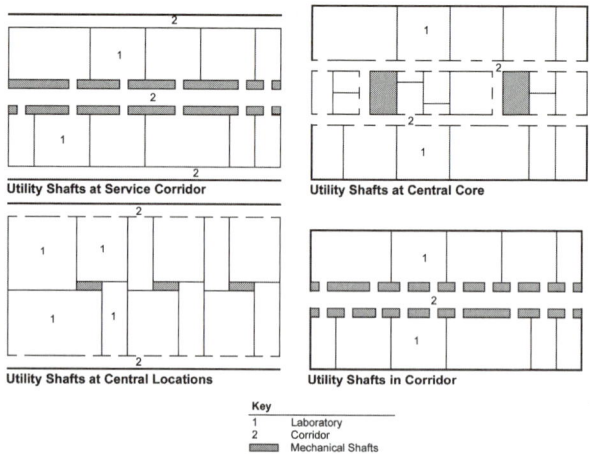

Figure 3-9 *Interior utility shaft configurations (diagrams modified with permission from DiBerardinis 1993).*

formance conditions for extended periods of time to protect the laboratory personnel and experiments and therefore need systems and equipment that can be sufficiently maintained. The continued reliability of the equipment and systems in a laboratory can be increased if they are selected and installed with consideration of accessibility and maintainability. Coordination between architects and engineers is needed to ensure that sufficient space is allowed for various HVAC and laboratory equipment and other building systems. Electrical, telecommunications, and fire protection should not interfere with the HVAC and laboratory equipment.

Utility Distribution

Various utilities, such as supply and exhaust ductwork, piping (HVAC, plumbing, gas), mechanical equipment technology, fire protection, and electrical conduits need to be distributed throughout the building to various laboratories, offices, and other spaces. During the planning phase, architects and engineers must coordinate the distribution of these services to fit within the available space yet remain maintainable. Methods of utility distribution for architects and engineers to choose from include utility corridors, multiple interior shafts, multiple exterior shafts, corridor ceiling distribution, and interstitial space.

Utility Corridors. A utility corridor is a separate hallway that is used to run piping, ductwork, conduit, and other systems to enable access to the systems between equipment rooms. Utility corridors are a horizontal layout method, with very few vertical shafts used for multi-story laboratory buildings. The use of utility corridors has the advantage of allowing O&M staffs to maintain laboratory equipment from an adjacent service corridor rather than routinely having to enter the laboratory. This also allows the utilities to be placed away from public access hallways so that they can be maintained without disrupting building occupants and when building occupant access to the critical systems that serve laboratories is limited. For additional information on utility distribution issues, see Chapter 6, "Primary Air Systems." Figure 8 contains a schematic of two possible utility corridor configurations.

Multiple Interior Shafts. Multiple interior shafts are a second method of utility distribution that use a primarily vertical layout with horizontal runs to individual equipment. Typically, multiple interior shafts only require a short horizontal run of ductwork, which is desirable for exhaust from fume hoods and other laboratory equipment. Depending on the building layout, interior shafts can be accessed either through a service corridor, a public corridor, or from within the laboratory. A possible downside of this layout is that multiple shafts can occupy a significant portion of the floor area in high-rise laboratories. Figure 3-9 illustrates several types of interior shaft configurations.

Multiple Exterior Shafts. Multiple exterior shafts are similar to interior shafts, except that they are located on the exterior wall of the building rather than in the interior of a building. As with interior shafts, exterior shafts typically only require short horizontal runs of ductwork. Exterior utility shafts are typically accessed from the laboratory or from the building exterior. Figure 3-10 contains a schematic of an exterior shaft configuration.

Corridor Ceiling Distribution. Corridor ceiling distribution uses the space above a drop ceiling (sometimes no drop ceiling) to contain ductwork and piping in

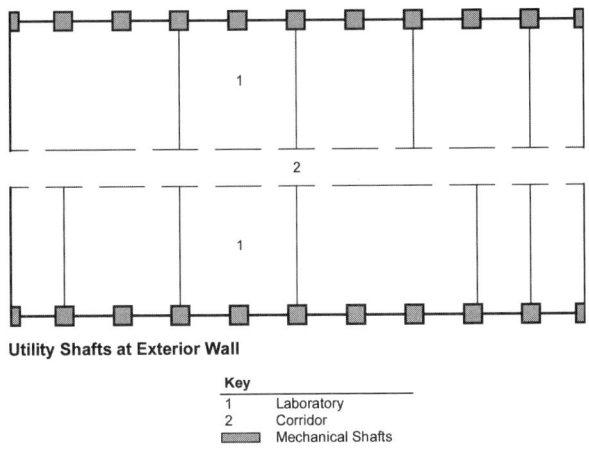

Figure 3-10 *Exterior utility shaft configuration (diagrams modified with permission from DiBerardinis 1993).*

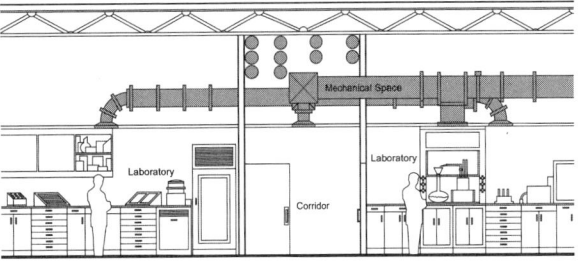

Figure 3-11 *Utility distribution in a corridor ceiling.*

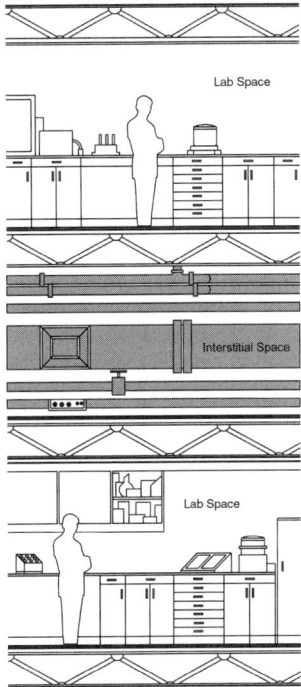

Figure 3-12 *Interstitial space utility distribution.*

a corridor just outside of a laboratory. This method of utility distribution requires a higher floor-to-floor height, as 3 to 4 feet (0.91 to 1.22 m) above the drop ceiling is required for the various utilities. While this type of system can be more difficult to maintain due to the need to move ceiling panels around to find equipment, it does provide increased usable floor area compared to multiple interior shafts or a separate corridor for utilities. Typically, ceiling distribution is confined to corridors and is not used above laboratory spaces, as the need to remove ceiling panels to maintain equipment could interfere with laboratory activities (e.g., cause dust to fall on sensitive laboratory equipment or experiments). Figure 3-11 is a schematic of utility distribution in a corridor ceiling.

Interstitial Space. Interstitial spaces are dedicated mechanical and electrical equipment floors located between each occupied floor. Ductwork, piping, and conduit are then either run up to the floor above and down to the floor below or only to the floor below. The benefit of placing systems in the interstitial space is ease of accessibility and ability to modify the layout of the various systems quickly without disturbing laboratory operations. However, this type of utility distribution is inherently expensive due to the need for additional floors in the laboratory building. Figure 3-12 is a schematic of utility distribution in an interstitial space.

Some laboratories and medical research facilities have been laid out with vertical utility spaces that are four to eight feet wide. This is not a conventional layout and requires careful planning and coordination by the owner and the design team. It does provide sidewall utilities versus ceiling and floor utility access. This is especially useful in laboratories that require frequent utility changes to meet use changes.

Laboratory Layout Approaches

Architects and engineers need to develop a plan for the internal organization of a laboratory and adjacent building spaces. The approaches to laboratory layout include modular design, workstations, and layout based on specific function.

Modular Design. A laboratory module is a basic work area, usually for one or two persons, which is used as a general building block in planning the complete floor plan. Each module can be contained in a separate room or combined into groups and used in larger rooms. The layout of modules should be carefully coordinated with the utility distribution layout. For laboratory buildings, which have different types of work that will be performed, multiple modules, one for each distinct type of laboratory work, may be chosen and used as appro-

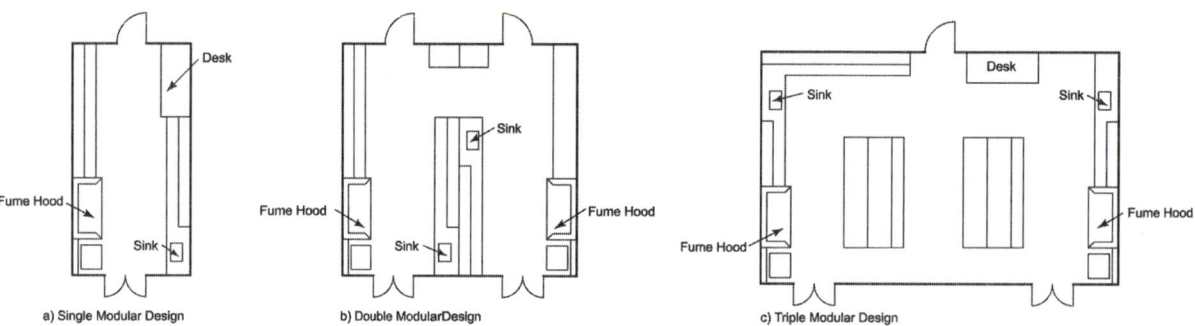

Figure 3-13 *Example laboratory modules (diagram modified with permission from DiBerardinis 1993).*

priate. Examples of a single, double, and triple module are shown in Figure 13.

Workstations and Specific Functions. The processes, materials, and waste products that are to be contained at a particular workstation needs to be considered with reference as to how they affect the function or pose a hazard to the functions of other workstations. As a result, their placement can be critical in order to avoid potential conflicts. Workstations should be located in low-traffic areas of the laboratory. Ventilated open face units should be at least 10 feet from the supply air if air is introduced in the ceiling and 20 feet from side wall diffusers. Recirculated biological safety cabinets and clean air devices should not be located less than 8 inches from a ceiling area. Island locations for workstations can be used provided visibility of the laboratory critical areas is maintained. An aisle clearance of a least 30 inches should be required in front of a workstation, and multiple fume hoods should be kept at least 12 inches apart, side wall to side wall.

The layout of internal spaces of a laboratory can be driven by the functions that are assigned to them. In the case of a renovation project, these spaces may already exist and appropriate modifications can be conducted in order to meet required goals in the most effective manner.

Specific Layout Issues

Once the general building concepts, method of utility distribution, and approach to laboratory layout have been planned, the specific layout issues for laboratory areas need to be determined. Specific layout issues for laboratories include the separation of laboratory and nonlaboratory areas, fire separations, access and egress, laboratory volumes, primary and secondary barriers, directional airflow, and workstation placement.

Separation of Laboratory and Nonlaboratory Areas. In order to address the layout of support offices, storage rooms, and mechanical rooms that are associated with laboratory buildings, the issue of physical separation must be determined. The following items need to be determined during planning:

- The separation distance needed between laboratory areas and nonlaboratory areas.
- The location of nonlaboratory areas, such as offices for laboratory workers, to determine if they should be placed within laboratories, across the hall, or in another part of the building.
- Whether or not the location of administrative offices should reflect the organization's hierarchical structure, if one is in place.
- The materials and construction methods required for the envelope that separates laboratory and nonlaboratory areas.

Fire Separations. Fire rated walls, floors, ceilings, and doors should be used according to the applicable codes and standards to separate laboratory spaces and contain a fire from spreading to the rest of the building. Planning between the architect and engineer is needed to determine fire containment zones and plan the HVAC system layout accordingly. *NFPA Standard 45,* Sections 3-1.3 through 3-1.5, stipulates that laboratory units should be separated from nonlaboratory areas or other laboratory areas (of equal or different hazard class) by construction equal to or greater than the fire resistance requirements specified in Table 3-19 and Table 3-20.

Access and Egress. Access and egress for laboratory facilities have several specific code and standard requirements, which are specified by the National Fire Protection Agency (NFPA 101 2000), U.S. Occupational Safety and Health Administration (OSHA 1990), and Building Officials and Code Administrators International, Inc. (BOCA 1991). These references provide regulations for the amount of exits, dimension of exits, door swing directions, and allowable door swings related to egress pathways. Additional information on codes and standards can be found earlier in this chapter.

Laboratory Volumes. Required ventilation rates for laboratories are commonly specified in air changes per hour. Designers have differing views on the importance of using air changes per hour as a measure as it places emphasis on the determination of ceiling heights. Often times, unjustified ceiling heights of 8 feet to 9 feet

Table 3-19: Construction and Fire Protection Requirements for Sprinklered Laboratory Units

		Sprinklered Laboratory Units	
		Any Construction Type	
Laboratory Unit Fire Hazard Class	Area of Laboratory Unit, ft² (m²)	Fire Separation[1] from Nonlaboratory Areas	Fire Separation[1] from Laboratory Units of Equal or Lower Hazard Classification
A	<1,000 (<93)	1 hour	NC;LC
	1,001-2,000 (93.1-186)	1 hour	NC;LC
	2,001-5,000 (186.1-465)	1 hour	NC;LC
	5,001-10,000 (465.1-930)	1 hour	NC;LC
	>10,001 (>930.1)	Not permitted	Not permitted
B	<20,000 (<1860)	NC;LC	NC;LC
	>20,000 (>1860)	Not permitted	Not permitted
C	<10,000 (<930)	NC;LC	NC;LC
	>10,000 (>930)	NC;LC	NC;LC
D	<10,000 (<930)	NC;LC; ½ C	NC;LC; ½ C; EC
	>10,000 (>930)	NC;LC; ½ C	NC;LC; ½ C; EC

Construction Type Designations:
NC = Noncombustible Construction
LC = Limited-Combustible Construction
½ C = ½-hr Combustible Construction
EC = Existing Combustible Construction

[1] Fire separation = from one level of rated construction to another
(Source: NFPA Standard 45-1996)

Table 3-20: Construction and Fire Protection Requirements for Non-Sprinklered Laboratory Units

		Non-Sprinklered Laboratory Units			
		Construction Types I and II		Construction Types III, IV, and V	
Laboratory Unit Fire Hazard Class	Area of Laboratory Unit, ft² (m²)	Fire Separation[1] from Nonlaboratory Areas	Fire Separation[1] from Laboratory Units of Equal or Lower Hazard Classification	Fire Separation[1] from Nonlaboratory Areas	Fire Separation[1] from Laboratory Units of Equal or Lower Hazard Classification
A	<1,000 (<93)	1 hour	1 hour	2 hours	1 hour
	1,001-2,000 (93.1-186)	1 hour	1 hour	Not permitted	Not permitted
	2,001-5,000 (186.1-465)	2 hours	1 hour	Not permitted	Not permitted
	5,001-10,000 (465.1-930)	Not permitted	Not permitted	Not permitted	Not permitted
	>10,001 (>930.1)	Not permitted	Not permitted	Not permitted	Not permitted
B	<20,000 (<1860)	1 hour	NC;LC	1 hour	1 hour
	>20,000 (>1860)	Not permitted	Not permitted	Not permitted	Not permitted
C	<10,000 (<930)	1 hour	NC;LC	1 hour	NC;LC
	>10,000 (>930)	1 hour	NC;LC	1 hour	1 hour
D	<10,000 (<930)	1 hour	NC;LC	1 hour	NC;LC
	>10,000 (>930)	1 hour	NC;LC	1 hour	1 hour

Construction Type Designations:
NC = Noncombustible Construction
LC = Limited-Combustible Construction
½ C = ½-hr Combustible Construction
EC = Existing Combustible Construction

[1] Fire separation = from one level of rated construction to another
(Source: NFPA Standard 45 - 1996)

are chosen to compensate. Therefore, during the planning phase, the volume of a laboratory needs to be discussed and properly coordinated with the selection of air-handling equipment to ensure that the required ventilation rates can be met.

Primary and Secondary Barriers. Laboratories typically use a primary barrier and a secondary barrier to control the spread of contaminants. The primary barrier is usually a piece of laboratory equipment, such as a fume hood or biological safety cabinet, which is intended to capture the hazardous materials used in the laboratory most of the time. A secondary barrier typically consists of a more passive measure, such as a negative pressure differential for room or a laboratory that is completely sealed off by an air lock. The secondary barrier then captures any hazardous materials such as fugitive emissions and spills and contaminants pulled out of the workstation by moving into and out of an open face station.

Directional Airflow. Maintaining pressure differentials between areas of different uses provides directional airflow. In doing so, the airflow direction is maintained from clean to dirty to ensure that contaminants from a laboratory are not spread throughout the building. Therefore, selection of pressure differentials is needed (i.e., creating a pressure map), which requires coordination between mechanical and architectural systems to ensure that the chosen pressure differential can be maintained. Also, the requirements for the materials necessary to maintain the proper pressure differentials (and thus directional airflows) need to be coordinated, such as any needs for tight sealing doors, anterooms, sealed plumbing penetrations, and electrical conduits. Pressure decay tests may be required to verify differential pressure.

Workstation Placement. As one moves within a laboratory space from the exterior wall toward the primary access door and public corridor, equipment should be placed according to its decreasing hazard potential. Therefore, fume hoods and biological safety cabinets for conducting hazardous experiments should be located away from the primary access doors and pathways. In addition, locating hoods and cabinets in this manner reduces occupant traffic past them, which can negatively affect the performance. Coordination of functionality and aesthetics is needed in selecting supply air diffusers, as laboratory areas need to have diffusers that can introduce large volumes of air at low speeds to prevent turbulence, which can disrupt airflow to open face workstations such as fume hoods, biological safety cabinets, and clean air stations.

Budgeting

When planning a laboratory, the following cost considerations should be made: the percent of construction budget available for HVAC equipment and systems, the

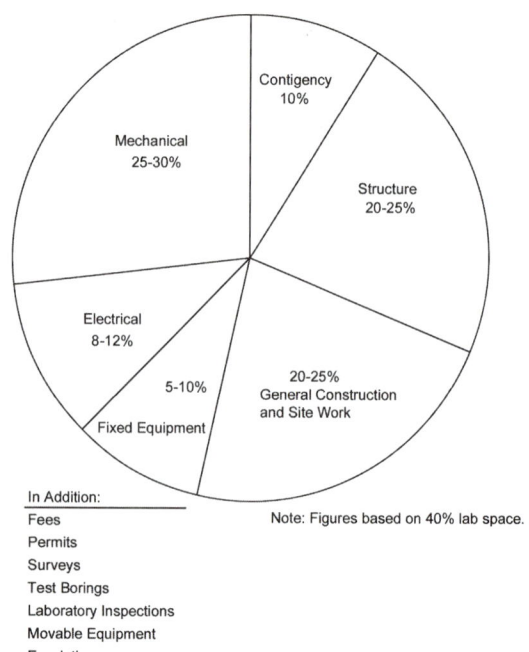

Figure 3-14 *Distribution of laboratory construction budget (diagram modified with permission from Ruys 1990).*

impact of system complexity on the budget, the impact of system redundancy on budget, and the differences between short-term (capital) cost and long-term (life-cycle) cost.

Percent of Construction Budget Available for HVAC. Due to the hazardous nature of the work performed in laboratories, extensive HVAC systems are needed to keep laboratory workers safe. Therefore, the mechanical costs of a laboratory can amount to a substantial portion of the total construction budget. Figure 3-14 shows the distribution of costs for a typical laboratory construction budget.

System Complexity Versus Budget. In many cases, features can be added to the mechanical or architectural systems of a laboratory to improve safety, reliability, or life-cycle costs. In general, however, more complex systems require additional capital cost. For example, variable air volume fume hoods with a complex control system can monitor the usage of each hood and reduce the operating costs for heating, cooling, and fans. However, this more complex system would require additional design work and additional equipment, which would add to the costs to construct the laboratory. During operation, it would require more regulation to maintain safety as well. Therefore, the design team should evaluate the benefits provided by more complex systems against the available budget and owner requirements.

System Redundancy Versus Budget. Laboratories may need to provide redundancy for various systems to ensure that workers and research are protected during unexpected failures, the most common of which is loss of power. The design team should evaluate the need for redundant systems in order to determine the probability of a failure, the hazards created, and the costs incurred when a failure occurs and compare this to the cost to provide a redundant system. Some laboratories will need to provide redundant systems regardless of cost to install, as serious injury or death may result if a system fails. For example, the exhaust system for a highly hazardous research laboratory may require the installation of two completely separate sets of exhaust fans and ductwork for fume hoods. In other laboratories, redundancy may simply provide additional peace of mind, convenience, or cost savings. Finally, in some cases, such as the selection of chillers and boilers, selecting two or three smaller units rather than one large unit can provide limited redundancy. Emergency power for limited areas should be provided to maintain critical systems to protect specimens that may be priceless.

Short-Term versus Long-Term Cost. Short-term, or capital, costs are those needed to initially plan, design, construct, and occupy a laboratory building. Long-term, or life-cycle, costs include the short-term costs (as they usually are financed over a long period) and the costs to operate and maintain the laboratory over its intended lifespan. When possible, alternative design possibilities should be evaluated based on life-cycle cost, as a small investment of money during the construction of a laboratory can save considerable costs over the life of a building. Coordination between architects and engineers is needed to accurately assess the impacts on the building as a whole and determine a life-cycle cost for various design options. For additional information on short-term and long-term costs, see Chapter 15, "HVAC System Economics."

DEVELOPMENT OF PLANNING DOCUMENTS

The Program of Requirements and the Owner's Design Intent are the two main planning documents used in the procurement of the laboratory facility. This section describes these programming documents and offers guidance for their development.

The Program of Requirements is not as detailed as the Owner's Design Intent and is used to record specific room types, their quantities and areas, number and type of personnel who will occupy the laboratory, the types of research, teaching, or industrial functions, and estimate of construction cost.

The Owner's Design Intent document provides a more detailed explanation of the information provided by the Program of Requirements by describing the laboratory's functional needs, intended levels and quality of environmental requirements and control, and description of basic mechanical, electrical, and plumbing systems. Producing a clear design intent is very critical to defining a benchmark to be used to judge the true success of the project of constructing a laboratory facility. This document is dynamic in nature and any changes to the design intent shall be documented, reviewed, and approved by the owner.

The following is a basic outline of the Owner's Design Intent:

- General project description
- Objectives
- Functional uses
- General quality of materials and construction
- Occupancy requirements
- Indoor environmental quality (IEQ) requirements
- Performance criteria
- Budget considerations and limitations

Chapter 14, "Laboratory Commissioning Process" discusses this significant document in the context of the commissioning process.

REFERENCES

American Conference of Governmental Industrial Hygienists (ACGIH). 1998. *Industrial Ventilation – A Manual of Recommended Practices*, 23d ed. Cincinnati, Ohio: ACGIH.

American Industrial Hygiene Association (AIHA). 1993. *Standard Z9.5-93, Laboratory Ventilation*. Fairfax, Va.: AIHA.

American Institute of Chemical Engineers (AIChE). 1985. *Guidelines for Hazard Evaluation Procedures*. New York: AIChE.

American National Standards Institute (ANSI). 1982. *Standard Z88.2-1980, American National Standard Practices for Respiratory Protection*. New York: ANSI.

American National Standards Institute (ANSI). 1993. *Standard Z9.5, Laboratory Ventilation. American National Standard Practices for Respiratory Protection*. New York: ANSI.

American Society of Heating, Refrigerating and Air-Conditioning Engineers, Inc. (ASHRAE). *1999. 1999 ASHRAE Handbook—HVAC Applications*. Atlanta: ASHRAE.

American Society of Heating, Refrigerating and Air-Conditioning Engineers, Inc. (ASHRAE). 1997. *1997 ASHRAE Handbook—Fundamentals*. Atlanta: ASHRAE.

American Society of Heating, Refrigerating and Air-Conditioning Engineers, Inc. (ASHRAE). 1995. *ANSI/ASHRAE Standard 110-1995, Method of Testing Performance of Laboratory Fume Hoods*. Atlanta: ASHRAE.

American Society of Heating, Refrigerating and Air-Conditioning Engineers, Inc. (ASHRAE). 1992. *ANSI/ASHRAE Standard 55-1992, Thermal Environmental Conditions for Human Occupancy.* Atlanta: ASHRAE.

American Society of Heating, Refrigerating and Air-Conditioning Engineers, Inc. (ASHRAE). 1989. *ANSI/ASHRAE Standard 62-1989, Ventilation for Acceptable Indoor Air Quality.* Atlanta: ASHRAE.

American Society of Heating, Refrigerating and Air-Conditioning Engineers, Inc. (ASHRAE). 1980. *ASHRAE Standard 90A-1980, Energy Conservation in New Building Design.* Atlanta: ASHRAE.

Berglund, B., and T. Lindvall. 1986. Sensory Reactions to Sick Building. *Environment International*, Vol. 12, pp. 147-159.

Bretherick, L. 1981. *Hazards in the Chemical Laboratory*, 3d ed. London: Royal Society of Chemistry, Burlington House.

Building Officials and Code Administrators International, Inc. (BOCA). 1987a. *The BOCA National Building Code*, 10th ed. Model building regulations for the protection of public health, safety and welfare. Country Club Hills, Ill.: BOCA Publications.

Building Officials and Code Administrators International, Inc. (BOCA). 1987b. *The BOCA National Mechanical Code*, 6th ed. Country Club Hills, Ill.: BOCA Publications.

Centers for Disease Control and Prevention (CDC)-National Institutes of Health (NIH). 1999. *Biosafety in Microbiological and Biomedical Laboratories*, 4th ed. Bethesda, Md.: U.S. Department of Health and Human Services.

Department of Justice. 1994. 28 CFR Part 36. ADA Standards for Accessible Design, July 1 Revision.

DiBerardinis, L.J., J.S. Baum, M.W. First, G.T. Gatwood, E. Groden, and A.K. Seth. 1993. *Guidelines for Laboratory Design: Health and Safety Considerations*, 2d ed. New York: John Wiley & Sons, Inc.

Hosni, M.H., B.W. Jones, and H. Xu. 1999. Measurement of Heat Gain and Radiant/Convective Split from Equipment in Buildings. Final Report, ASHRAE 1055-RP, American Society of Heating, Refrigerating and Air-Conditioning Engineers, Inc. Atlanta, Ga.

Huchingson, R.D. 1981. *New Horizons for Human Factors in Design.* New York: McGraw-Hill Book Company.

Illumination Engineering Society of North America (IESNA). 1989. *IES Lighting Ready Reference.* New York: IESNA.

Illumination Engineering Society of North America (IESNA). 1993. *IESNA Lighting Handbook*, 9th ed. New York: IESNA.

Institute of Laboratory Animal Resources. 1985. *Guide for the Care and Use of Laboratory Animals.* Washington, D.C.: National Academy of Sciences.

Lawley, H.G. 1974. Operability Studies and Hazard Analysis. *Loss Prevention,* Vol. 8, pp. 105-116.

Mayer, L. 1995. *Design and Planning of Research and Clinical Laboratory Facilities.* New York: John Wiley and Sons, Inc.

National Fire Protection Association (NFPA). 1998. *NFPA 801, Facilities Handling Radioactive Materials.* Quincy, Mass.: NFPA Publications.

National Fire Protection Association (NFPA). 2000. *NFPA 101, Life Safety Code*, Chapter 5, Section 13-2, Section 14-2. Quincy, Mass.: NFPA Publications.

National Fire Protection Association (NFPA). 2000. *NFPA 30, Flammable and Combustible Liquids Code.* Quincy, Mass.: NFPA Publications.

National Fire Protection Association (NFPA). 2000. *NFPA 45, Fire Protection for Laboratories Using Chemicals.* Quincy, Mass.: NFPA Publications.

National Research Council (NRC). 1995. *Prudent Practices in the Laboratory: Handling and Disposal of Chemicals.* Washington, D.C.: National Academy Press.

National Research Council (NRC). 1996. *Guide for the Care and Use of Laboratory Animals.* Washington, D.C.: Institute of Laboratory Animal Resources, Commission on Life Science.

National Sanitation Foundation International (NSF). 1992. *Class II (Laminar Flow) Biohazard Cabinetry.* Ann Arbor, Mich.: NSF Publications.

Nielson, D.S., O. Platz and B. Runge. 1975. A Cause-Consequence Chart of a Redundant Protection System. *IEEE Transactions on Reliability*, Vol. R-24, pp. 8-13.

Occupational Safety and Health Administration (OSHA). 1990a. *OSHA Regulated Hazardous Substances–Health, Toxicity, Economic and Technological Data*, Vols. 1 and 2. New Jersey: Noyes Data Corporation.

Occupational Safety and Health Administration (OSHA). 1990b. *OSHA 29 CFR Part 1910.1450, Occupational Exposure to Hazardous Chemicals in Laboratories.* Washington, D.C.: U.S. Government Printing Office.

Occupational Safety and Health Administration (OSHA). 1990c. *OSHA 29 CFR Part 1910.1330, Occupational Exposure to Bloodborne Pathogens.* Washington, D.C.: U.S. Government Printing Office.

Occupational Safety and Health Administration (OSHA). 1991. *OSHA 29 CFR Part 1910.35, Health and Safety Standards*, Chapter 12, Subpart

E. Washington, D.C.: U.S. Government Printing Office.

Occupational Safety and Health Administration (OSHA). 1991b. *OSHA 29 CFR Part 1990—Identification, Classification, and Regulation of Potential Occupational Carcinogens*. Washington, D.C.: U.S. Government Printing Office.

Ruys, T. 1990. *Handbook of Facilities Planning, Volume 1, Laboratory Facilities*. New York: Van Nostrand Reinhold.

Scientific Equipment and Furniture Association (SEFA). 1996. *SEFA 1.2, Laboratory Fume Hoods, Recommended Practices*. Maclean, Va.: SEFA.

Scientific Equipment and Furniture Association (SEFA). 1997. *SEFA 2.3, Installation of Scientific Furniture and Equipment, Recommended Practices*. Maclean, Va.: SEFA.

Scientific Equipment and Furniture Association (SEFA). 1999. *SEFA 8, Laboratory Furniture, Recommended Practices*. Maclean, Va.: SEFA.

Simmons, C.G., and R. Dvoodpour. 1994. Design Considerations for Laboratory Facilities Using Molecular Biology Techniques. *ASHRAE Transactions*, Vol. 100, Part 1, pp. 1266-1274.

U.S. Department of Justice (DOJ). 1994. Americans with Disabilities Act (ADA) Regulation for Title III, Non-discrimination on the basis of disability by public accommodations and in commercial facilities. Washington, D.C.; DOJ.

Varley, J.O. 1998. Applying Process Hazard Analysis to Laboratory HVAC Design. *ASHRAE Journal*, Vol. 40, No. 2, pP. 54-57.

Vesley, W.E., et. al. 1981. *Fault Tree Handbook*. Report (NUREG-0492), National Technical Information Service, Springfield, Va.

Wisconsin Department of Commerce (WI-DOC). 1998. *Commercial Building Energy Conservation/HVAC Code Handbook*. Madison, Wisc.: WI-DOC.

BIBLIOGRAPHY

Lee, F.P. 1980. Loss Prevention in the Process Industries, Volume I. London: Butterworths & Co. Ltd.

Rankin, J.E., and G.O. Tolley. 1978. *Safety Manual No. 8, Fault Tree Analysis*. Washington, D.C.: National Mine and Safety Academy, National Fire Protection Association.

Siemens. 1999. *Laboratory Ventilation Standards and Codes*. Buffalo Grove, Ill.:.Siemens Building Technologies, Inc., Landis Division.

Chapter 4
Design Process

OVERVIEW

During the planning phase of the project, the general layout of the building and system are determined along with the specific exhaust airflow required by codes and standards to maintain a safe and healthy environment. In this chapter the planning information, such as ventilation requirements, indoor air quality (IAQ), environmental conditions, pressure relationships, occupancy and appliance loads, and special conditions for various laboratory functions and types, is transformed from criteria into a system design. It is intended that the general design process be detailed in this chapter with references to later chapters on specific topics and guidance. The design process for laboratory systems includes the following steps:

- Step 1 – Determine exhaust/supply requirements
- Step 2 – Load calculation
- Step 3 – Pressure mapping
- Step 4 – Evaluate system options (Chapters 5, 6, and 7)
- Step 5 – Layout of ducts and rooms (Chapter 6)
- Step 6 – Size primary air systems (Chapter 6)
- Step 7 – Evaluate air treatment (Chapter 8)
- Step 8 – Design exhaust stack (Chapter 9)
- Step 9 – Evaluate energy recovery options (Chapter 10)
- Step 10 – Develop control strategies (Chapter 11)
- Step 11 – Determine TAB and certification requirements (Chapter 12)
- Step 12 – Document O&M requirements (Chapter 13)
- Step 13 – Commissioning integration (Chapter 14)
- Step 14 – Economic evaluation (Chapter 15)

While presented as a linear process, the design of a laboratory system is very iterative due to the interrelationships between the different systems and their respective abilities to meet the owner's requirements. Therefore, it is important to evaluate how the assumptions being made impact the owner's design intent at each step in the design process.

Also included in this chapter is a review of special space design considerations for the different types of laboratories.

DESIGN PROCESS

Step 1 — Determine Exhaust/Supply Requirements

During the risk analysis for the laboratory, the chemicals and toxins to be used in the laboratory are determined along with the required exhaust airflow pressurization (positive or negative). The first step in the design process is to take this information and summarize it in a table form. A sample format that can be used is shown in Table 4-1.

The recirculated air column in Table 4-1 is included for nonlaboratory spaces and for those laboratories (e.g., cleanrooms) where recirculation of the air is acceptable and allowable by the governing codes and standards. For most laboratories, 100% ventilation air will be used, with the recirculated air being zero.

Step 2 — Load Calculation

Using the assumptions and guidelines on occupancy rates, loads, and construction materials developed during the planning phase, the second step of the design process is to calculate the design cooling and heating loads of the space. The actual load calculation can be divided into three steps:

- Document assumptions
- Calculate loads for comfort
- Calculate loads for safety

Document Assumptions. A critical step in preparing for calculating loads is to gather and document the key system criteria and assumptions made in transforming the owner's requirements into a physical design. This documentation not only provides a checklist for the designer to ensure key items are not missed but also provides the O&M staff with key information on the limitations of these systems and their intended operation. It is recommended that the assumptions be documented according to a format similar to that shown in Table 4-2, where the assumptions include not only the occupancy rates, equipment loading, and diversities but also include references to codes and standards that were used as the basis for the assumptions. Table 4-2 is referred to as the "basis of design" as it clearly documents all of the assumptions a designer made during the design of a system. This document is critical to the long-term success of a laboratory as it provides the operator information on the limits of the system (i.e., peak outdoor air temperature), which is invaluable during system changes and renovations. Since most laboratories have multiple types of spaces, Table 4-2 can easily be modified to contain a column for each use type for information that changes from space to space.

In addition to the example shown in Table 4-2, the designer should clearly document the loads in each space. This includes:

- Envelope
- Occupants
- Lighting
- Computers
- Equipment
- Process loads

Calculate Loads for Comfort. As with estimating the load for a typical building, it is no different for estimating the envelope, occupant, lighting, computers, and equipment load for a laboratory. If the load is in the space or is transferred across the envelope, then the load is included in the calculation. However, if the load does not directly impact the space (i.e., contained within an exhaust hood), then it is not included as a load on the central system. Figure 4-1 details the various loads within a laboratory space.

The grayed items in Figure 4-1 are considered internal loads that must be handled by a central or local cooling/heating system. The other items' loads are removed

Table 4-1: Exhaust/Supply Requirements

Room Data				Exhaust, cfm (L/s)			Pressure Relationship		Supply, cfm (L/s)		Recirculated, cfm (L/s)
No.	Use	Air Exchange Rates, ACH	Room Volume, $ft^3 (m^3)$	Required	Hood	General	+/-	%	Source*	Required	Required

*Source is either clean (e.g., hallway or anteroom) or dirty (e.g., adjacent laboratory space).

Table 4-2: Basis of Design Documentation

Project Information		
Designer/Engineer:		
Company:		
Project:		
Date:		
Unit system: English SI		
Design Criteria	**Value**	**Reference/Comments**
Latitude/longitude		
Elevation		
Clearness value		
Summer outdoor air design dry bulb/wet bulb		
Winter outdoor air design dry bulb		
Ground reflectance		
Cooling load methodology		
Cooling setpoint, dry bulb/relative humidity		
Heating setpoint, dry bulb/relative humidity		
Cooling/heating setback temperatures		
Roof U-factor		
Roof construction type		
Wall U-factor		
Wall construction type		
Ceiling U-factor		
Overall roof U-factor		
Glass U-factor (summer/winter)		
Glass shading coefficient (interior/exterior)		
Overall wall U-factor		
Overall building U-factor		
People sensible/latent heat generation		
Lighting density		
Outdoor air ventilation rate		
Infiltration rate		
Cooling air change rate		
Reheat minimum value		
Supply air temperature		
Duct heat gain		
Duct leakage		
Maximum duct noise level/ceiling effect		
Air distribution system diversity		
Number of occupants at peak load		
Fan heat gain		
Equipment loads		
Expected utility costs		
Hazard types and level		

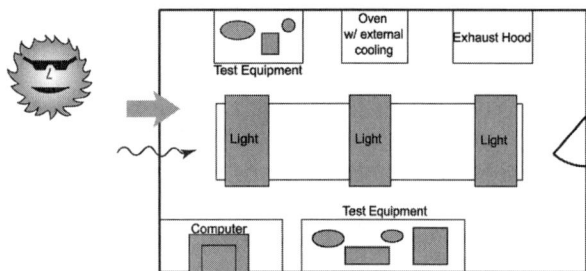

Figure 4-1 *Laboratory equipment loads.*

from the space and thus do not impact the conditioning system for the space. Guidance determining the typical load was provided in the planning chapter. The one unique load in laboratory spaces is a process load caused by laboratory equipment and processes.

Within the laboratory there is a mix of equipment available for use by the occupants depending upon the specific experiment being conducted. Since it is rare, if not impossible, to use all of the equipment at one time, care must be taken in estimating the loads and diversity of specified equipment for the design of the systems.

One of the first tasks in estimation of equipment loads is a thorough understanding of the energy consumption of each type of equipment. Typically, this will require contacting manufacturers for the rated energy consumption for each piece of equipment or, if necessary, researching and testing of the actual equipment. With these data, standard load calculations as described in *ASHRAE Applications*, can be used (AHSRAE 1993, Ch. 26). Proper estimation of loads is crucial in creating acceptable conditions for personnel and experiments.

In calculating the specific load for a piece of equipment, it is critical to designate where the load is being rejected to. The rejection media can be the surrounding air, domestic water, process water, or the outdoor air. Depending upon the source of rejection, the cooling load may or may not affect the central system's size requirements.

In terms of the diversity factor for the laboratory and equipment, this can be developed by consulting current and future personnel to determine usage patterns. However, caution in the use of this factor should be stressed, particularly in its use for hoods. Although the use of this factor can greatly reduce unnecessary energy consumption and costs, careful design of the interaction of the type of controls this system uses and the main system controls must be accomplished. Also, diversity factors that lower system exhaust volumes below 75% of the sum of peak zone exhaust (without diversification) should be extensively checked and verified.

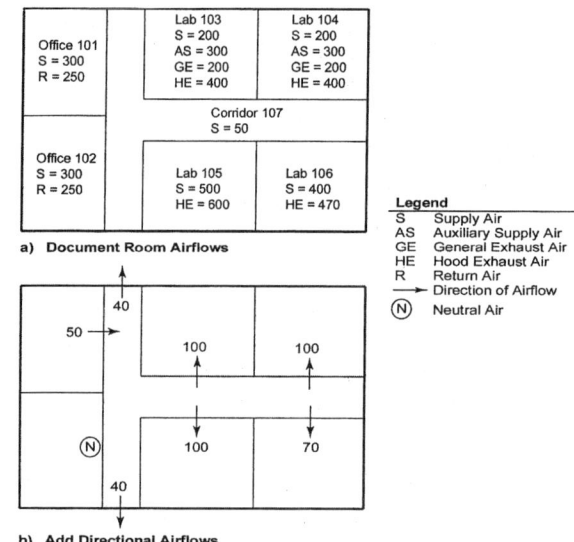

Figure 4-2 *Pressure mapping process.*

Calculate Loads for Safety. The comfort loads represent the minimum supply air required to maintain the space at acceptable conditions. These minimum airflows must be compared to the exhaust required for safety. Therefore, the exhaust airflow requirements for each space are compared to the supply airflow required for comfort to determine the increase in supply airflow required to maintain the required pressure relationship. Table 4-1 is ideal for this purpose.

Step 3 – Pressure Mapping

With the supply, return, and exhaust airflows determined for each space, the next step is the development of a pressure map for the entire facility. A pressure map graphically depicts the airflows within and between each room. The intent is to ensure that the pressure relationships documented during the planning phase are achieved.

For each room/space, the supply (primary and auxiliary), return (if any), and exhaust (general and hood) airflows are listed. Once all room airflows are detailed, the directional airflow between the rooms is shown with arrows and values. Figure 4-2 details these two steps.

While Figure 4-2 indicates air transfer between rooms occurring at discrete points, in reality, the air transfer will occur at the lowest pressure loss. This can be at an open door, exterior wall/window, or interior partition. Therefore, close attention to detail on wall, window, and door construction must be taken to minimize unwanted air movement. As the toxicity of the substances within a space increases, the greater attention to detail is required.

Step 4 – Evaluate System Options

With the design load and airflow values calculated for each space, the next step in the design process is to evaluate the different system options to meet these loads. Chapters 5, 6, and 7 cover the selection, design, and layout of systems in detail. In general, the decisions to be made during this phase of the design process include:

- Type of exhaust system
  - General/hoods
  - Constant/variable volume
- Type of supply system
  - Constant volume
  - Variable volume
  - Auxiliary
- Handling special process loads
- Meeting space heating needs
- Evaluate division of systems/areas
- Central vs. local air systems

Chapter 5, "Exhaust Hoods" details the options available for exhaust hoods and provides guidance on their selection and application. Chapter 6, "Primary Air Systems" encompasses the selection of the central exhaust, supply, and auxiliary air systems, along with space heating system. Finally, Chapter 7, "Process Cooling" provides details on process system layout and application. Division of systems/areas is similar to a normal HVAC design process where the division is done based on functionality of the area, location of areas for mechanical equipment, room for ductwork, etc.

Central Air System. By sizing the central air system to handle the worst case scenario for cooling, it is possible to eliminate the need for any local systems. Variable speed fans in conjunction with variable air volume boxes are required to handle the large variability in space load. The control of this system is complicated if room pressurization is to be maintained. This can result in higher capital costs for the central system than for the local air systems. However, big central systems may not be possible for some laboratory types that require the air system to be completely separated from other air systems.

Local Air Systems. Local air systems have individual control of each unit, and the control systems are less complicated than those with one central air system and can in many cases be designed to have a higher performance than comparable central air systems. These systems typically take less building space for piping and ducting than a central air system. The disadvantages of these systems are that they require more extensive maintenance due to the many components, can cause more noise due to the relatively closer location of the fans, compressors, pumps, and motors to the conditioned space, and will require many penetrations of the building envelope.

Step 5 – Layout of Ducts and Rooms

This step in the design process, while often taken for granted, is one of the most critical. If done improperly, pollutants will not be properly contained, occupants' health will be at risk, thermal loads will be unmanageable, and the system can be very costly to construct and operate.

The recommended procedure to follow for the layout of the rooms and ductwork is:

- Locate exhaust points and flows
- Determine placement of supply diffusers
  - From clean to dirty
  - Avoid disturbing hood face velocity profile
- Identify manifolding of exhaust systems
- Identify supply/exhaust main locations
- Layout supply/exhaust duct run outs

Chapter 6, "Primary Air Systems" provides details on accomplishing these tasks.

Step 6 – Size Primary Air Systems

Using the space airflow requirements, the central systems can be selected and sized. This includes the supply and exhaust air systems and cooling and heating plants. Due to the critical nature of the exhaust systems, these typically take precedence when determining the type and location of central systems. This is followed by the design of the supply system and finally the auxiliary air systems. The complexity of interrelationships between the different systems and rooms makes this a critical step in the design process. Chapter 6, "Primary Air Systems" provides details for this step.

Step 7 – Evaluate Air Treatment

An important aspect of the air systems is their impact on the indoor and outdoor air quality. Therefore, the treatment of the supply and exhaust airstreams must be determined to ensure that the indoor air quality is adequate for the comfort and safety of the occupants and the experiments and that the exhaust air is not a danger or nuisance to nearby people and buildings. Chapter 8, "Air Treatment" provides details on the technology available and requirements for air treatment. In general, the following should be accomplished:

- Document acceptable ambient pollutant levels
- Identify treatment options
- Select and size treatment system

It is important to recognize, depending upon the quality of the ambient outdoor air, that the supply air in some instances will require extensive treatment prior to use.

Step 8 – Design Exhaust Stack

There is a strong correlation between the level of pollutants in the exhaust air and the design of the exhaust stack. As the level of pollutants increases, the exhaust stack must discharge the air further from the building to ensure the pollutants are properly dispersed before they come into contact with any people. Therefore, exhaust stacks are designed to reduce the concentration of pollutants below recommended or allowable levels. This can be a challenging process due to aesthetics considerations, climatic conditions (wind), and location of stacks away from air intakes. However, measures taken in Step 7 can significantly simplify the stack design. The design of exhaust stacks is detailed in Chapter 9, "Exhaust Stack Design."

Step 9 – Evaluate Energy Recovery Options

Laboratory facilities are high consumers of energy due to the large volume of outdoor air that has been conditioned only to be exhausted. To minimize the energy required to condition the supply air, energy can be recovered from the exhaust air and transferred to the supply air. The savings from employing an energy recovery system can be as high as 80% for some circumstances. Chapter 10, "Energy Recovery" details the options and their application for laboratory energy recovery systems.

Step 10 – Develop Control Strategies

The main systems are sized and selected for peak design condition. However, this condition only occurs once or twice a year. During the other periods, the system must be actively controlled to efficiently maintain space comfort and safety conditions. Therefore, control strategies must be developed for non-peak conditions and must address the following:

- Space temperature control
- Space pressurization
- Building pressurization
- Occupied/unoccupied periods
- Emergency situations

Chapter 11, "Controls" provides detailed guidance on control strategies for laboratory systems.

Step 11 – Determine TAB and Certification Requirements

To ensure the systems can be properly started and operated, it is important to identify and ensure during the design phase that sufficient locations throughout the system can be accessed for the purpose of balancing the systems. This includes measurement ports on both sides of the valves and coils, straight runs of ductwork and piping, accessibility to dampers, and proper documentation on what is expected of the TAB firm.

Therefore, the design developed by the architect and engineer should, as a minimum, include:

- Six duct diameters of straight duct for measuring supply airflow and test instrument ports
- Six duct diameters of straight duct for measuring exhaust airflow for exhaust fans and ventilated workstations
- Measurement ports on both sides of valves, AHU coils, energy recovery coils, and terminal reheat coils
- Fan inlet and outlet ports for static pressure measurements

As a minimum the TAB and certification should:

- Balance all systems to design conditions
- Verify proper airflow direction/pressurization
- Verify system operation under part-load conditions (75%, 50%, 25%, and minimum)
- Perform face velocity tests
- Calibrate airflow and direction monitors for all fume hoods
- Certify all clean air and biological safety cabinet workstations

Details on determining the TAB requirements for laboratories are contained in Chapter 12, "Airflow Patterns and Air Balance."

Step 12 – Document O&M Requirements

During the development of the construction documents (drawings and specification) it is critical to identify those items that are necessary for the proper lifetime operation of the laboratory systems. These include:

- Training requirements, including a video
- System documentation, including CAD drawings and all control software
- Involvement of O&M personnel throughout the construction

Chapter 13, "Operation and Maintenance" provides guidelines for the development of these O&M requirements, with specific emphasis on the O&M manual.

Step 13 – Commissioning Integration

Commissioning is a quality process an owner elects to use as the process for planning, designing, constructing, and operating the facility. The key characteristics of the commissioning process, as with any quality process, are:

- Work is accomplished correctly the first time
- The individual worker determines the level of quality
- What constitutes a "successful" project is clearly defined

Several commissioning tools/tasks can be integrated into an owner's current process. Specifically, the following should be completed:

- Develop owner's design intent
- Develop commissioning plan
- Accomplish design reviews
- Develop basis of design
- Integrate quality requirements into specifications
- Verify installation and calibration
- Accomplish functional performance test on installation
- Ensure warranties maintained
- Have a "Lessons Learned" meeting

Chapter 14, "Laboratory Commissioning Process" provides a detailed overview of the commissioning process relative to laboratory facilities.

Step 14 – Economic Evaluation

The final step in the design process is an economic evaluation of the system. This evaluation should be based on a life-cycle cost for a typical 20- or 30-year period. Details on accomplishing a life-cycle cost analysis for a laboratory are contained in Chapter 15, "HVAC System Economics."

SPECIAL SPACE CONSIDERATIONS

During the design of laboratories, each type of laboratory has unique requirements. These requirements are typically due to pollutants generated in the space by the experiments and their potential adverse effects on the occupants. The types of laboratories covered in this section include:

- General chemistry laboratory
- Radiation laboratory
- Physics laboratory
- Teaching laboratory
- Research laboratory
- Hospital and clinical laboratory
- Biological containment laboratory
- Animal laboratory
- Isolation/cleanrooms
- Materials testing laboratory
- Electronics/instrumentation laboratory
- Support spaces

It is important that the users, safety personnel, and industrial hygienists who are helping select the laboratory types be consulted throughout the design phase.

In each section, specific information that influences the HVAC design is provided. In addition, a sample layout for each laboratory is shown to aid in the duct layout and diffuser and exhaust locations.

General Chemistry Laboratory

General chemistry laboratories are considered to be the basic type of laboratory. A variety of chemicals and heat-generating equipment is used. This laboratory is preferably located in a building with similar laboratories and rooms with activities related to these laboratories. A sample layout for this type of laboratory is shown in Figure 4-3

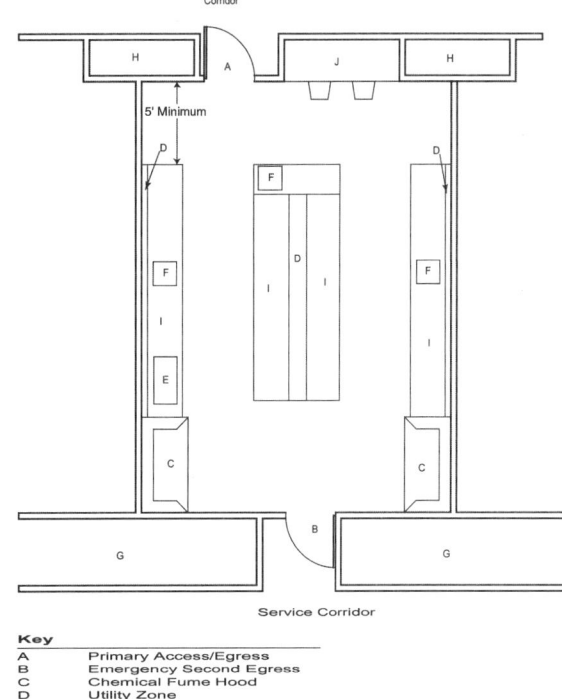

Figure 4-3 *Sample layout of general chemistry laboratory. (Diagram modified with permission, DiBerardinis et al. 1993.)*

Radiation Laboratory

Radiation laboratories are designed to provide a safe and efficient workplace for activities associated with materials that emit ionizing radiation that can be harmful. This radiation can cause harm either directly or indirectly by ingestion or inhalation. The materials handled have the same general chemical properties but in addition emit radiation. These types of laboratories must for that reason be able to handle the requirements for a general chemistry laboratory in addition to the requirements added by the radioactivity of the chemicals. The safety measures required will depend on the chemical and radioactive properties of the substances used. DiBerardinis et al. (1993) suggest as a design guideline that laboratories that handle materials with a total activity greater than 1 µCi should be considered radiation laboratories. Consultation with an industrial hygienist and health physicist is recommended if the radiation level is expected to be above 25% of the recommended maximum exposure.

Depending on the substances used, a shower may be required as a part of the laboratory or should be available in the building (according to OSHA and NRC). A shower may not be needed if the substances handled only require gloves and a laboratory coat, but handwashing facilities should be available in both the laboratory and the shower rooms. The work surfaces should be smooth, easily cleaned, with impervious materials (such as steel) (DiBerardinis et al. 1993). Floor coverings should be constructed of monolithic materials, while cracks in floors, walls, and ceilings may need to be sealed. Lighting should be flush with the ceiling and sealed to be vapor- and waterproof. Radiation-producing equipment should not point toward the entrance to the laboratory. Access to radio-chemistry laboratories should be restricted and labeling may be required. It is recommended that entrance doors have windows that do not compromise the fire rating required of the door.

Variable air volume is not recommended for radiation laboratories due to the need for dilution of the radioactive emissions. Isotope fume hoods with cleanable surfaces should be used. Glove boxes may be necessary if the work is with powdery or volatile chemicals or other chemicals that can produce high amounts of airborne radioactive materials. The glove box must have a negative pressure difference to the laboratory of 0.25 in. w.g. while the laboratory must have 0.05 in. w.g. compared to the atmospheric pressure. All storage areas should be vented to prevent radioactive contaminant buildup.

The design of the exhaust system from radiation laboratories should consider future provisions for HEPA filtration capabilities of at least 99.97% efficiency for particle sizes of 0.3 μm. If the exhaust contains gaseous radioactive materials, an activated carbon filter may be required. The air-cleaning elements should be replaceable without causing any hazards to the maintenance staff (e.g., bag-in, bag-out procedures). NRC requires continuous monitoring of the exhaust for some radioisotopes if the expected concentration levels are above regulated levels. If HEPA filters must be installed, a written authorization of intent by the owner should be provided and the unit(s) certified annually.

Waste handling of the radioactive waste generated from these laboratories must be considered. Shielding of rooms can be required with certain radioactive materials (cobalt and cesium). These shields (lead and heavy concrete) add a significant amount of weight to the structure. It may be desirable to designate iodinization activities to one laboratory per floor (or an appropriate number of floors) to better control this activity.

The layout of radiation laboratories can resemble the chemistry laboratory, but many layouts are designed for a specific function. Walters (1980) describes several layouts for radiation laboratories. It is advantageous to design radiation laboratories as small as practical due to the possible need for decontamination.

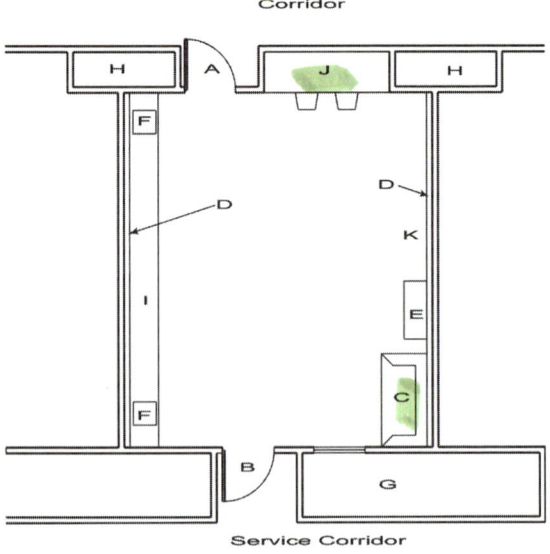

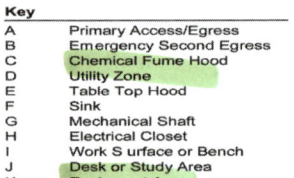

Figure 4-4 *Sample layout of a physics laboratory. (Diagram modified with permission, DeBerardinis et al. 1993.)*

Physics Laboratory

Physics laboratories are categorized by activities that often require electricity. These laboratories should for that reason have multiple electrical outlets easily available throughout the laboratory. In case chemicals are used, sinks and eyewash stations should be placed in hallways or other areas away from the electrical outlets but close enough for easy access in case of an emergency. In addition, physics laboratories with radioactive materials to produce iodizing radiation should meet the requirements of radiation laboratories.

The fire suppression system is a very costly system in a physics laboratory and should be considered from the initial phases of the design. A water-based fire suppression system can be used if the electrical system is disconnected during a fire. Otherwise, CO_2 based (ventilation requirements must follow *NFPA 12* (NFPA 2000) or total flooding dry chemical systems must be used. An automatic sprinkler system cannot be used if water-sensitive equipment or high voltage or high current electronics are present. A handheld fire extinguisher should be located in the laboratory to help the personnel exit the laboratory.

A grounding system must be in place for all the equipment, and the electrical supply system must be adequate to provide electric outlets with only a small potential difference between the closest electrical outlet, and the one farthest away. Often a great quantity of cooling

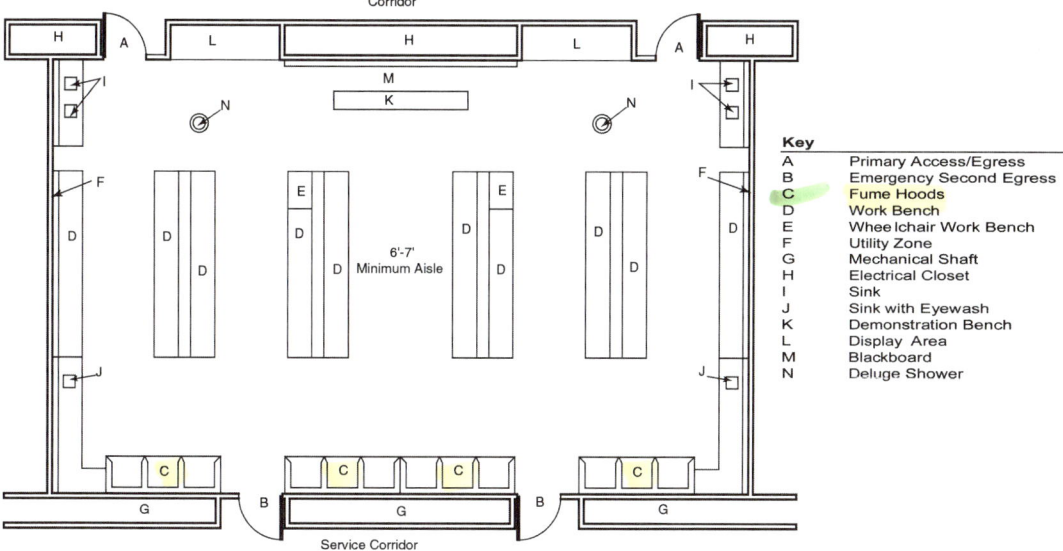

Figure 4-5 *Sample layout of a teaching laboratory. (Diagram modified with permission, DiBerardinis et al. 1993.)*

water is required, and a closed loop cooling system should be evaluated for this purpose. The laboratory must be designed to reduce the risk from high-intensity energy forms such as strong laser beams, high electrical energy demands, and cryogenics. A sample layout of a physics laboratory is shown in Figure 4-4.

Teaching Laboratory

Teaching laboratories should be designed for a maximum of 30 students in high schools, while undergraduate laboratories may be designed for more than 30 students. Sound control to an RC of 40 decibels should be a design goal since communication is critical. These laboratories should also be designed to encourage safe laboratory practices, as the student will bring experiences from these laboratories into their professional work. Microscale chemistry teaching laboratories conduct the same experiments as general chemistry laboratories, but the amount of chemicals used are 10 to 100 times less than the amounts normally used in a typical general chemistry laboratory (DiBerardinis et al. 1993). Teaching laboratories can be placed into two categories, dry laboratories (typically physics, biology, and engineering activities) and wet laboratories (activities that use liquid, solid, and gaseous chemicals including heating devices).

The space requirements for each student are shown by experience to be an absolute minimum of 32 ft^2 (3.0 m^2). The instructors and students must be able to easily move around and the benches should be at least 6 ft (1.8 m) apart when students are working back to back. Wall type fume hoods work well for groups of 12 students or less, while island type fume hoods work well for larger groups. The instructor must be able to see all parts of the laboratory and easily be able to access all areas of the laboratory to provide a quick response to emergency situations. A sample layout of a teaching laboratory is shown in Figure 4-5 (care must be taken to reduce traffic in the area in front of the fume hoods for this layout). A microscale laboratory can have a layout similar to that shown in Figure 4-5 but often with the students seated rather than standing. This requires lower workbenches and a recommended minimum of 7 ft (2.1 m) aisles. At least two exit doors swinging out of the room are required in teaching laboratories. These doors must be not require more than 50 ft (15 m) of travel from any point in the laboratory.

An adequate ventilation rate of teaching laboratories is considered to be between 0.75 cfm/ft^2 (10.8 (m^3/h)/m^2) or 5 air changes per hour when fume hoods are not in operation. Fume hoods should be placed as far from entrances and walkways as possible with easy access for the students but must not block escape routes in case of emergencies. At least one fume hood should be installed in dry chemistry laboratories. Local exhaust is desirable for each bench workplace (organic chemistry), but for general chemistry, one for each four workspaces is sufficient. The local exhaust should be located on the bench closest to the most hazardous area of the laboratory. Microscale chemistry laboratories may not require any fume hoods, and the room ventilation may be reduced due to the reduced amount of chemicals involved. A slot exhaust in the rear of the laboratory bench or a modified downdraft bench should be considered for microscale chemistry laboratories. *Industrial Ventilation* (ACGIH 1998) provides design guidance for microscale chemistry laboratories.

Emergency showers must be placed within 25 ft (7.5 m) of any location of the laboratory and an eyewash facility should be available at each bench if the laboratory contains more than four benches. A laboratory with less than four benches requires one eyewash facility within 3-4 seconds reach. Highly reactive or flammable materials must be stored in separate rooms. Waste areas should be close to fume hoods, and one fire extinguisher per bench with a clean agent such as CO_2 is appropriate. Larger backup fire extinguishers with a dry agent should be located in the hallway.

A preparation room is required for a teaching laboratory or a group of teaching laboratories that use chemicals or hazardous materials. The chemicals stored in these rooms must be stored in approved chemical storage cabinets in adequate numbers. The preparation room should be provided with approximately ten air changes per hour of general ventilation to prevent buildup of hazardous gases and a well-planned fire-suppression system.

Research Laboratory

Research laboratories can be chemistry, physics, clean room, controlled environment, and radiation laboratories. There are no general design issues for research laboratories, and the design issues should be evaluated based on what type of laboratory is required.

However, team research laboratories, where more than one discipline of activities is performed, can save money and improve cooperation and interaction within the facility. The large size of team laboratories is the biggest difference between this laboratory and other specialized laboratories. Team laboratories must have a simple layout without dead ends and cul-de-sacs to ensure easy evacuation. Dead end configurations must not exceed 20 ft (6 m) and an exit should be within 50 ft (15 m) of any location in the laboratory. Emergency showers must be available within 25 ft (7.5 m) of travel, and an eyewash station must be at every two-module equivalent at sinks distributed evenly in the laboratory.

Hospital and Clinical Laboratory

Hospital and clinical laboratories can apply procedures that require almost any type of laboratories, such as radiation, biological, cleanrooms, chemical, etc. However, these laboratories are not designed to handle large amounts of hazardous materials, conduct dangerous procedures, or do research. Many of the procedures are automated. The access to hospital and clinical laboratories does not need to be restricted, and the layout resembles chemical or biological laboratories.

Chemical fume hoods are used on a limited basis in hospital and clinical laboratories, while Class II, Type A and Type B biological safety cabinets usually are needed. Class II, Type B biohoods should be used in pharmaceutical areas. Local exhaust should be considered for equipment or processes that give off obnoxious fumes, hazardous fumes, or a large amount of heat rather than to try to ventilate the complete space. The location of large equipment must be considered when placing the diffusers because they can disturb the airflow and create drafts. A separate HVAC system is advantageous for hospital and clinical laboratories due to the need for a high degree of temperature control ($\pm 1.5°F$) and sometimes a strict control of humidity (DiBerardinis et al. 1993). The heat gain from some equipment typically used in hospital and clinical laboratories is given in Table 4-3 (Alereza and Breen 1984).

Safety precautions depend on the type of processes conducted in the hospital and clinical laboratory. It is an advantage to separate the types of activities into different laboratories to avoid large and cluttered laboratories that can restrict access to exits and safety devices.

Table 4-3: Hospital Equipment Heat Gain

Appliance Type	Size	Maximum Input Rating, Btu/h (W)	Recommended Rate of Heat Gain[a]
Autoclave	0.7 ft^3 (0.0198 m^3)	4270 (1250)	480 Btuh (140.64 W)
Bath, hot or cold circulating, small	1.0-9.7 gal (3.785-36.7145 L) −22°F-212°F (−30°C-100°C)	2560-6140 (750-1800)	Sensible: 440-1060 Btu/h (130-310)
			Latent: 850-2010 Btu/h (249.05-588.93 W)
Blood analyzer	120 samples/h	2510 Btu/h (735.43 W)	2510 Btu/h (735.43 W)
Blood analyzer with CRT screen	115 samples/h	2510 Btu/h (735.43 W)	5120 Btu/h (1500.16 W)
Centrifuge (large)	8-24 places	3750 Btu/h (1098.75 W)	3580 Btu/h (1048.94 W)
Centrifuge (small)	4-12 places	510 Btu/h (149.43 W)	480 Btu/h (140.64 W)

Table 4-3: Hospital Equipment Heat Gain

Appliance Type	Size	Maximum Input Rating, Btu/h (W)	Recommended Rate of Heat Gain[a]
Chromatograph	—	6820 Btu/h (1998.26 W)	6820 Btu/h (1998.26 W)
Cytometer (cell sorter/analyzer)	1000 cells/s	73,230 Btu/h (21.46 kW)	73,230 Btu/h (21.46 kW)
Electrophoresis power supply	—	1360 Btu/h (398.48 W)	850 Btu/h (249.05 W)
Freezer, blood plasma, medium	13 ft^3 (0.368 m^3), down to –40°F (–40°C)	340[b] Btu/h (99.62 W)	136[b] Btu/h (39.85 W)
Hotplate, concentric ring	4 holes, 212°F (100°C)	3750 Btu/h (1098.75 W)	2970 Btu/h (870.21 W)
Incubator, CO$_2$	5-10 ft^3 (0.142-0.283 m^3), up to 130°F (54.4°C)	9660 Btu/h (2830.38 W)	4810 Btu/h (1409.33 W)
Incubator, forced draft	10 ft^3 (0.283 m^3), 80-140°F (26.7-60°C)	2460 Btu/h (720.78 W)	1230 Btu/h (360.39 W)
Incubator, general application	1.4-11 ft^3 (0.040-0.311 m^3), up to 160°F (71.1°C)	160-220[b] Btu/h (46.88-64.46 W)	80-110[b] Btu/h (23.44-32.23 W)
Magnetic stirrer	—	2050 Btu/h (600.65 W)	2050 Btu/h (600.65 W)
Microcomputer and monitor	—	341-2047 Btu/h (99.91-599.77 W)	300-1800 Btu/h (87.9-527.4 W)
Minicomputer	—	7500-15,000 Btu/h (2197.5-4395 W)	7500-15,000 Btu/h (2197.5-4395 W)
Oven, general purpose, small	1.4-2.8 ft^3 (0.040-0.079 m^3), 460°F (237.8°C)	2120[b] Btu/h (621.16 W)	290[b] Btu/h (84.97 W)
Refrigerator, laboratory	22-106 ft^3 (0.623-3.002 m^3), 39°F (3.9°C)	80[c] Btu/h (23.44 W)	34[c] Btu/h (9.96 W)
Refrigerator, blood, small	7-20 ft^3 (0.198-0.566 m^3), 39°F (3.9°C)	260[b] Btu/h (76.18 W)	102[b] Btu/h (29.89 W)
Spectrophometer	—	1710 Btu/h (501.03 W)	1710 Btu/h (501.03 W)
Sterilizer, free-standing	3.9 ft^3 (0.110 m^3), 212-270°F (100-132.2°C)	71,400 Btu/h (20.92 kW)	8100 Btu/h (2373.3 W)
Ultrasonic cleaner, small	1.4 ft^3 (0.040 m^3)	410 Btu/h (120.13 W)	410 Btu/h (120.13 W)
Washer, glassware	7.8 ft^3 (0.221 m^3) load area	15,220 Btu/h (4459.46 W)	10,000 Btu/h (2930 W)
Water still	5-15 gal (18.925-56.775 L)	14,500[d] Btu/h (4248.5 W)	320[d] Btu/h (93.76 W)

[a] For hospital equipment installed under a hood, heat gain is assumed to be zero.
[b] Heat gain per cubic foot of interior space.
[c] Heat gain per 10 ft^3 of interior space.
[d] Heat gain per gallon of capacity.
(Source: Alereza 1984)

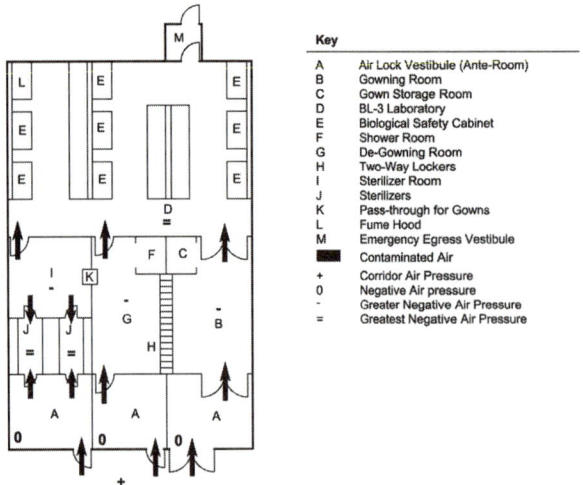

Figure 4-6 *Sample layout of a biological containment laboratory. (Diagram modified with permission, DiBerardinis et al. 1993.)*

Biological Containment Laboratory

Biological containment laboratories (BCLs) are laboratories that handle biological substances within all four safety levels. These levels of hazard range from level 1, where there is no known or only minimal potential hazard to the laboratory personnel, to level 4, where the microbiological agents are known to cause life-threatening diseases (described in greater detail in Chapter 16). The minimum size of a BCL for a single worker with one Class II biological safety cabinet (see Chapter 5 for details) is 7.5 ft × 10 ft. The walls and floor must be impervious and vermin proof and be easy to clean and decontaminate. Each laboratory must contain one autoclave for decontamination as well as an automatic elbow- or foot-operated sink for handwashing (DiBerardinis et al. 1993).

Often the exhaust air from BCLs must be HEPA filtered before the air is released either in the laboratory or outdoors. A gas filter may also be required if especially hazardous gaseous substances are used. An alarm must warn the users when the filters need to be replaced. Pressure control of the spaces is important for BCLs and to maintain the correct pressure relationships, and constant volume hoods and cabinets are preferable to ensure that the needed pressure relation is always present. Figure 4-6 shows a sample layout of a biological containment laboratory with pressure relationships.

Due to the need to decontaminate biological containment laboratories, usually with formaldehyde (a suspected carcinogen), suspended ceilings are not acceptable. The laboratory must be designed as gas tight as possible with only minimal work needed to prepare and decontaminate the laboratory. Waste must be treated similarly to waste at hospital and clinical laboratories (autoclaving).

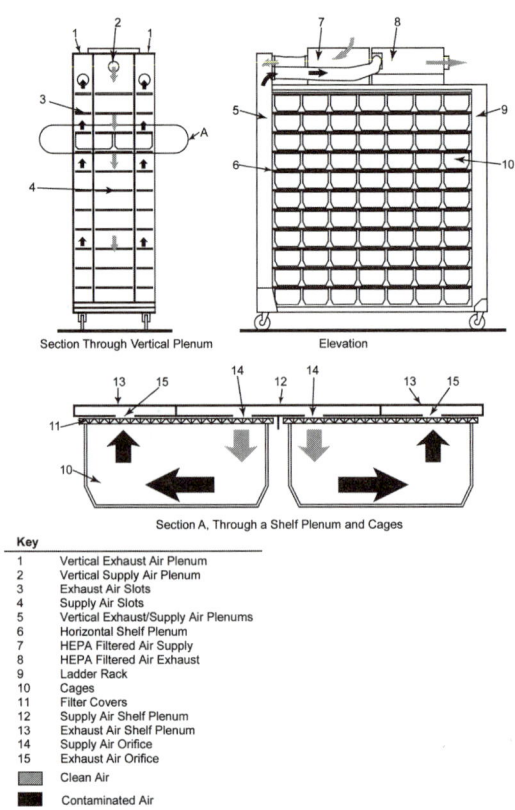

Figure 4-7 *Individual cage ventilation. (Diagram modified with permission, DiBerardinis et al. 1993.)*

Animal Laboratory

Due to the complexity of the needs for different species, animal laboratories should be designed with a veterinarian on the design team. Animal laboratories have also been strongly criticized and this can cause changes in regulations and the requirements. It is therefore suggested that animal laboratories be constructed to be flexible and beyond the current requirements to ensure that they don't become obsolete after only a few years of operation (DiBerardinis et al. 1993).

The minimum areas required for an animal laboratory are an animal reception and quarantine area, animal holding rooms, sanitation facilities (dirty areas), vermin proof storage rooms, experimental treatment, surgery, and autopsy room, and freezer for dead animals. To be able to properly care for the animals, animal laboratories must have impervious floors, ceilings, and walls that can withstand rough treatment. They should also have metal window frames that are resistant to moisture and that cannot be opened, doors that are animal proof (keep wild animals out), carefully sealed walls to prevent vermin infestation, and, possibly, boric acid in wall spaces to reduce cockroach problems.

The HVAC system should be designed to maintain the ideal temperature and humidity for the species con-

Design Process

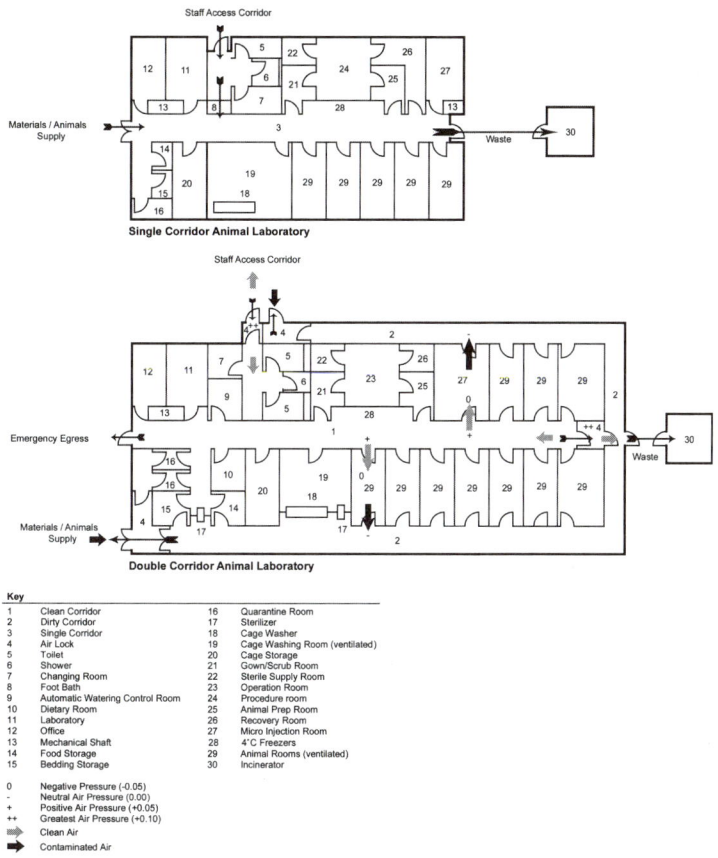

Figure 4-8 *Sample layout of an animal laboratory. (Diagram modified with permission, DiBerardinis et al. 1993.)*

tained with the appropriate number of air changes to reduce stress and thereby improve the accuracy of the experiments. The capacity of the HVAC system should be sufficient to maintain an adjustable temperature between 65°F and 84°F (18°C and 29°C) and humidity between 30% and 50%. Cage racks may create difficulties in maintaining a constant temperature with good ventilation in all cages. However, a study by Manning et al. (2000) shows that with careful diffuser selection and diffuser and exhaust location, an acceptable environment can be achieved. Figure 4-7 shows a system that ventilates each cage, keeping positive pressure in each cage. It may be necessary to filter the exhaust air, preferably at the exhaust grille, to protect the equipment from hair, dander, bedding fragments, and feces. Recirculation of the air is not recommended. The air change rate needed depends on the density of animals in the rooms (typically 10 to 25 air changes per hour).

Correct pressurization of animal laboratories is important to ensure that odors and pests are not spread to adjacent rooms. Animal laboratories should be built in separate parts of a building to be able to pressurize the animal areas correctly and to keep odors from spreading to other spaces. A sample layout of an animal laboratory is shown in Figure 4-8.

The ductwork from cage washers and sterilizers should be made in stainless steel. Survival surgery rooms require HEPA filtration of the supply air and positive pressure. In case workers are sensitive to the animal products, disposable respirators will generally relieve the symptoms and should be available. Smoke detectors are not recommended for animal laboratories as they very quickly get dirty and give false alarms. Chemical fire extinguishers can be a hazard and also are not recommended for animal laboratories. Sprinkler systems must consider high cages that will interfere with spray pattern, and a denser than normal sprinkler head location may be needed. Animal laboratories may require decontamination with formaldehyde to get rid of diseases.

Animals are very sensitive to illumination, and an automatic timer should control the lighting. A veterinarian should be consulted regarding the color of the light and illumination level. Animals are sensitive to noise and vibration, and this should be considered throughout the design. Emergency power is recommended in case of power failure to keep the temperature within a reasonable range.

Further details about animal laboratories are available in Chapter 16, "Microbiological and Biomedical Laboratories."

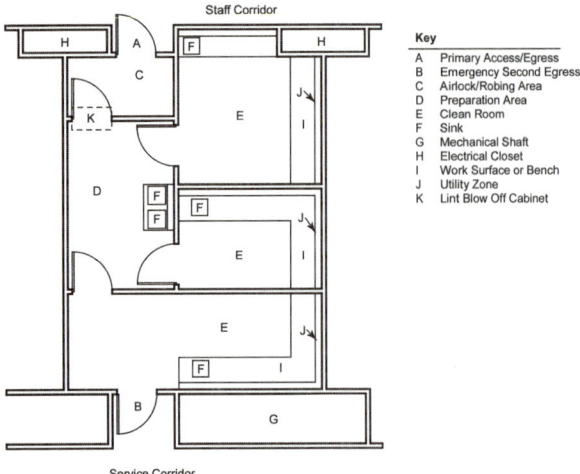

Figure 4-9 *Sample layout of a clean room. (Diagram modified with permission, DiBerardinis et al. 1993.)*

Isolation/Cleanrooms

Isolation/cleanrooms are designed to maintain a low concentration of airborne particles, constant temperature, humidity, and air pressure, and well-defined airflow patterns. There are six levels of cleanrooms depending on particle count: 100,000; 10,000; 1,000; 100; 10; and 1 particles at 0.5 μm particle size with designated limits at 5.0, 0.3, 0.2, or 0.1 μm depending on level of cleanliness. Cleanrooms are usually required, for example, for development and experimentation with microchips, miniature gyroscopes, pharmaceuticals, and photographic films. These rooms will have access restrictions and require the use of special lint-free clothing. The static pressure of the rooms should be positive to prevent infiltration of air from areas with more airborne particles. The surfaces in cleanrooms should be easy to clean, without seams and with smooth monolithic surfaces that are resistant to chipping and flaking.

Laminar flow workstations may be used for processes that require a higher level of cleanliness than the rest of the laboratory. These flow benches may be self-contained units with a fan blowing air through either HEPA (high efficiency particulate air) or ULPA (ultra low penetration air) filters. The air may be induced through the rear wall or the ceiling of the bench. The process in the laminar flow bench must not emit any hazardous materials as the laboratory worker breathes the air exiting the flow bench (DiBerardinis et al. 1993). Hazardous materials are limited to laminar flow workstations such as vented vertical flow stations or Class II, Type B biological safety cabinets.

A sample layout of a clean room is shown in Figure 4-9.

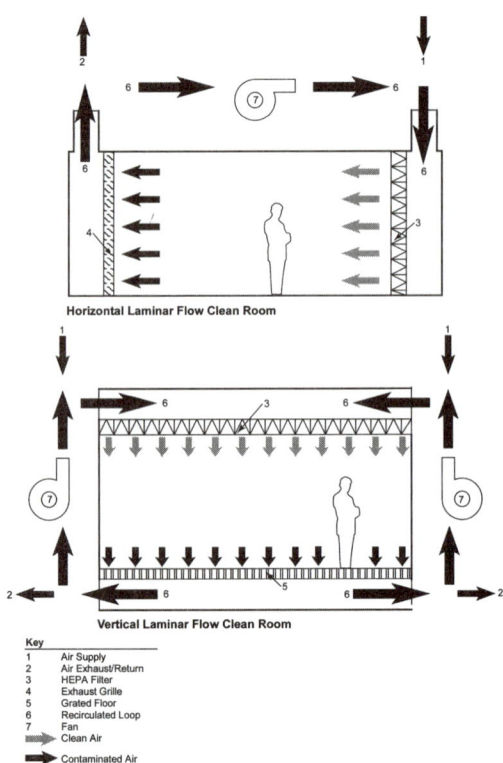

Figure 4-10 *Horizontal and vertical laminar flow cleanrooms. (Diagram modified with permission, DiBerardinis et al. 1993.)*

A minimum of two fire exits spaced as far apart as possible leading to different fire zones is recommended. Double egress doors cannot be interlocking due to prevention of an easy emergency exit. Options such as alarms should be used to prevent simultaneous opening of both doors and thereby introducing contaminants into the clean room. The traffic flow can be a two-way flow with the exception of laboratories where a contaminant generated in the clean room must be prevented from leaving the laboratory.

A good ventilation system that is independent of the main HVAC system is essential to cleanrooms. The makeup supply air to a clean room is often taken from areas surrounding the clean room, but outside air can also be used. Large amounts of particle free air that moves in one direction at a uniform velocity, typically 60-90 fpm (0.30-0.46 m/s), is needed to achieve the required cleanliness. The air should be filtered through roughing filters to remove large particles, then a pre-filter of 85-95% atmospheric dust efficiency, and finally a HEPA or ULPA filter (99.97% efficiency or higher). HEPA or ULPA filters can be placed in the wall or the ceiling as shown in Figure 10.

The pressure difference between contiguous pressure control zones should be 0.05 in. w.g. (12 Pa). A greater pressure can make it hard to open and close doors

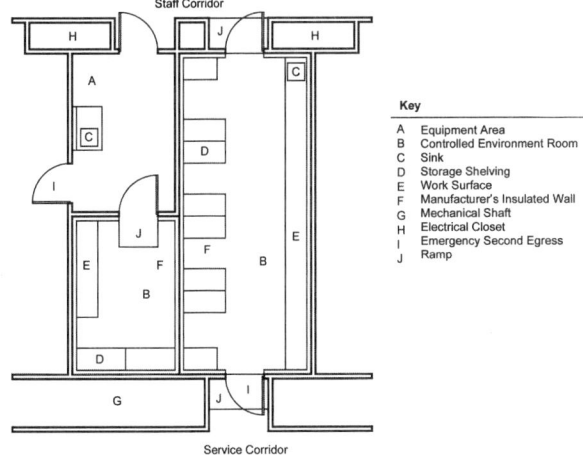

Figure 4-11 *Sample layout of a materials testing laboratory. (Diagram modified with permission, DiBerardinis et al. 1993.)*

and a lower pressure will not be effective in separating the zones. Fume hoods and local exhaust points should be operating continuously and there should be air locks between adjoining areas. Alarms should be provided to tell when the filters need to be replaced and if the directional airflow malfunctions. Humidity control is important for corrosion and condensation control, static electricity elimination, and personal comfort. In general, the temperature and humidity should be between 68°F and 73°F ±1°F (20°C and 23°C ±0.5°C) and 30%-50% ±5%RH. In critical applications the tolerances can be lower. If the humidity is supposed to be maintained at a lower level than 35%, special considerations should be given to static electricity.

Materials Testing Laboratory

Materials testing laboratories are here considered laboratories that are in need of very strict temperature and humidity control to ensure repeatable and consistent measurements of the materials that are tested and for storage of sensitive materials. These rooms can be maintained at a temperature ranging from 35°F to 120°F (2°C to 49°C) with a temperature variation of ±1°F (0.5°C) and ±0.5% of the humidity span.

Due to the strict temperature and humidity control for these rooms, it is advisable that one contractor be responsible for the completion of the rooms and that the room be pre-built and tested by the manufacturer before installation. Some building components may deteriorate at a faster rate in hot laboratories. An example is electrical wires and the National Electrical Code should be consulted. Cold laboratories may have condensation problems due to people entering and working in the laboratory,

and these rooms should be constructed with materials that are not affected by condensation. High humidity rooms (over 60% relative humidity) should have easily cleanable surfaces, and the use of HEPA filtration should be considered for the recirculating air due to the possibility of biological growth.

The amount of outside air should be as small as possible; 50 cfm (85 m^3/h) is minimum when people need to work in the laboratory for an extended period of time. It may be an advantage to precondition the outside air to achieve sufficient control of the room conditions. Air can be exhausted to the building ventilation system. The temperature and humidity controller must have the accuracy, stability, and sensitivity to provide the temperature and humidity required. A direct expansion cooling system designed to operate continuously is preferred if the temperature has to be low. Insulation of supply duct, airflow in laboratory, location of equipment and tables, and heat (longwave) radiation should be investigated.

A steam injector usually controls the room humidity. Steam from central steam plants may contain chemicals that are hazardous and should be carefully evaluated before being used directly. Steam injection does not increase the temperature of the air, but the injector is hot and may transfer a significant amount of heat. It may be necessary to both heat and cool the air at the same time to achieve the necessary control. Each laboratory should be supplied with a temperature and humidity controller that displays current conditions. A sample layout of a material testing laboratory is shown in Figure 4-11.

Electronics/Instrumentation Laboratory

Electronics/instrumentation laboratories are similar to cleanrooms and require strict control of airborne particulate matter, temperature, and humidity. Any contamination caused by hazardous materials used in the laboratory must in addition be controlled by local and general exhaust.

The major difference between cleanrooms and electronics/instrumentation rooms is the need to control contamination as well as maintaining a low amount of airborne particles (DiBerardinis et al. 1993). To achieve this, strict control of the airflow and a sealed work area are required. Anterooms or corridors surrounding the laboratory, which draw air from both the clean room and surrounding areas, will prevent hazardous materials from entering the surrounding areas as well as prevent air with high particle counts to enter the clean room. Two anterooms or a double shell of corridors may be required.

Some electronics/instrumentation laboratories are sensitive to vibration. These laboratories should have corridors that are supported separately from the labora-

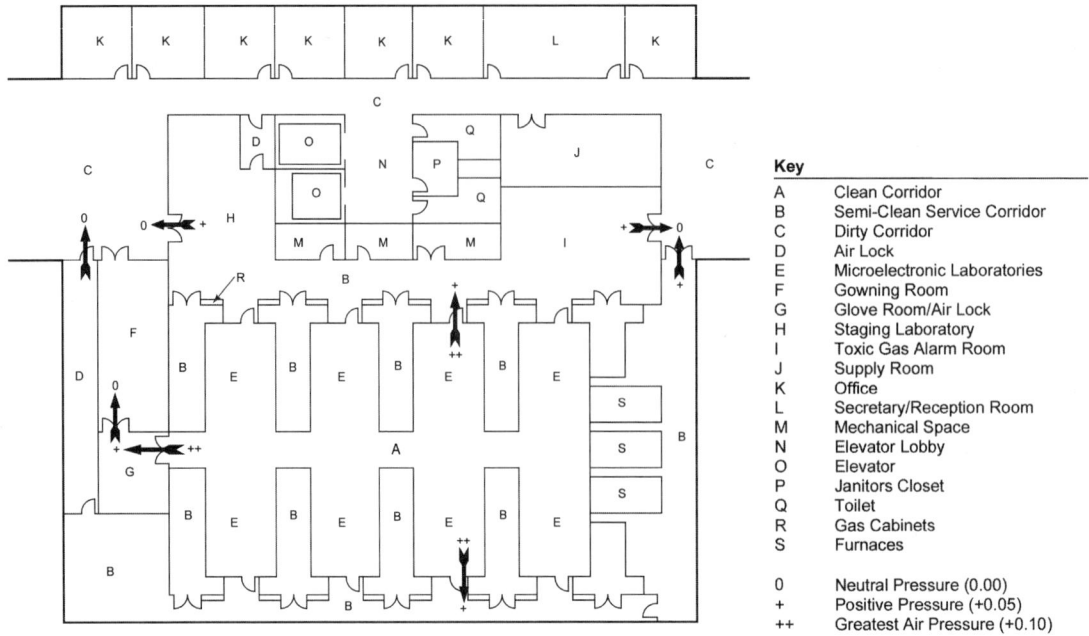

Figure 4-12 *Sample layout of an electronics-instrumentation laboratory. (Diagram modified with permission, DiBerardinis et al. 1993.)*

tory. The separation of the corridor will prevent vibrations from footfalls of people walking by. Isolation of vibrations from machines, motors, and motor vehicles should also be considered.

A sample layout of an electronics/instrumentation laboratory is shown in Figure 4-12.

Equipment used in microelectronic laboratories is often required to be operated under vacuum and is provided with an exhaust outlet. The ventilation system must provide the exhaust capability to vent these gases out. This exhaust should not be recirculated but can be combined with other exhaust if mixing the effluent does not create increased hazards. The exhaust may need effluent treatment to remove specific particulate and toxic gases.

A constant volume HVAC system is recommended for electronics/instrumentation laboratories to ensure correct pressurization at all times. The air distribution equipment must be designed and selected carefully to prevent cross-drafts and minimize turbulence.

Piping for toxic materials should be of stainless steel or other compatible material, have sufficient strength and durability, and be welded. Fittings, if any, should be in an exhaust enclosure. Highly toxic gases should be piped through double-walled pipes. It is an advantage to place gas tanks and cylinders outside the cleanroom to more easily replace empty tanks and cylinders. These rooms should be exhausted and the portable tanks should be provided with a flow control and be marked with maximum design flow rate. Some gases are acutely toxic and the exhaust may need to be treated instantaneously in case of an accidental release if the contamination is above safe levels to exhaust. A system that can handle the decontamination of these gases must be in place (typically a scrubber).

Some activities require respirators, and air-supplied respirators should be evaluated since they are usually easier to use than portable respirators.

In applications where metal ducts cannot be used, plastic may be a usable material. However, polypropylene and polyethylene have performed disastrously, in fires and are not recommended. PVC and FRP are recommended, and other fire resistant and fire spread retardation treated plastic can also be used (DiBerardinis et al. 1993).

Support Spaces

Support spaces are spaces for functions that require special equipment or conditions normally not found in the laboratory itself. The support spaces may be required for several types of laboratories. The spaces discussed in this chapter are:

- Photographic darkroom
- Support shop
- Waste-handling facility

Photographic Darkroom. Photographic darkrooms are facilities designed for filming, developing, printing, enlarging, and cassette loading and are mainly needed for support of clinical and research laboratories. Where hazardous materials are used, storage and processing must be done in exhaust vented areas. If manual operations are required, eyewash and showers must be available. The HVAC system required for photographic

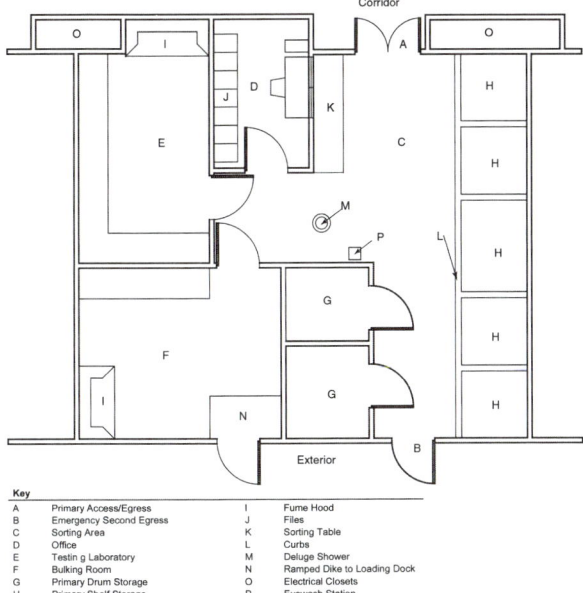

Figure 4-13 *Sample layout of hazardous chemical waste storage. (Diagram modified with permission, DiBerardinis et al. 1993.)*

darkrooms is very complex due to the high humidity caused by the use of chemical solutions in open tanks. Relative humidity is recommended to be between 40% and 50% for optimal quality and effectiveness of the processes, and the temperature should be maintained between 68°F and 78°F (20°C and 25.5°C) (DiBerardinis et al. 1993).

The minimum outdoor air requirements are 0.5 cfm per square foot (9 (m^3/h)/m^2) of darkroom area, while 10-15 air changes per hour is considered sufficient. A darkroom should be kept under negative pressure to prevent odors from spreading, and the exhaust air from the darkroom is not recommended for recirculation.

Support Shop. Support shops handle the supplies and equipment needed by the users of laboratories. These spaces usually have high heating loads due to welding, use of blowtorch, and heavy machinery. Air conditioning is recommended. Some processes, such as glass blowing, degreasing, and cleaning, use or emit hazardous gases and local exhaust scoops. Fume extraction devices are recommended. Dust can be generated in large amounts especially with woodworking. Dust collectors at sanders, grinders, and saws are recommended.

Waste-Handling Facility. Wastes that are generated in laboratories and have to be handled include hazardous chemicals, radioactive materials, and biological materials. These types of waste cannot be disposed of through the sewers or the landfill, and the design of the laboratories has to consider handling these wastes in an environmentally safe method. Radioactive waste is highly regulated and needs special attention during the design

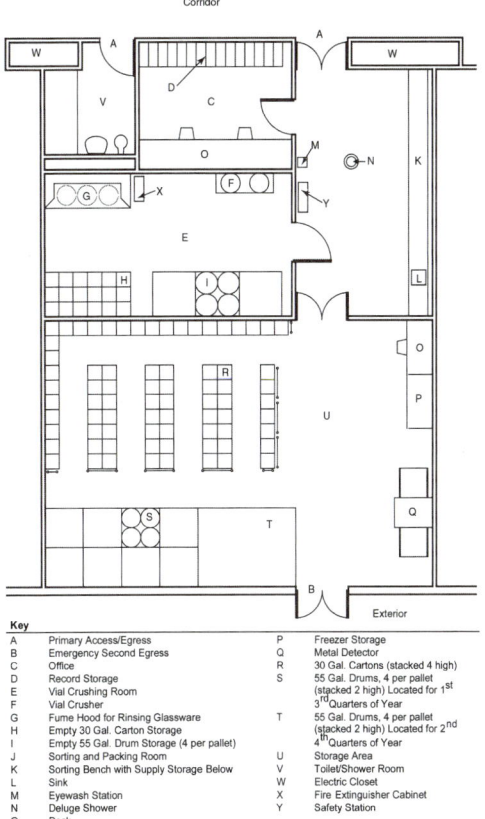

Figure 4-14 *Sample layout of radioisotope waste storage. (Diagram modified with permission, DiBerardinis et al. 1993.)*

phase. Biological waste that is infectious or pathological is also in need of special attention during the design. Chemical, radioactive, and biological wastes have different characteristics and need to be handled differently and individually. Hazardous waste handling is expensive, labor intensive, and heavily regulated, and it requires storage areas as well as sufficient workspace. Wastes that are radioactive and chemically or biological hazardous are considered mixed wastes and are very difficult to dispose of due to the many mandatory regulations from both the Nuclear Regulatory Commission (NRC) and Environmental Protection Agency (EPA). The standard that regulate chemical waste handling is U.S. EPA *Code of Federal Regulations 40 CFR 260* (EPA 1992b), biological waste handling is covered by *40 CFR 259* (EPA 1992a), and radioactive waste handling is covered by *10 CFR 20* (NRC 1991).

Waste generated in laboratories must be stored and later either be treated in house or transported to commercial waste facilities. The type of storage facility needed depends on the type of waste generated. Generally, chemical, radioactive, and biological waste should not be handled in the same room. A sample hazardous chemical waste storage is shown in Figure 4-13, a radioisotope

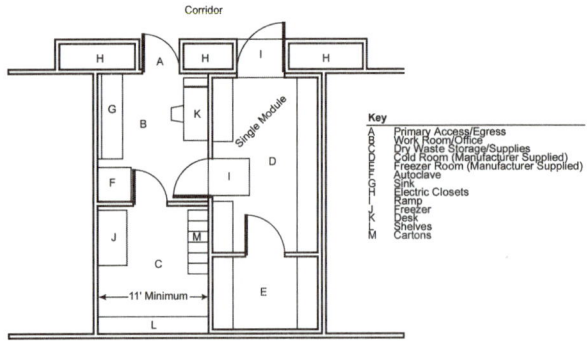

Figure 4-15 *Sample layout of biological waste storage. (Diagram modified with permission, DiBerardinis et al. 1993.)*

waste storage is shown in Figure 4-14, and a biological waste storage is shown in Figure 4-15.

The ventilation requirements for the waste facilities are 8 to 10 air changes per hour. This might be achieved by fume hoods or exhausts that are continuously operating, a general exhaust system, or a combination. Recirculation is not permitted, and the waste facility ventilation systems must be separate from all other ventilation systems. At least one flexible barrel exhaust is needed with a minimum of 150 cfm (250 m^3/h).

Flammable materials must be handled and stored in rooms with electrical services that comply with Class I Group C&D Division II installations, and a static electricity bonding and grounding system must be in place where flammable materials are handled. A non-water-based fire suppression system is required in areas where water reactive chemicals make up a substantial portion of the chemicals. An emergency and personal safety equipment cabinet should be located close to the waste handling room. The water supply should have capacity for the sprinkler system, and the potable water system should be able to handle a deluge shower of 30 gpm (6.8 m^3/h) and an eyewash of 7 gpm (1.6 m^3/h), and there should be a sink for handwashing.

Chemical waste is usually not treated in the laboratory waste facility, but activities related to recycling and bulking, such as mixing chemicals together into bigger volumes for cheaper disposal, as well as sink disposal of nonhazardous materials, will most likely take place. Activities that take place in the hazardous chemical waste facility are filling laboratory packs, consolidating chemically compatible wastes into larger volumes, transfer of chemicals from one container to another, sampling, sorting for recycling, decontamination, inventory sheet preparation, transportation documentation, and data entry.

The hazardous chemical waste facility should be separated from the other rooms and facilities by not less than a 2 hour fire rated wall and a 1 ½ hour fire rated door. Wastes should be protected from weather, large temperature swings, and excessive temperatures (50°F to 100°F (10°C to 38°C) is an acceptable range). At least one wall should be an exterior wall above grade for pressure relief.

The size of the waste handling facility depends on the amount of waste generated, but a facility containing 700 to 800 teaching and research laboratories requires 200 ft^2 (19 m^2) bulking room and 800-1000 ft^2 (74-93 m^2) space for the other activities. Bulking should be done in a separate room.

Radioactive material is usually separated by activity level and nuclide type. The activities in a radioactive waste handling facility include receiving, shipping, drum handling, bulking of liquids and dry material, record keeping, labeling, surveying, volume reduction, and documentation. Some of the radioactive material with short half-life isotopes may be kept in storage until the activity level is below acceptable limits. Radioactive waste facilities have the same fire, environmental, and pressure relief requirements as chemical waste facilities. The floor should be of low porosity cement with a continuous, impervious surface and have no drains. The walls may need to be constructed of solid cement or lined with lead to protect the surrounding areas from radiation. The size of a radioactive waste facility for a 500-1,000-bed teaching hospital is about 600 ft^2 (56 m^2) for storage, plus, if a vial crushing room is required, a separate room for highly flammable solvent activities is required. An office and a shower and toilet room may be desirable. Spill control berms or dikes should be installed on openings in each room.

Biological waste facilities should have good odor control through ventilation. The exhaust must be located away from sensitive areas such as air intakes, public areas, and entrances. The size of the biological waste facility should be based on past experience or experience with similar facilities. A typical biological research laboratory requires about 500 ft^2 (46 m^2) of floor space of waste handling facilities.

Biological wastes include:

- Blood and blood products
- Pathogenic wastes
- Cultures of infectious agents
- Contaminated animal carcasses and wastes
- Needles, scalpels, broken medical glass
- Biotechnology byproducts
- Dialysis wastes
- Isolation wastes

These biologicals can be disposed of at the facility or sent to commercial waste handling facilities. Disposal techniques include:

- Incineration
- Steam or gas sterilization

- Ionizing radiation and non-ionizing radiation followed by burial
- Maceration and discharge to sewers

Other activities in biological waste facilities are:
- Autoclaving
- Bagging and boxing
- Freezing or refrigeration
- Record keeping
- Labeling
- Spill decontamination
- Clean up

REFERENCES

American Conference of Governmental Industrial Hygienists (ACGIH). 1998. *Industrial Ventilation – A Manual for Recommended Practices*, 23rd ed. Cincinnati, Ohio: ACGIH.

Alereza, T., and J. Breen. 1984. Estimates of Recommended Heat Gains Due to Commercial Appliances and Equipment. *ASHRAE Transactions* 90 (2A): 25-58.

DiBerardinis, L.J., J.S. Baum, M.W. First, G.T. Gatwood, E. Groden, and A.K. Seth. 1993. *Guidelines for Laboratory Design: Health and Safety Considerations*, 2nd ed. New York: John Wiley & Sons, Inc.

Manning, A., F. Memarzadek, and G.L. Riskowski. 2000. Analysis of air supply type and exhaust location in laboratory animal research facilities using CFD. *ASHRAE Transactions* 106(1): 877-883.

National Fire Protection Association (NFPA). 2000. *NFPA 12, Carbon Dioxide Extinguishing Systems*. Quincy, Mass.: NFPA Publications.

National Research Council (NRC). 1991. *Standards for Protection against Radiation*. U.S. Nuclear Regulatory Commission, 10 CFR 20.

U.S. Environmental Protection Agency (EPA). 1992a. *Standards for the Tracking and Management of Medical Waste., 40 CFR 259*.

U.S. Environmental Protection Agency (EPA). 1992b. *General Regulations for Hazardous Waste Management., 40 CFR 260*.

Walters, D. 1980. Safe Handling of Chemical Carcinogens, Mutagens, Teratogens, and Highly Toxic Substances. Vol. 1, *Ann Arbor Science*.

Chapter 5
Exhaust Hoods

OVERVIEW

Exhaust hoods are generally the dominant type of unique HVAC equipment used in laboratory applications. Detailed knowledge of exhaust hoods, including the types of hoods available, their features, limitations, and design and operational requirements, are crucial to ensure the proper application and design of laboratory HVAC systems. Therefore, the different types of exhaust hoods are discussed in this chapter to provide information for the proper application of exhaust hoods in the later chapters of this guide. Also included in this chapter is a discussion on proper certification, monitoring, and selection of hoods. The types of hoods covered in this chapter are:

- Chemical fume hoods
- Perchloric acid hoods
- Biological safety cabinets
- Other hood types

CHEMICAL FUME HOODS

A common type of exhaust hood found in almost all facilities is the chemical fume hood. These hoods are designed to contain general odorous, toxic, or otherwise harmful chemical substances that are used in a wide range of laboratory activities, including research, demonstration, and numerous laboratory processes and experiments. In laboratories where the primary work of the occupants involves handling chemicals, it is recommended by the American Chemical Society and others that there be at least one fume hood per two workers and each worker should have at least 2.5 linear feet (0.76 meters) of working space at the face of the fume hood (ACS 1991).

The type of chemical fume hood chosen depends upon the process being accomplished within the hood, safety requirements for the intended processes, usage patterns, and energy efficiency. Due to the range of requirements, many configurations of the chemical fume hood have been developed, including:

- Conventional fume hood
- Bypass fume hood
- Baffle fume hood
- Variable air volume fume hood
- Auxiliary fume hood

Conventional Fume Hood

The conventional fume hood is the most basic type of chemical fume hood. Since its core characteristics are typical to other types of chemical fume hoods, the information in this section provides the basic terminology and features that are common for all chemical fume hoods.

Conventional fume hoods consist of an enclosed box with a movable sash, either vertical, horizontal, or a combination of the two, which is connected to a constant speed exhaust fan. Due to the air being drawn into the sash opening, fumes are contained inside the hood so they do not affect the laboratory workers.

Since a conventional fume hood only draws air through the sash opening, the air velocity through the sash opening changes, based on the position of the sash. With a design face velocity of 100 fpm at the fully open sash position, the velocity can increase to 400 to 500 fpm (2.0 to 2.5 m/s) at 25% open and can increase to 1,000 to 2,000 fpm (5.1 to 10.2 m/s) before becoming completely closed. See Figure 5-1. A concern with conventional fume hoods is that the high velocities encountered as the sash is closed cause turbulent airflow in the hood and can allow fumes to escape the fume hood. Further, once the sash is completely closed, the exhaust airflow is significantly reduced to air, which leaks around the sash and airfoil sill. Once the sash is closed, the reduced exhaust airflow can cause fumes to build up within the hood and the room pressurization (direction of airflow) can change, unless the supply airflow is reduced.

Sash stops are used to prevent the fume hood from being closed all the way to help limit high face velocities as the sash is closed and to ensure the ventilation rate is achieved when the hood is not in use. Also, sash stops can be used to prevent sashes from being opened all of the way and ensure the face velocity doesn't decrease too much when the sash is open. While the primary function of a fume hood is to contain fumes and vapors, it can also protect laboratory personnel from small explosions within the hood since the sash is typically made of shatterproof material.

The simplicity of conventional fume hoods results in their being less expensive than other hoods. However, due to the variable face velocity as the sash is moved and the potential for affecting room pressurization, conventional fume hoods are used only for chemicals that present a low-to-moderate risk to laboratory occupants and hood users.

The key to the successful containment of chemical fumes within any fume hood is the barrier at the intersection of the interior of the hood and the occupied space. This barrier for conventional fume hoods, and most others, is a combination of inward airflow to the hood and the use of a sash as a physical barrier.

As shown in Figure 5-1, the face velocity increases as the sash on the hood is closed since conventional fume hoods are at constant volume and must exhaust the same volume of air through a smaller area. The recommended minimum face velocity for chemical fume hoods is 100 fpm (0.51 m/s). These values are based on SEFA 1-2, *Laboratory Fume Hoods-Recommended Practices* (SEFA 1996). However, it is possible to use face velocities as low as 80 fpm (0.41 m/s) and as high as 150 fpm (0.76 m/s), depending on the application and whether the laboratory is occupied or not.

Sash Movement. Sash movement for conventional fume hoods is typically vertical, but it can also be horizontal or a combination of vertical and horizontal. The sash movement choice is usually dictated according to occupant input and the requirements of experiments to be performed within the hood. For instance, a vertically

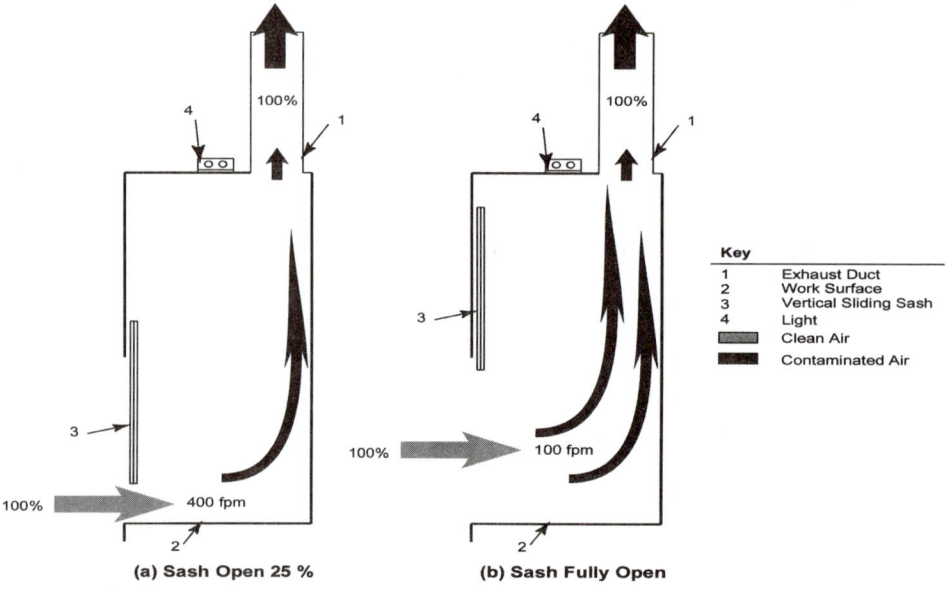

Figure 5-1 *Conventional fume hood. (Diagram modified with permission, DiBerardinis et al. 1993.)*

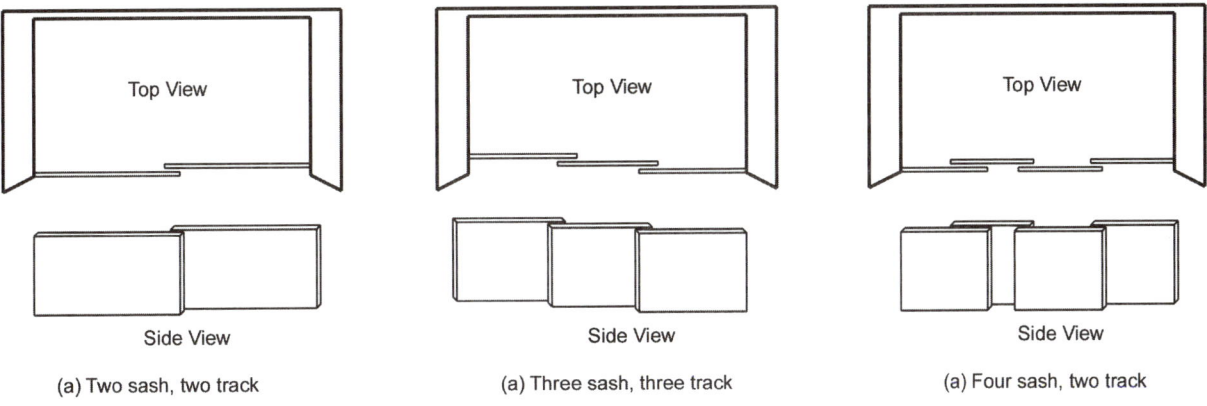

Figure 5-2 *Horizontal sash configurations.*

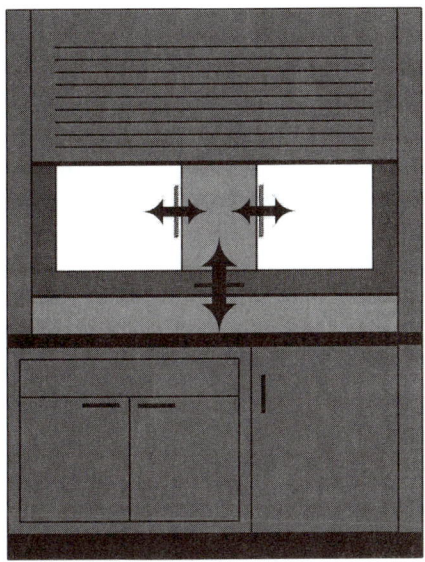

Figure 5-3 *Combination sash configuration.*

moving sash limits access to the upper or back ends of the fume hood at operating heights. Therefore, the sash must be opened to the full position to place tall test equipment into the fume hood, potentially resulting in insufficient face velocities to capture chemicals if the fume hood is not properly designed, balanced, and maintained.

Conversely, horizontal sashes allow the user access to the full height of a portion of the hood area at a single time, but not to both sides for wide equipment. Depending on the number of horizontal panels, the portion accessible can be either 50% for two and four panels or 66% for three panels. These horizontal sash configurations provide convenience for the user while maintaining sufficient face velocities. In addition, horizontal sash hoods can conserve energy relatively due to vertical sash hoods decreasing exhaust volume resulting from a smaller design face area. Figure 5-2 shows possible horizontal sash arrangements.

A final sash movement option is a combination of vertical and horizontal sashes as shown in Figure 5-3. This type of sash opening offers the greatest flexibility for access by the user while maintaining containment of experiments and protecting laboratory occupants. This option typically costs more but can pay for itself due to its energy-saving potential and flexibility for the type of laboratory work that can be performed.

Sash Stops. A key component of any sash is the use of physical stops and monitoring systems to ensure that the user cannot open or close the sashes to positions where containment of hood contents is compromised. Typically, these stops are coupled with alarms for notification of a potential hazard. The alarm systems are required according to ANSI Standard Z9.5, which states that all hoods be equipped with some type of flow-monitoring device, which is typically connected to a display for the user and an alarm for high/low limits (ANSI 1992).

Bypass Fume Hood

A bypass fume hood uses an air bypass opening, typically in an area that is covered by the sash when it is open, which allows air to enter the hood when the sash is partially (>75%) or fully closed. This prevents the face velocity from varying widely, minimizes turbulence, and reduces or eliminates the potential for the fume hood to affect room pressurization. Bypass fume hoods are the most common type of chemical fume hoods due to their improved safety performance compared to conventional fume hoods and their lower initial cost compared to variable air volume hoods.

The operating principle of the bypass hood is that a constant volume of air enters the hood through either the sash opening or the bypass opening, with the face velocity of the hood varying little with regard to sash position. As the sash is opened and closed, varying degrees

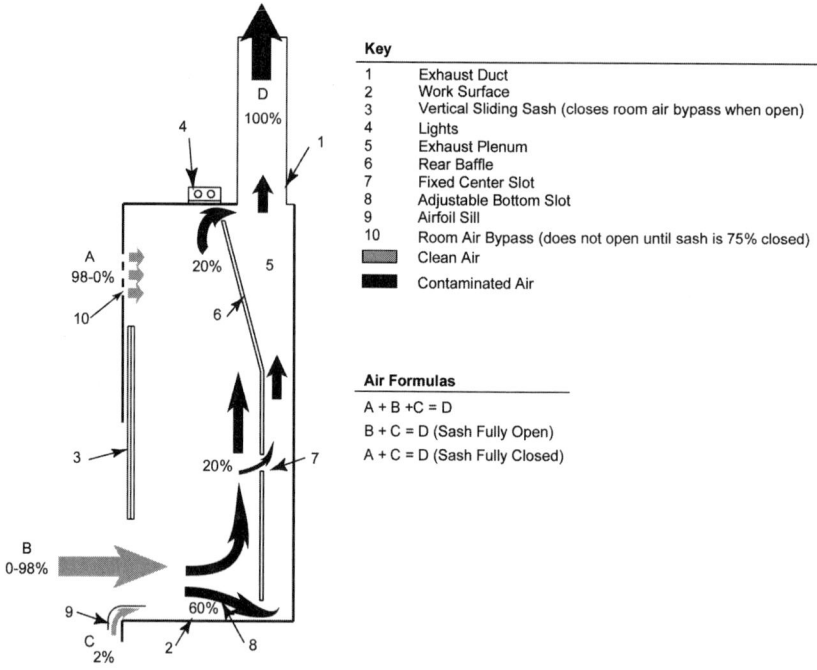

Figure 5-4 *Bypass fume hood. (Diagram modified with permission, DiBerardinis et al. 1993.)*

of air enter the hood through the face of the hood or the bypass opening. The intent of this design is to maintain a constant face velocity at all sash positions. Figure 5-4 shows a bypass hood at the partially open position.

There are two additional features of the bypass fume hood that improve its performance over a conventional fume hood. First is the addition of the airfoil sill. The airfoil sill helps minimize turbulence and ensures all fumes are exhausted within the hood when the sash is fully closed by always allowing 2% of the exhaust flow to enter at the very bottom of the hood. Second, the addition of the rear baffle, center slot, and adjustable bottom slot helps ensure that regardless of where the room air is entering the hood, all of the fumes are exhausted.

Baffle Fume Hood

The baffle fume hood uses a baffle control system that varies the position of a rear baffle in accordance with the sash position. When the sash is mostly closed, the baffle is moved to the rear of the fume hood so that the airflow is mostly horizontal to the back of the hood and above the work zone. As the sash is opened, the baffle is moved forward so that the turbulence at the rear of the hood is reduced, creating a floor sweep effect in the work zone. By varying the position of the baffle in the back of the fume hood, the turbulence created in a conventional fume hood at low sash heights is minimized, which reduces spillage of fumes and vapors from the hood. Figure 5-5 is a schematic of a baffle fume hood. Current research suggests that this type of hood can con-

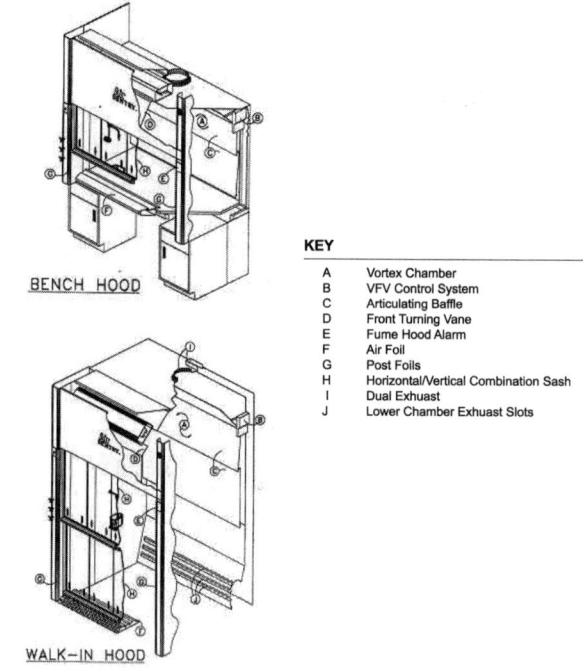

Figure 5-5 *Baffle fume hood. (Diagram modified with permission from Lab Crafters Inc.)*

tain at lower face velocities, down to as low as 51 fpm (0.254 m/s) (WI DOA 2000).

Also, fixed baffles can be added to the back of many types of fume hoods (auxiliary, VAV, bypass) to improve the flow characteristics through the sash open-

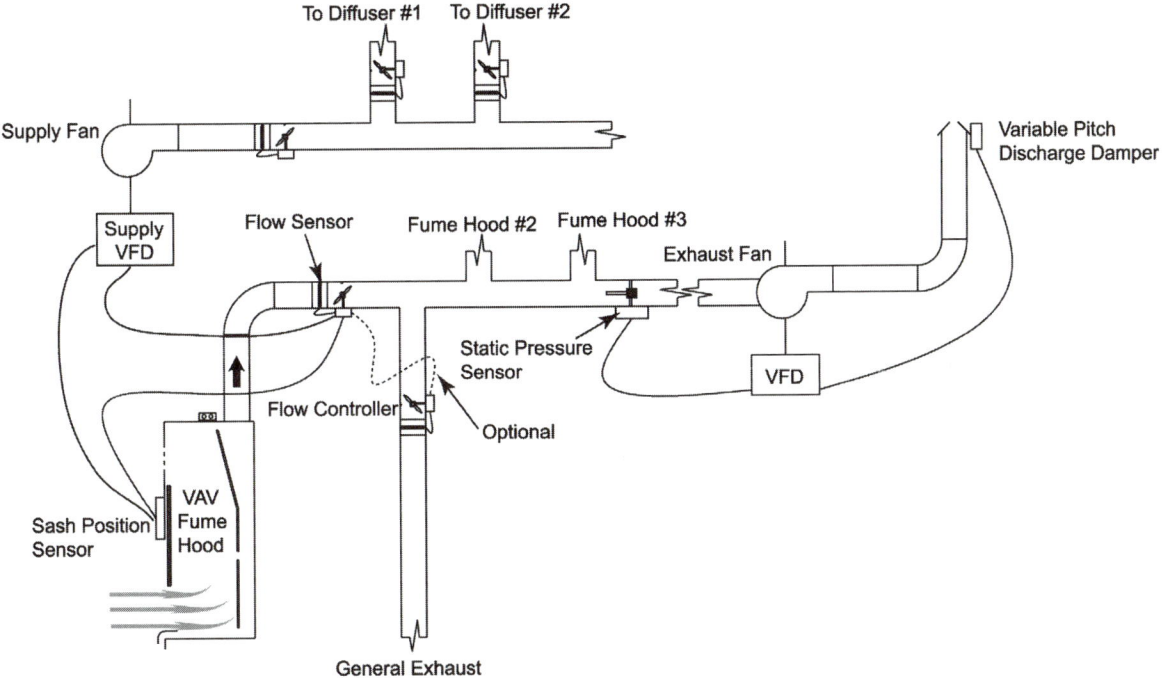

Figure 5-6 *VAV fume hood.*

ing by minimizing turbulence and reducing the potential for spillage from the hood.

Variable Air Volume Fume Hood

The variable air volume (VAV) fume hood is an energy-saving adaptation of the conventional fume hood that varies the exhaust air volume according to the sash position to maintain a constant face velocity. The energy savings are a result of reduced energy for conditioning the supply air, and reduced fan energy for both the supply and exhaust air when the fume hood sash is partially or fully closed. In order to achieve energy savings with VAV fume hoods, there must be times when either the laboratory is unoccupied or the fume hoods are not being used, and the laboratory occupants must be educated to keep fume hoods closed when they are not in use. VAV fume hoods typically make use of a small bypass opening (airfoil sill) to ensure that a minimum amount of air continues to enter the fume hood, even when the sash is completely closed. This bypass ensures that any fumes that accumulate in the hood due to stored chemicals are exhausted.

Due to the varying exhaust volume from VAV fume hoods, the supply air volume must also vary in order to maintain pressure relationships in the laboratory. There are two common control strategies used for VAV hoods: sash position measurement and direct air velocity measurement. Sash position measurement uses a potentiometer or other method to determine the sash position. The required exhaust airflow is then determined by multiplying the open face area by the minimum face velocity. Exhaust volume is controlled to this value using a damper and an airflow measurement sensor in the hood exhaust duct. This method ensures an average face velocity over the entire face area and does not account for the location of the fume hood operator or the location of the equipment in the fume hood.

The second control method is to directly measure the face velocity along the outer edge of the fume hood just inside the sash. An exhaust duct damper is then controlled to maintain a constant preset face velocity. This method accounts for the presence of the fume hood operators and the location of equipment in the fume hood. Figure 5-6 is a schematic of a VAV fume hood and some of its associated controls.

When a damper is used to control the exhaust airflow, a central exhaust fan is often used to exhaust multiple VAV fume hoods. This central exhaust fan maintains a constant negative pressure in the ductwork by varying its speed. A second means of controlling the exhaust air volume is to have a variable speed exhaust fan dedicated to each VAV fume hood. This is commonly used in laboratories when there are independent exhaust stacks for each exhaust device.

The energy savings for VAV fume hoods compared to the constant volume conventional fume hoods is an average 80% reduction in exhaust fan energy due to a 20% reduction in exhaust air volumes. These savings assume that the hoods are closed when not in use and are due to the cubic relationship between fan power and airflow or speed. (See Equation 5-1.)

$$P_1 = P_2\left(\frac{N_1}{N_2}\right)^3 = P_2\left(\frac{Q_1}{Q_2}\right)^3 \quad (5\text{-}1)$$

where

P	=	power, hp (kW)
N	=	speed, rpm
Q	=	airflow rate, cfm (L/s)
1	=	state 1
2	=	state 2

While there are significant operating cost savings, VAV hoods generally have a higher first cost and require a more complex control system for the hood and main system to ensure space pressurization and contaminant containment. VAV fume hoods, like other fume hoods in general, require periodic maintenance and testing to ensure that the fume hood is operating correctly and maintaining the required face velocity. The O&M personnel should be knowledgeable about the uniqueness of VAV systems compared to their CAV counterpart. Therefore, the ability to provide maintenance to a VAV fume hood through adequate training should be an important hood selection criterion.

Auxiliary Air Fume Hood

Auxiliary air fume hoods are also referred to as add air hoods, makeup air hoods, and supplementary air hoods. They often use less energy than conventional or bypass hoods, not by reducing airflow, but by reducing the amount of conditioned air that is exhausted through the hood. To accomplish this, between 50% and 70% of the air supplied to the fume hood is from a separate unconditioned or partially conditioned supply system. This auxiliary air is introduced to the hood using a canopy or plenum above and on the sides of the fume hood's face. Figure 5-7 details an auxiliary air fume hood.

Since the energy savings of an auxiliary air fume hood comes from reducing the conditioning of the air that is exhausted, the climate where the laboratory is located may have a significant impact on the amount of energy that can be saved. The amount of air needed to meet the cooling and heating loads of a laboratory can also determine the feasibility of using auxiliary air fume hoods. If the volume of air needed to satisfy the comfort heating or cooling requirements is greater than the volume of air that is exhausted through the fume hoods, there is little energy-saving potential for using an auxiliary air fume hood.

ASHRAE 110-1995 Testing. Given the potential for turbulence from the unconditioned air entering the fume hood face, containment "as installed" should be verified in all auxiliary fume hoods. Unless containment

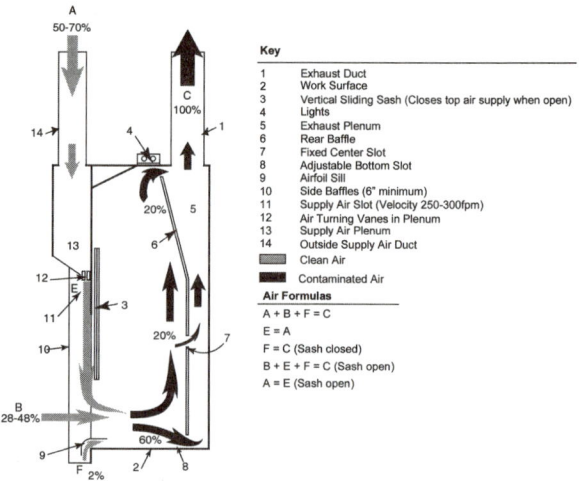

Figure 5-7 *Auxiliary air fume hood. (Diagram modified with permission, DiBerardinis et al. 1993.)*

testing verifies that each fume hood complies with a leakage rate of less than 0.1 ppm at a release rate of 508.6 cfm (4.0 L/min), the hoods should not be used until a hood operating protocol can be found that verifies containment and is incorporated into the laboratory's chemical hygiene plan (CHP).

Application Disadvantages. Although auxiliary air fume hoods can save energy, there are several significant disadvantages in their application that must be considered when selecting the type of fume hood to use including:

- Supply air vortexes at sash face
- Effects on laboratory sensible and latent loads
- Drafts on hood user
- Difficulty maintaining experimental conditions
- Two air supply systems

Supply Air Vortexes at Sash Face. By supplying air at the hood face, turbulence and disruption of the inward airflow to the fume hood can occur, resulting in fumes from the hood escaping into the occupied space. To avoid this problem, the exit velocity of the auxiliary air must be evenly distributed across the front of the fume hood and have a velocity less than 50 fpm (0.25 m/s). Due to the increased potential for fumes to escape, auxiliary air fume hoods are usually only used for handling low to moderately hazardous materials.

Effects on Laboratory Sensible and Latent Loads. A second disadvantage to auxiliary air fume hoods is their effect on the conditioning of the laboratory space. Auxiliary air hoods present a challenge in maintaining a consistent temperature and humidity within the laboratory, since the unconditioned or partially conditioned (heated) air can mix into the laboratory rather than be exhausted by the fume hood. This increases the load on the central system that supplies the laboratory. The problem is heightened when the fume hood sash is mostly closed.

It has been shown that the use of auxiliary air fume hoods can have a large effect on the control of room pressurization (Neuman 1989). Finally, the energy savings quoted for this type of hood are typically based on the use of fully unconditioned air, which is not possible in colder climates.

Drafts on Hood User. A primary complaint of auxiliary fume hoods is the drafts they create on the user, which result from the user needing to stand in the vicinity of the unconditioned or partially conditioned supply airstream. This draft of supply air creates uncomfortable conditions for hood users and can lead to users attempting to defeat the drafts by plugging the supply canopy, deflecting the supply air, or disconnecting the auxiliary air supply altogether. Each of these attempts to defeat the auxiliary air supply can result in an imbalance of room pressurization and decreased containment performance of the auxiliary air fume hood.

Difficulty Maintaining Experimental Conditions. Along with creating uncomfortable drafts on the hood users, the introduction of unconditioned supply air into an auxiliary fume hood can result in poor or failed experiments. Unsuccessful experiments can be the result of not providing constant temperature and humidity levels in the fume hood and from temperature and humidity extremes. For example, in experiments using desiccants, maintaining a constant and controlled humidity level is crucial, since desiccants readily absorb water from the air. With the possibility of varying temperature and humidity levels in the supply air from an auxiliary air fume hood, there is a possibility of the experimental results being influenced.

Two Air Supply Systems. A final concern with auxiliary hoods is the requirement for two supply systems. This can increase the capital costs of constructing a laboratory, create construction problems associated with needing to coordinate the location of additional ductwork, and increase the time needed to troubleshoot or fine tune the associated controls. Also, building configurations that require extensive use of small branch ductwork to move auxiliary air to the individual fume hoods can substantially increase the energy used by auxiliary fume hood fans and in some cases eliminate the savings from reduced conditioning of the supply air.

PERCHLORIC ACID HOOD

Perchloric acid fume hoods deserve special consideration due to the hazards accompanying this chemical and the subsequent special construction requirements.

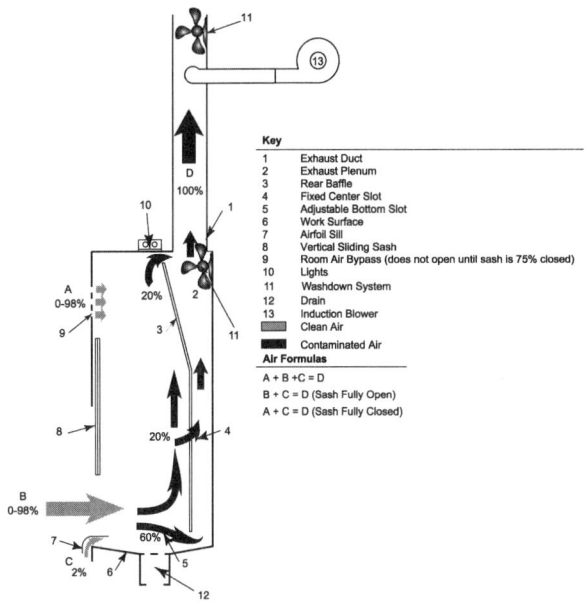

Figure 5-8 *Perchloric acid fume hood. (Diagram modified with permission, DiBerardinis et al. 1993.)*

Hazards

Perchloric acid is a strong oxidizing agent that reacts with organic materials, is highly volatile in its solidified form, and is highly acidic in its solution form. In its perchlorate, or salt state, this substance is highly unstable and can easily cause a flash explosion if sparked or agitated. Using a generic fume hood can allow perchloric acid to condense on ductwork and the inside of the hood, especially on horizontal surfaces, and create an explosion potential or damage the materials used in the construction of the exhaust hood. Therefore, special attention is required in the design of perchloric acid hoods to properly handle the liquid and solid phases. In terms of maintenance, perchloric acid hoods require careful consideration in cleaning and repair of hoods that handle the acid. Due to the dangerous nature of perchloric acid and the specialized construction requirements of perchloric acid fume hoods, experiments conducted in perchloric acid fume hoods should be strictly limited to work involving perchloric acid.

Hood Construction Requirements

Perchloric acid fume hoods should be composed of nonorganic materials, such as stainless steel, which will not degrade due to the high acidic environment. The interior of the hood should be smooth and have rounded corners to minimize areas where perchloric acid and its salts can accumulate and allow the wash down system to freely drain. All joints must be welded or sealed with neoprene gaskets, and all surfaces and joints should be as smooth and vertical as possible to make accumulation of the acid difficult. These hoods must be frequently or continually washed with a water spray that covers all of the hood interior, thereby preventing the buildup of perchlorate deposits. Treatment of the rinse water to balance the pH level is recommended prior to disposal. Figure 5-8 is a schematic of a perchloric acid fume hood.

Duct Construction Requirements

For the ductwork of perchloric acid hoods, many of the same design considerations as for the hood must be followed. As in hood construction, duct materials should be nonorganic and able to withstand the corrosive effects of the acid (e.g., stainless steel), be welded or sealed with neoprene gaskets, and be as smooth as possible. Also, duct runs should be as vertical as possible, with as few horizontal runs or elbows as possible. For this reason, perchloric acid fume hoods should be located near the outside walls of the laboratory or where they can be exhausted straight up to the roof. Due to the need to minimize horizontal duct runs, manifolded perchloric acid exhaust systems are not acceptable.

The exhaust fans, like other components of the perchloric acid exhaust system should be made of a nonorganic material that can withstand exposure to acid, although it is better to locate these fans outside the airstream. Also, as an additional safety measure, a redundant exhaust fan system may be installed to ensure the buildup of perchlorate does not occur in the event of the breakdown of the primary fan.

As with the fume hood itself, a wash down system for the ductwork associated with perchloric acid fume hoods should be used, with the rinse water being treated to balance the pH level. The washdown system is crucial to ensure that any deposition of the acid is removed before perchlorates can form and create dangerous conditions and it must be incorporated in a manner that washes down the entire exhaust system and not just the hood. In northern climates, the ductwork and water system should be winterized on all exterior locations.

BIOLOGICAL SAFETY CABINET

In biological laboratories and clinics, the typical exhaust hood, which simply exhausts its contents to the atmosphere for dilution, is not sufficient to properly handle the materials used within the laboratory. In these instances, biological safety cabinets are needed to provide a more secure environment. These hoods are designed to specifically handle organisms, infectious agents, processes, and particles that can potentially contaminate and harm other projects, personnel, or the environment. Therefore, all biological safety cabinets require filtration of exhaust, and some require filtration of supply air, depending on the classification type of the hood. Because HEPA filters and prescribed airflows provide the protective envelope, these devices should be annually certified according to field test protocols described in *National Sanitation Foundation Standard, NSF 49 – 1992* (NSF 1992).

Biological safety cabinets are divided into Classes I, 2, and 3, with Class II hoods being divided further into

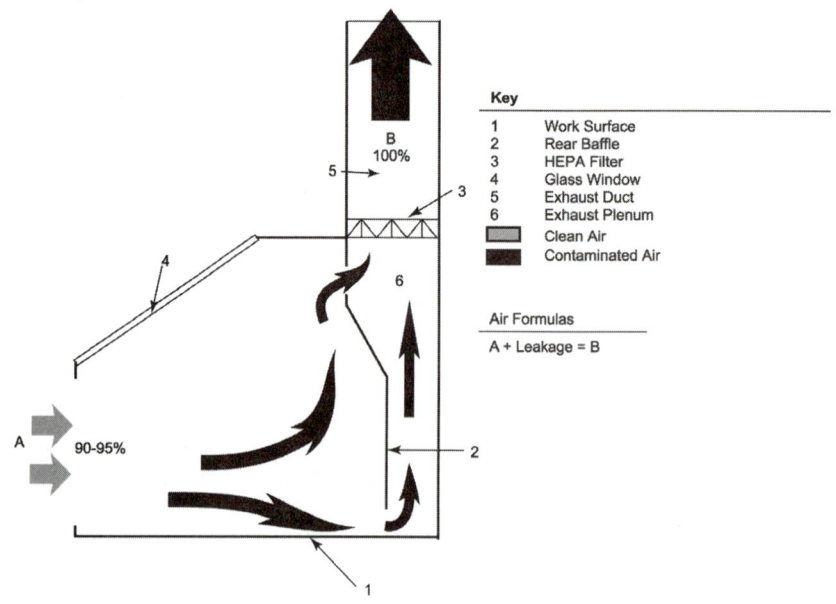

Figure 5-9 *Class I biological safety cabinet. (Diagram modified with permission, Diberardinis et al. 1993).*

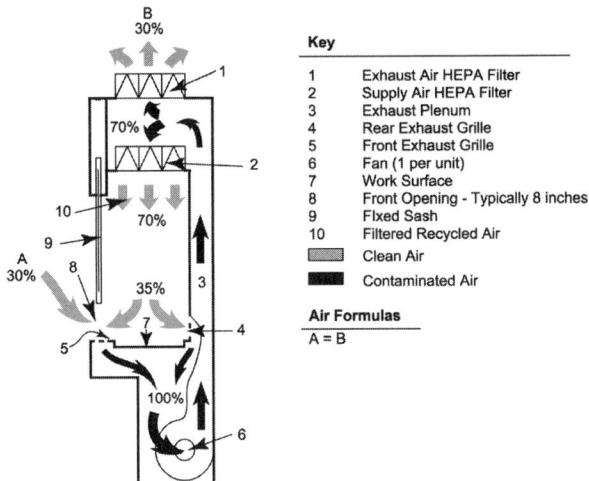

Figure 5-10 *Class II Type A biological safety cabinet. (Diagram modified with permission, DiBerardinis et al. 1993.)*

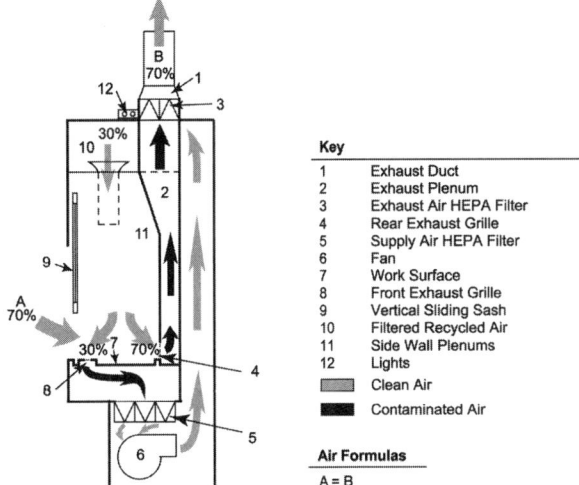

Figure 5-11 *Class II Type B1 biological safety cabinet. (Diagram modified with permission, DiBerardinis et al. 1993.)*

four types depending on the percentage of exhaust air recirculation within the hood, typical face velocities used, and exhaust air vented out of doors. This rating system is according to the National Sanitation Foundation standard, NSF 49-1992 (NSF 1992). The three classes of biological safety cabinets should not be confused with the four biosafety levels, which deal with the general level of infection and risk associated with a laboratory.

Class I

The first class of safety cabinet is the most general and is used for agents that pose little or moderate risk to users and the environment. Class I cabinets do not use supply filtration or recirculation of supply air, but all of the exhaust air is HEPA filtered. This type of biosafety cabinet has the airflow inward past the operator. Due to this airflow arrangement, Class I biosafety cabinets do not provide protection for the products contained in the hood, as the inward flow can move past multiple items on the work surface. The typical minimum face velocity for a Class I cabinet is between 75 and 100 fpm (0.38 and 0.51 m/s). Figure 5-9 is a schematic of a Class I biological safety cabinet. Class I safety cabinets are typically vented out of doors with negative pressure ductwork inside the building.

Class II

The Class II safety cabinets protect laboratory workers by maintaining inward airflow with the addition of having vertical laminar flow and the use of HEPA filters for all of the supply air and all of the exhaust air. Due to the vertical laminar flow arrangement, this type of biological safety cabinet protects the

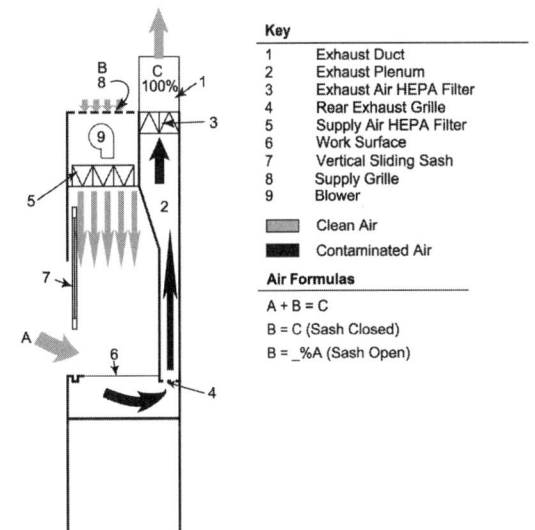

Figure 5-12 *Class II Type B2 biological safety cabinet. (Diagram modified with permission, DiBerardinis et al. 1993.)*

cabinet contents from cross-contamination or product protection.

This requirement for filtration of both the supply and exhaust and the specific type of filtration is due to the processes intended for this class of hood requiring a clean and controlled environment. Further, Class II cabinets contain differences within this classification, such as face velocities and exhaust recirculation percentages. The subtypes of Class II cabinets are Type A, Type B1, Type B2, and Type B3.

Type A. The Class II Type A safety cabinet is capable of handling low to moderate risk agents associated with Biosafety Level 2 and, if properly designed and annually certified, can meet Biosafety Level 3 requirements. The vertical laminar airflow is provided from a common plenum, where a portion of the air is exhausted and a portion is supplied to the work area of the cabinet. Since Class II Type A cabinets use approximately 70% recirculated air from this common plenum, they are not intended for experiments using toxic chemicals, flammables, or radionuclides. This cabinet has a minimum required face velocity of 75 fpm (0.38 m/s) and has a fixed opening (no movable sash).

This type of cabinet is configured for post-HEPA filtered exhaust air to be exhausted to the laboratory space. When the exhaust air from Type A cabinets is configured to be exhausted from the building and not returned to the laboratory space, negative pressure ductwork must be used; otherwise the air balance inside the hood can not be maintained. Figure 5-10 details a Class II Type A biological safety cabinet.

Type B1. The Type B1 safety cabinet requires a face velocity of 100 fpm (0.51 m/s) and allows up to 30% of the air within the hood to be recirculated, but exhausting to the laboratory space is not permitted. Unlike Type A cabinets, Type B1 cabinets have a vertically sliding sash. Type B1 cabinets are suitable for agents up to the Biosafety Level 3 classification, and they do allow for limited amounts of toxic chemicals and trace amounts of radionuclides to be used. This type of cabinet requires that all exhaust ductwork and plenums be maintained at negative pressure or be contained in a negatively pressurized duct or plenum.

Additional features and requirements of Type B1 biological safety cabinets are that they

- provide unidirectional airflow in the work zone,
- double the supply HEPA filtration,
- can be used with microgram quantities of toxic material,
- require an exhaust system designed to maintain cabinet exhaust airflow under increased static pressure of up to 50%,
- require an airflow monitor and sash alarm, and
- require annual field certification.

FIgure 5-11 is a schematic of a Class II Type B1 biological safety cabinet.

Type B2. Type B2 safety cabinets are referred to as "total exhaust" cabinets, since they exhaust 100% of the air and do not recirculate exhaust air within the hood or exhaust air back to the laboratory once it has been HEPA filtered. All of the vertical supply air is HEPA filtered prior to being introduced to the work area, but some air can enter through the sash without being filtered. Type B2 cabinets can be used with milligram quantities of toxic materials and trace quantities of radioisotopes. A certified industrial hygienist should be consulted and a chemical hygiene plan developed for the use and maintenance of this unit.

Additional features and requirements of Type B2 biological safety cabinets are that they

- provide vertical laminar flow through inward airflow to the cabinet,
- require negative pressurization of exhaust ducts and plenums,
- are suitable for Biosafety level 3,
- can use small amounts of toxic materials and radionuclides,
- require a minimum face velocity of 100 fpm (0.51 m/s),
- require their exhaust system to be designed for a 50% increase in static pressure for less than a 5% decrease in airflow,
- provide a switch for turning off the internal fan in the event of an exhaust system failure,
- require an airflow monitor and sash alarm, and
- require annual field certification.

Figure 5-12 is a schematic of a Class II Type B2 biological safety cabinet.

Type B3. The final category for the Class II biological safety cabinet is Type B3. This type of biological safety cabinet has a minimum face velocity of 100 fpm (0.51 m/s) and is similar in performance to the Type A cabinet. They exhaust 30% of the air after it has been HEPA filtered. All exterior plenums are under negative pressure.

Class III

The Class III biological safety cabinet is used for processes and biological agents that pose the greatest hazard risk, namely, for Biosafety Level 3 and Level 4. These cabinets are of airtight construction and have attached rubber gloves for handling the materials within the hood. Double door autoclaves and chemical dunk tanks may be used, depending on the owner's requirements, to pass materials in and out of a Class III cabinet. In order to contain all contaminates within the hood, a minimum negative pressure of 0.5 in. w.g. (120 Pa) must be maintained, at the required airflow rate. These biological safety cabinets require pressure monitors, decontamination ports, gas-tight shutoff dampers, and like the other classes of cabinets, must be field certified annually.

Class III cabinets use single HEPA filters for supply air, use double HEPA filters in series or single HEPA filtration and incineration for all exhaust air from the cabinet, and do not recirculate any air within the cabinet. This class of cabinet requires that exhaust ductwork be under negative pressure with the exhaust typically being separate from other exhaust systems for the laboratory. Class III safety cabinets are only required in very specialized laboratories and are not commonly used, compared to the other classes and types of biological safety cabinets. However, this type of cabinet is the preferred

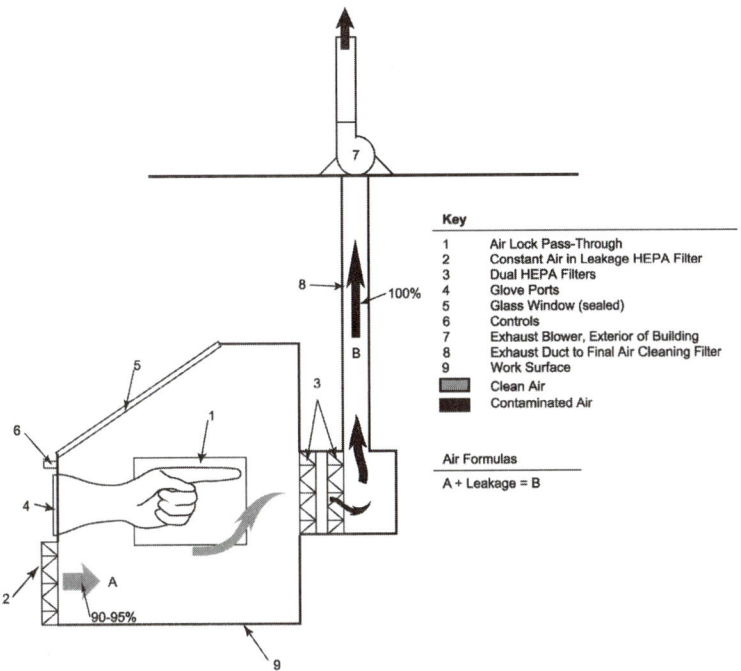

Figure 5-13 *Class III biological safety cabinet. (Diagram modified with permission, DiBerardinis et al. 1993.)*

method of containment when a high level of personnel and community protection is needed. Figure 5-13 shows an example of a Class III biological safety cabinet.

OTHER HOOD TYPES

In addition to chemical fume hoods and biological safety cabinets, there are numerous other types of hoods that are available to serve the exhaust requirements for laboratories, including:

- Radioisotope hood
- Special purpose hood
- Slot hood
- Canopy hood and dedicated equipment exhaust
- Local exhaust scoop
- Laminar flow clean air stations

Radioisotope Fume Hood

Radioisotope fume hoods, while similar to chemical fume hoods, have additional requirements for handling radioactive materials, namely, different recommended face velocities, different construction materials, and the potential need for filtration of exhaust air.

The recommended face velocity for radioisotope fume hoods is between 125 fpm (0.635 m/s) and 150 fpm (0.762 m/s), since radioisotopes usually have a low permissible exposure limit. A radioisotope fume hood is typically made from a smooth, nonporous material such as stainless steel so that radioactive liquids cannot soak in and the surfaces can be easily cleaned. It is recommended that the exhaust duct material be 18 gauge (1.27 mm) 316 stainless steel (DiBerardinis et al. 1993). In some instances, NRC regulations require further safety features, such as HEPA filtration, activated charcoal absorption for exhaust air, and continuous radioactivity monitoring of the exhaust airstream.

Variable air volume control is not recommended for radioisotope fume hoods, due to the potential need for air filtration and because emissions from radioisotope hoods are based on the concentration of radioactive materials, which will increase with the decreased airflow of a VAV hood.

Special Purpose Hood

There are three types of special purpose hoods for laboratories, including the glove box, walk-in hood, and weighing station.

Glove Box. The glove box is used in cases where the toxicity or radioactivity is too high to be handled by a chemical fume hood and in cases where the substance being handled reacts with air. In these instances, a glove box is used to completely isolate the experiment from the surroundings of the laboratory. To accomplish this, the glove box is usually of airtight construction and is made of stainless steel and safety glass with smooth finishes. Also, as the name suggests, full-length rubber gloves are attached to the side of the box and are used for handling the materials within the box. The glove box

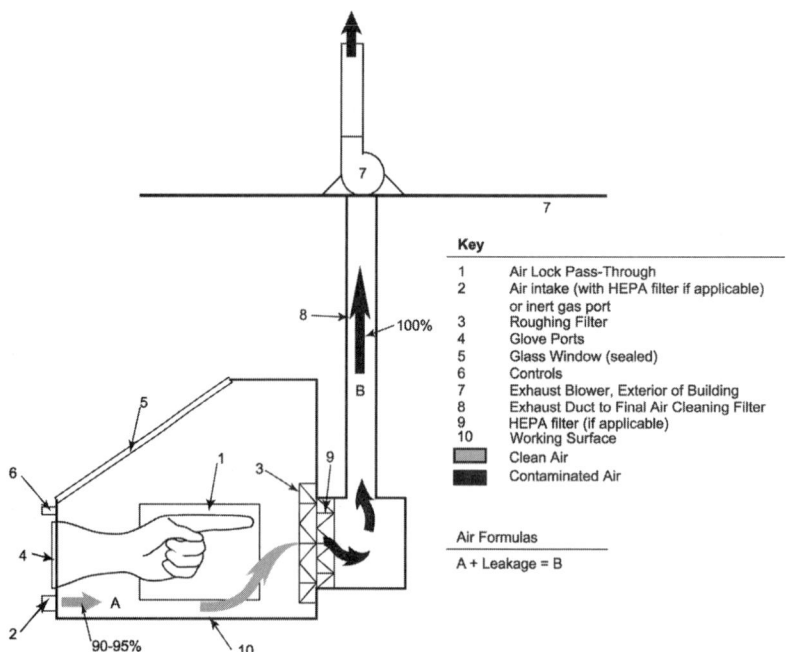

Figure 5-14 *Glove box. (Diagram modified with permission, DiBerardinis et al. 1993.)*

depth should be 24 inches or less to prevent glove damage.

Glove boxes are maintained under a relatively high negative pressure of 0.25 to 0.5 in. w.g. (62 to 124 Pa). Depending on the material handling requirements of the glove box, the following may also be used: HEPA filtration of supply air, HEPA filtration of the exhaust air for toxic and infectious materials, activated charcoal absorption for volatile chemicals, or airlocks for introduction of items into the box. When HEPA or activated charcoal filtration is used, a pre-filter for removing particulate matter is recommended. Figure 5-14 is a schematic of a glove box.

Glove boxes are also used to contain substances that react with air. An inert gas or one that does not react with the substance is used. When inert gas is used, the bypass opening for the glove box is replaced with a connection to an inert gas supply. Inert gas glove boxes may or may not be ducted to the outside, depending on the hazards to laboratory personnel associated with the materials that will be used in the glove box.

Walk-In Hood. The walk-in hood is a larger version of the chemical fume hood and is used in instances where the size of the experiment exceeds the size of typical hoods. With walk-in fume hoods, the worker is only inside the hood to set up the experiment and not actually in the hood while the experiment is being conducted. This type of hood has a horizontal or vertical sash that extends to the floor level and may use either or both a dedicated supply and dedicated exhaust system with filtration (HEPA, activated charcoal, etc.), depending upon the needs of the experiment being conducted in the hood. Figure 5-15 is a picture of a walk-in fume hood.

Figure 5-15 *Walk-in fume hood (courtesy of Labconco Corp.)*

Weighing Station. A weighing station hood is designed to contain highly toxic chemicals and instrumentation during measurement and weighing operations. Due to the toxic nature of some materials, it is necessary to measure and weigh them in a ventilated space. However, typical fume hoods can make accurate measurements difficult to obtain because of the high airflow velocities within the fume hood. Therefore, weighing stations are used to provide a ventilated space with low airflow velocities for measuring and weighing toxic materials.

Figure 5-16 *Slot hood (courtesy of New-Tech.).*

Figure 5-17 *Canopy hood (courtesy of Labconco Corp.).*

Slot Hood

Slot hoods are used to contain fumes and toxic materials close to their source when the use of chemical fume hoods or canopy hoods is not practical or necessary. The primary advantages of a slot hood is that it uses significantly less exhaust air volume than a chemical fume hood and that it can be used in general laboratory work areas.

The slot hood works by drawing odors and vapors through small, high-velocity exhaust slots in the rear of an open basin. This hood is not as effective in protecting laboratory personnel as chemical fume hoods and should not be used as a substitute when materials need to be handled in a fume hood. Typically the laboratory operations where slot hoods are applicable include specimen preparation, mixing, and weighing operations. Exhaust velocities for slot hoods are in the range of 500 to 2,000 fpm (2.5 to 10.2 m/s), with the slots not being located more than 12 in. (30 cm) from the generation source. Figure 5-16 is a picture of a slot hood.

Canopy Hood and Dedicated Equipment Exhaust

Canopy hoods are used to exhaust low hazard gases, vapors, and aerosols created by permanent laboratory equipment, such as ovens, gas chromatographs, autoclaves, and atomic absorption spectrophotometers. This type of hood is usually used in cases where effluents and exhaust are at high temperatures or are directed upward and are thus suspended over the equipment for which they are intended to provide exhaust. Exhaust velocities for canopy hoods are in the range of 500 to 2,000 fpm (2.5 to 10.2 m/s), and the air intake should not be located more than 12 in. (30.5 cm) from the generation source. The effectiveness of canopy hoods is dependent largely on the contour and structure of the opening used. For more information on the contour and structure impacts of canopy hoods, and exhaust hoods in general, consult Chapter 3, "Local Exhaust Hoods" of *Industrial Ventilation* from the American Conference of Governmental Industrial Hygienists (ACGIH 1998).

As with slot hoods, canopy hoods should be placed as close to the pollutant source as possible, use significantly less air volume than a chemical fume hood, and should not be used as a substitute when laboratory processes require chemical fume hoods. Figure 5-17 is a picture of a canopy hood.

Local Exhaust Scoop

Local exhaust scoops perform similarly to canopy and slot hoods, yet with a more focused role. The local exhaust scoops are designed to remove heat, humidity, and smoke from specific areas and equipment, and can be used to exhaust chemical odors, vapors, and gases located very close to the source.

Laminar Flow Clean Air Stations

These workstations provide product protection by design. Unidirectional air can be introduced vertically or horizontally depending on design. Some stations are designed to be vented either partially or 100% depending on the requirements of the experimental protocol. These units produce class 100 conditions in the work zone.

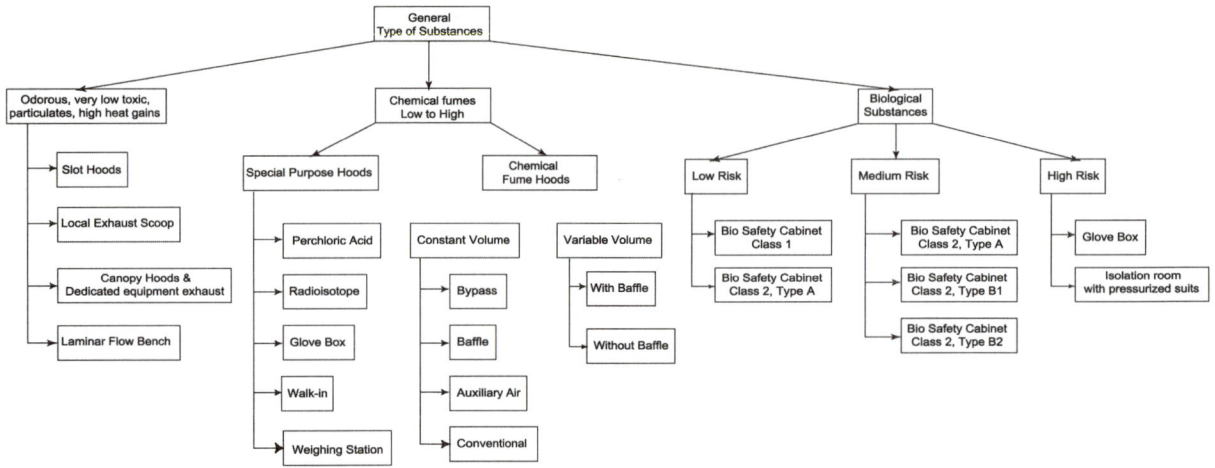

Figure 5-18 *Exhaust hood selection matrix.*

HOOD SAFETY CERTIFICATION AND CONTINUOUS MONITORING

One of the key issues in ensuring proper protection for laboratory workers is the certification and continuous monitoring of all types of exhaust hoods. Usually, biological safety cabinets are initially certified by the NSF and include proof of certification (NRC 1995). Then, upon installation, the cabinets are field certified according to NSF 49-1992, Annex F (NSF 1992). This certification should be repeated on a regular, ongoing basis (typically yearly), whenever the hood is relocated and if occupant complaints arise about the performance of the hood.

Fume hoods are certified at the factory according to *ASHRAE Standard 110-1995* and *ANSI/AIHA Standard Z9.5* (ASHRAE 1995; ANSI 1992). Once installed, the fume hood should be certified for containment to meet criteria as defined in ANSI/AIHA Standard Z9.5. Once the fume hood has passed certification, standard air velocity testing and directional airflow (smoke tests) can be run thereafter unless conditions change. The certification tests are designed to verify the face velocity and ensure that proper leakage limits from the fume hood are maintained. ANSI Standard Z9.5-1992 recommends leakage limits of 0.05 (for manufacturer's ideal conditions) to 0.1 ppm (for an operating laboratory) at a rate of 4 liters per minute of tracer gas diffused inside the fume hood. However, acceptable leakage limits may vary with the type of fume hood and with the materials being used in the hood. To accomplish a hood leakage test, a tracer gas is emitted within the hood via a gas diffuser. Sensors placed at the hood face at a defined height of a typical user provide readings for tracer gas emissions at various user positions. If at any position there is a tracer gas amount greater than the ANSI limit, the hood fails and must be rebalanced and retested for certification.

Another safety requirement for hoods is the use of appropriate continuous monitoring devices. This can include alarms for the sash position traveling beyond sash stops and for excess or insufficient face velocity. These types of monitoring devices and alarms can aid in informing personnel of possible hazardous operating conditions of the hood.

SELECTION OF EXHAUST HOODS

The selection of exhaust hoods is an important step in the design of a laboratory and can greatly impact the health and safety of the laboratory workers, the results of the laboratory experiments, and the life-cycle cost of owning and operating the laboratory. To begin the selection process, a clear understanding of the current and future types of work that are to be conducted in the laboratory needs to be researched and documented. The hazards associated with the current and future work in the laboratory are then identified, and the applicable codes, standards, and laws that regulate the anticipated types of work and associated hazards need to be located and followed. Once the hazards and applicable regulations have been identified, the appropriate type of exhaust hoods for the laboratory processes can be chosen. Figure 5-18 shows a matrix of the exhaust hood selection process.

In some cases, there are no options for the type of hoods to use. If perchloric acid is to be used, a perchloric acid fume hood is required and other types of hoods cannot be substituted. The same is true for regular work with radioisotopes; only a radioisotope fume hood should be used. Biological safety cabinets allow some flexibility in the type of cabinet to use, as long as the appropriate level of containment is achieved. Chemical fume hoods and some of the hoods mentioned elsewhere in this chapter offer additional choices of variable air volume or constant air volume systems.

The choice of constant air volume or variable air volume fume hood is made in conjunction with the selection of the type of primary air system that is to be used in the laboratory. (Primary air systems are discussed in Chapter 6.) Also made in conjunction with the choice of constant or variable air volume is the choice of dedicated or manifolded exhaust air systems. Generally, manifolded exhaust systems need to be grouped into areas of similar hazards.

Constant volume fume hoods have the benefit of being less expensive for first cost, require a less complex supply and exhaust system, and are generally easier to maintain. However, constant volume fume hoods can waste substantial amounts of energy if the laboratory is occupied infrequently. Variable volume fume hoods have the benefit of saving substantial amounts of energy over their lifespan if the laboratory is used periodically, and they can provide continuous monitoring to document that the required face velocity of the hood is being met. However, the first cost of variable volume fume hoods is greater than that of constant volume hoods. Variable volume hoods also require a more complex supply and exhaust system, require additional controls, and can be more difficult to maintain, as it is essential that their performance be checked regularly.

The compatibility of materials used, and the layout of the laboratory building influence choosing dedicated or manifolded exhausts systems. The choice of dedicated or manifolded exhaust systems is also closely tied to whether constant or variable volume fume hoods are selected. Variable volume systems are generally more economical when a manifolded exhaust system is used, as a dedicated system would require a separate, costly VFD for each fume hood fan motor. A manifolded system would only require a less expensive damper for each fume hood and one VFD per manifolded system. The choice of dedicated or manifolded exhaust for constant volume fume hoods is generally determined by the compatibility of the materials used and the cost difference between a few long runs of ductwork to create a manifolded system and numerous individual runs of ductwork for a dedicated system.

The choice of constant or variable volume fume and dedicated or manifolded systems can be an iterative process. Also, a laboratory building does not need to be exclusively constant volume or exclusively variable volume, exclusively dedicated exhaust or exclusively manifolded exhaust. Similar areas of work and associated hazards can be grouped together to use one or the other type of system without the whole building necessarily being the same.

REFERENCES

American Conference of Governmental Industrial Hygienists (ACGIH). 1998. *Industrial Ventilation – A Manual for Recommended Practices*, 23rd ed. Cincinnati, Ohio: ACGIH.

American Chemical Society (ACS). 1991. *Design of Safe Chemical Laboratories: Suggested References*, 2nd ed. Washington, D.C.: Committee on Chemical Safety.

American National Standards Institute (ANSI). 1992. Standard Z9.5, *Laboratory Ventilation—American National Standard Practices for Respiratory Protection*. New York: ANSI.

American Society of Heating, Refrigerating and Air-Conditioning Engineers (ASHRAE). 1995. *ASHRAE Standard 110-1995, Method for Testing Performance of Laboratory Fume Hoods*. Atlanta: ASHRAE.

DiBerardinis, L.J., J.S. Baum, M.W. First, G.T. Gatwood, E. Groden, and A.K. Seth. 1993. *Guidelines for Laboratory Design: Health and Safety Considerations*, 2nd ed. New York: John Wiley & Sons, Inc.

National Research Council (NRC). 1995. *Prudent Practices in the Laboratory, Handling and Disposal of Chemicals*. Washington, D.C.: National Academy Press.

National Sanitation Foundation International (NSF). 1992. *Class II (Laminar Flow) Biohazard Cabinetry*. Ann Arbor, Mich.: NSF Publications.

Neuman, V.A. 1989. Disadvantages of Auxiliary Air Fume Hoods. *ASHRAE Transactions* 95(1): 73.

Scientific Equipment and Furniture Association (SEFA). 1996. *SEFA 1.2, Laboratory Fume Hoods, Recommended Practices*. Maclean, VA: SEFA.

Wisconsin Department of Administration (WI-DOA), Division of Facilities Development. 2000. Fume Hood Performance Test and Life Cycle Cost Analysis, p. 7 and 19. Milwaukee, Wisc.: WI-DOA.

Chapter 6
Primary Air Systems

OVERVIEW

In the development of laboratory systems, the proper layout and design of the primary supply and exhaust air systems is critical in ensuring the performance and safety of the laboratory environment. When designing the primary systems, the complex and often conflicting system options must be evaluated. These include the type of hoods used, special requirements for exhaust and supply duct systems, the location and types of distribution devices, and the first and operating cost of any proposed system. A key step in the layout and design of the laboratory system is the selection and sizing of the primary air systems. However, prior to the design of these components (air-handling units, etc.), the zone air distribution and heating must be determined.

The main reason the design of the primary systems is started at the zone level is that for the primary system to perform properly and efficiently, the conditions in the occupied space must first be met. Therefore, for the selection and design of the primary systems, the following, in order, should be evaluated:

- Zone air distribution
- Zone heating
- Exhaust air system
- Supply air system
- Duct construction
- Energy efficiencies

ZONE AIR DISTRIBUTION

The distribution (supply) of air into a laboratory, relative to the exhaust air requirements (based on hoods, etc.), determines the integrity of hood containment, as well as ensuring proper pressure control (direction of airflow) with respect to adjacent nonlaboratory (administrative) zones and spaces differing in cleanliness. Therefore, the location of diffusers, discharge velocities, and volume control of the exhaust and supply for a room are all critical to maintaining a safe and comfortable work environment. As was covered in Steps 1 through Step 3 of Chapter 4, "Design Process," the quantities of supply and exhaust air, as well as the integrity of the room envelop and the influence of outside forces (wind and stack effect), determine the air movement between spaces.

Therefore, with the total supply (including auxiliary air) and exhaust air volumes known for a space, the next question to answer is how to properly supply and exhaust the air to maintain safety and comfort within the space.

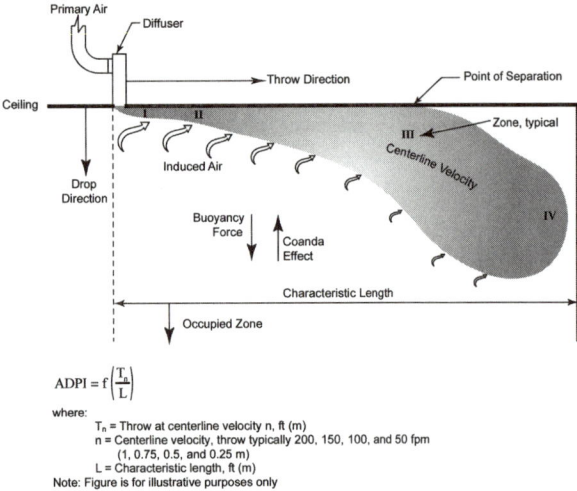

Figure 6-1 *Generic air jet characteristics.*

The key to maintaining proper air movement in a laboratory is the proper selection and location of supply, return, and exhaust devices. When designing laboratory air systems, it is important to ensure that the following are evaluated and their interaction with one another thoroughly understood:

- Room air velocities
- Relative device location
- Device types
- Off-peak loads and balancing

Room Air Velocities

To ensure proper containment of pollutants within a laboratory hood, it is necessary to maintain proper velocities at the entrance (face) of the hood. Disturbances in the room air motion near the face of a hood can result in entraining pollutants out of the laboratory hood into the occupied space. Therefore, it is crucial that the supply, return (if any), and general exhaust systems do not adversely affect the air velocities at the laboratory hood faces.

The greatest adverse influence on air motion near a laboratory hood is from the supply air system, through the supply air diffusers. Due to the typical requirement of 100 fpm (0.508 m/s) face velocity for a laboratory hood, it is recommended that the maximum supply air velocity within the occupied space, below approximately 7 ft (2.134 m), be 50 fpm (0.254 m/s) (DiBerardinis et al. 1993). In general, a supply air velocity of 50% of the required hood velocity should be maintained within the occupied space. However, some experts recommend that within the frontal area of a fume hood or biological safety cabinet, the room supply air velocity should be no more than 20% of the required hood velocity to maintain containment in the hood (Wunder 2000).

Relative Device Location

The placement of supply air diffusers relative to the laboratory hoods and to other diffusers determines the airflow within the room. Typically, it is best practice to place diffusers as far from the hoods as possible. There are two primary reasons for this general rule. First, when a diffuser is located close to a hood, it is likely that there will be insufficient distance between the diffuser discharge and hood face for the air velocity to reduce to an acceptable value, thus creating drafts and compromising the containment of the hoods. Second, if diffusers are located near hoods, it is possible that the ventilation of the general laboratory space will be hampered, as the supply air could short-circuit and be exhausted out of the hood.

Device Types

The space velocity and diffuser location requirements make the selection and application of the room supply air system the key to the safety and comfort of the occupants. To understand what type of distribution systems (diffusers) are commonly used in laboratory systems, a basic knowledge of air distribution fundamentals is required.

With the introduction of a jet of air into a large, relatively stagnant space, the disturbance of the air jet is well defined. As shown in Figure 6-1, the jet enters at a high velocity and dissipates with distance. This is a result of the induction of room air into the jet.

The higher the jet velocity, the greater the induced air and turbulence of the space. Conversely, the lower the velocity, the less the induced air and turbulence. However, there is still the concern of occupant comfort. If the air velocity is too high, the occupants will complain of drafts, and if too low, they complain of stuffiness. Therefore, the proper selection of diffusers must address the velocity near hood faces, the general mixing of room air and supply air, and the local velocity of room air. It is the combination of these requirements with the number and type of diffusers that determines the final room air motion.

Observing Figure 6-1, it is obvious that a diffuser producing the represented characteristic of nondirectional flow is poor for laboratories due to its relatively turbulent nature. To minimize the turbulence within the space, the diffuser supply velocity needs to be reduced. This can be accomplished through the addition of multiple diffusers or the use of a diffuser with multiple orifices. While each orifice has the same general performance characteristics, by having multiple orifices, the jet velocity is reduced and a more unidirectional (laminar) flow is achieved. Figure 6-2 shows this principle.

To achieve the low-velocity, high-volume air requirements typical in most laboratories, several laminar type systems have been developed:

- Perforated duct
- Perforated diffusers
- Perforated ceiling panels
- High-capacity radial

These four options are shown in Figure 6-3.

It is important to understand that the systems shown in Figure 6-3 are specialized for critical environments and are separate product lines from diffuser manufacturers. While these types of supply systems offer the best characteristics for safety and comfort in laboratories, they are not always practical due to cost limitations.

Table 6-1: T_{50}/L Ranges

Diffuser Type	T_{50}/L Range
Side Wall	1.3 - 2.0
Ceiling-round Pattern	0.6 - 1.2
Ceiling-cross Pattern	1.0 - 2.0
Slot	0.5 - 3.3

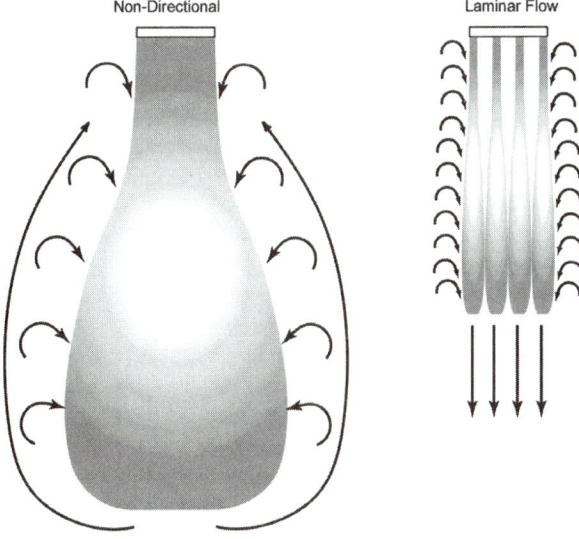

Figure 6-2 *Nondirectional vs. laminar flow.*

To select a system under these circumstances requires returning to basic air distribution principles. Figure 6-4 shows a typical sectional view of a laboratory, and Figure 6-5 has a plan view with airflow shown.

The key to diffuser selection is that the throw of the diffuser where the velocity falls below 50 fpm (0.254 m/s), designated as T_{50}, does not enter the area near the hoods. Since most air distribution systems are laid out symmetrically, the T_{50} envelope should not enter the occupied space. While this criterion appears simple to meet, it is complicated by the fact that sufficient velocities are required to thoroughly mix the room air for comfort. Table 6-1 lists the throw divided by diffuser characteristic length, L, required for comfort.

For any diffuser with a T_{50}/L greater than unity will result in the throw being greater than 50 fpm (0.254 m/s) at walls and within the occupied space. Therefore, the use of sidewall and ceiling cross-pattern diffusers must be considered and evaluated carefully. Detailed data on the specific diffusers to be used must be obtained.

While the T_{50}/L addresses the horizontal direction, of more concern is how far down the 50 fpm (0.254 m/s) velocity profile extends into the occupied space. Unfortunately, most manufacturer's catalogs do not provide drop information for diffusers and this must either be obtained from the manufacturer or physically verified prior to selecting a diffuser. However, even if the T_{50} falls within the occupied space, the diffuser may still work. Figure 6-6 shows an example of one such instance.

As Figure 6-6 illustrates, the diffuser can be used if the T_{50} line does not reach the hood face. Care must be taken in these instances. Other situations, as shown in Figure 6-7, must be avoided. In this instance the sidewall diffuser application results in higher than acceptable velocities and turbulence near the hood face due to the vane position and characteristics of sidewall diffuser.

Regardless of the type of diffuser or air distribution system chosen, DO NOT allow substitutions during construction without thoroughly analyzing and approving the

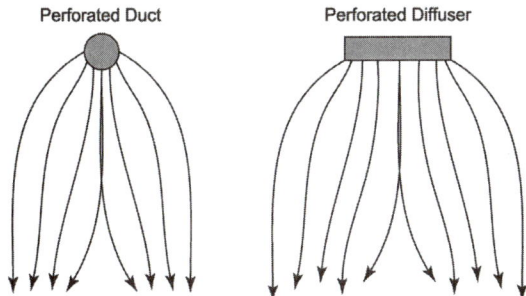
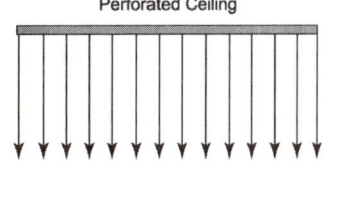
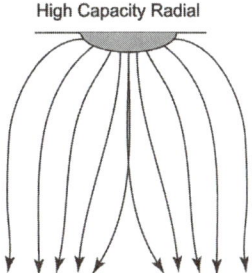

Figure 6-3 *Perforated supply systems.*

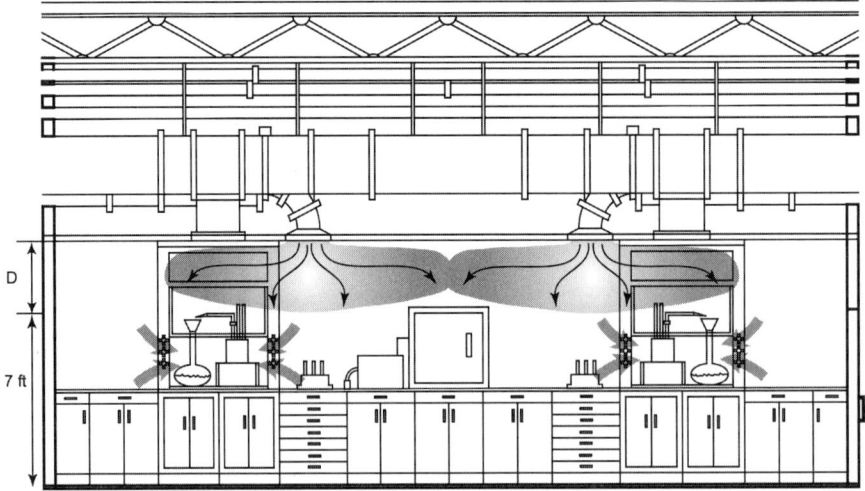

Figure 6-4 *Sectional view (A-A).*

change. This must be accomplished for every room and every unique diffuser.

Finally, it is recommended that non-locking adjustable louvers not be allowed on diffusers. When designed properly, the louvers typically will only provide proper air distribution and safety in one position. Therefore, if used, the position of the adjustable louvers will change due to occupant or O&M personnel intervention, usually resulting in more problems, not less.

Off-Peak Loads and Balancing

The final step in the design of a zone distribution system is to ensure that the room air motion is maintained under all system conditions. The primary concern is with variable air volume (VAV) systems. However, there are also fluctuations in constant volume systems due to changes in processes and equipment used. Therefore, the supply diffuser performance (throw and drop) must be verified for minimum flows.

The last consideration in the design is to ensure adequate dampers and test ports are available for balancing of the system. Of critical importance is the requirement for straight exhaust duct from each laboratory hood at least ten duct diameters downstream and five duct diameters upstream of any obstruction. This is required to ensure accurate exhaust flow measurements are possible during balancing.

Once installation of all air distribution systems has been completed, the system must be properly balanced. This work should be accomplished by a separate contractor with the aid of a commissioning authority to ensure that the balancing is done properly and that any errors in the system are resolved. Air balancing must be completed to ensure that the volume difference between supply and exhaust volumes is maintained at peak and non-peak con-

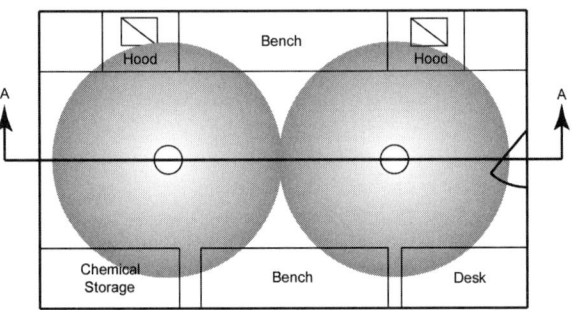

Figure 6-5 *Plan view.*

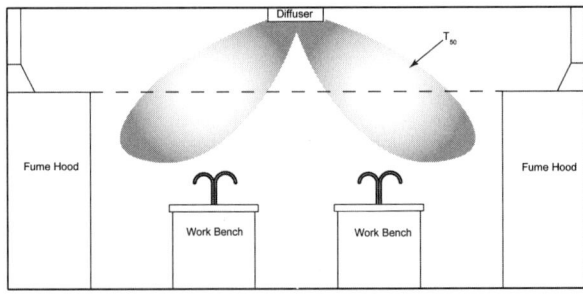

Figure 6-6 *Layout of laboratory and diffuser selection.*

ditions, that the specified airflow from diffusers is within acceptable volumes and velocities, and that comfort levels are maintained.

ZONE HEATING

Due to the fact that the exhaust requirement often exceeds the volume of supply air required for comfort, heating is often required all year. However, the need for heat during cooling periods focuses on not overcooling the space and during winter avoiding comfort problems form cold exterior walls. It is important to understand two different heating loads. The first one, due to heat loss through the building envelope, is identical to any other building and is not unique to laboratories. In addition, it is important to recognize that this heating load only affects the perimeter and not the interior zones. To meet this load, sufficient heat must be provided to the room to offset the loss. This heat can be added to the supply airstream or provided separately through a separate system.

The second heating load found in many VAV systems due to the fact that excess supply air is required is magnified in laboratory systems due to the higher air exchange rates. Although the heat can be provided via a second system to meet this load, heating the supply system is often most economical and sensible.

Three main options for providing heat to individual spaces are:

- Baseboard
- Radiant panel
- Supply air

The source of heat for all of these options is hot water, steam, or electricity. Figure 6-8 details these systems.

Baseboard

Baseboard heating is typically applied along exterior walls to offset the heat loss to the outdoors during cooler periods. The primary advantages of baseboard heating are that it is easily zoned and has few moving parts to maintain. However, there are several disadvantages of using perimeter baseboards within a laboratory. First, control of space conditions may not be suitable for laboratory applications due to humidification and temperature control issues. For instance, many laboratories require that space environment be maintained in a consistent state for experimentation consistency. Unfortunately, the response time of perimeter baseboard heating is typically much slower than is acceptable for laboratories. Also, this type of heating system has no humidification option and requires the addition of such a system to control humidity at acceptable levels. Finally, maintenance and cleaning of these systems can be difficult and expensive when decontamination is necessary.

Relationship to Casework. A consideration of perimeter baseboard heating systems is the relationship they must have to laboratory casework and equipment. Generally baseboard heating protrudes approximately 2 inches (5.08 cm) from the wall and cannot be blocked by any objects to ensure proper heating and performance. Thus, use of this type of system requires careful planning

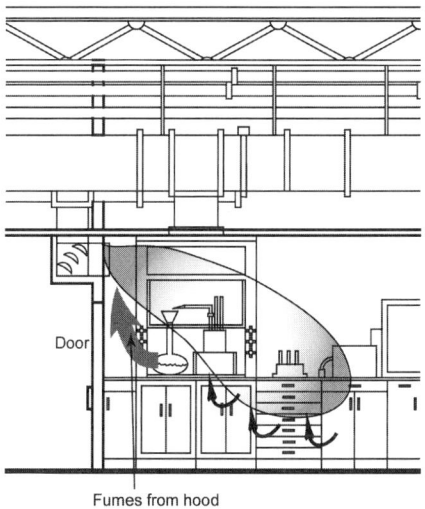

Figure 6-7 *Misapplication of diffuser layout.*

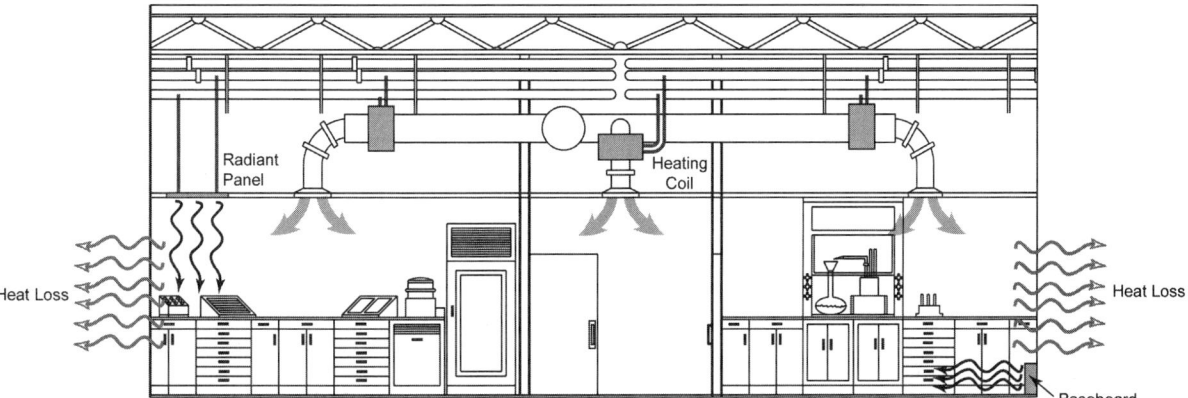

Figure 6-8 *Zone heating options.*

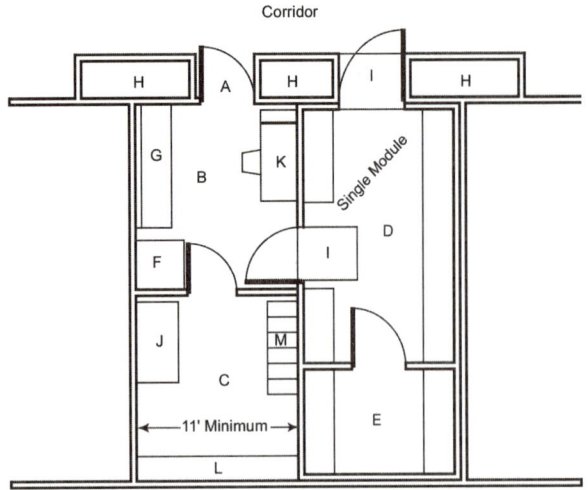

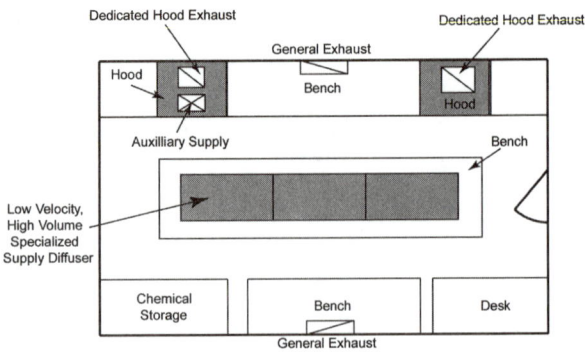

Figure 6-10 *Air systems in typical laboratory.*

Figure 6-9 *Plan view of typical laboratory. (Diagram modified with permission from DiBerardinis et al. 1993.)*

and involves loss of perimeter space, which is very valuable in many laboratories.

Radiant Panel

The second option for zone heating is the use of radiant panels. Radiant panels generally come in five configurations:

- Metal ceiling panels
- Embedded piping
- Electric ceiling panels
- Electrically heated ceiling or floor
- Air-heated floors

Similar to the perimeter baseboard systems, all of these configurations use the principle of free convection, with the addition of radiant heating. These systems do not require any laboratory floor space since they are integrated into the infrastructure of the laboratory.

The radiant panels are made of various materials with steam, hot water piping, or electrical elements passed through them. These panels can then be located within floors, walls, or ceilings. Unfortunately, with radiant panel systems, temperature control is often slow when heavier materials such as concrete are used. Another temperature-related problem for radiant panels is the uneven distribution of heat around objects. Finally, radiant panels, like perimeter baseboard systems, do not account for humidification control issues.

Supply Air

The final heating option involves the use of a central air distribution system to heat the zones. Typically this option will be the better choice in laboratory applications due to its quick response times in temperature control and even heating distribution.

EXHAUST AIR SYSTEM

With the room layouts completed and the exhaust airflow requirements known and the exhaust hoods chosen (see Chapter 5, "Exhaust Hoods"), the primary exhaust air systems can be selected and sized. Since the exhaust system in a laboratory facility is often the most critical system for ensuring the safety and health of the occupants, its design is accomplished first. Further, as most laboratories do not allow recirculation of air (100% ventilation air), the focus of the following section is on 100% exhausted spaces. However, in cases where recirculation is allowed (e.g., adjacent non-laboratory spaces), the guidance for general exhaust can be followed.

A typical plan view of a laboratory is shown in Figure 6-9.

Figure 6-10 shows the supply and exhaust systems for the typical laboratory.

The air systems within the typical laboratory are:

- Hood exhaust
- General exhaust
- General supply
- Auxiliary supply

The exhaust systems are addressed in this section and the supply systems in the next section.

A laboratory space can have any combination of general and specialized (hood) exhaust:

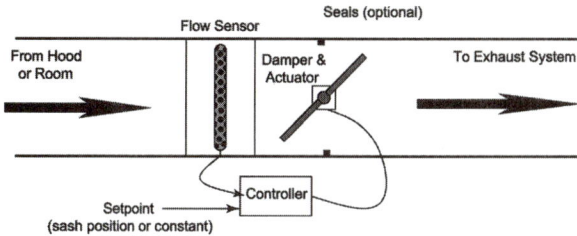

Figure 6-11 *Generic flow controller.*

- Only general exhaust
- Only hood exhaust
- General and hood exhaust

The application of the various exhaust points within a laboratory is dictated by the safety and health codes and standards detailed in Chapter 3, "Laboratory Planning." In review, hood exhaust is typically required when dangerous, toxic, or carcinogenic materials are being handled or used in such a manner as would be dangerous for human contact, whereas, general exhaust is used for less toxic, less odorous substances and procedures, such as chemical storage, experiment preparation, and cleanup. General exhaust is also used to minimize the impact that heat-producing equipment (e.g., ovens) has on the occupant by removing the heat prior to its introduction into the occupied space.

The design and layout of an exhaust system for a laboratory facility is accomplished through a series of steps.
1. Determine type of exhaust system
2. Determine separation of exhaust systems
3. Accomplish duct design and layout
4. Accomplish exhaust stack design
5. Select exhaust fan(s)

Determine Type of Exhaust System

As introduced at the end of Chapter 5, "Exhaust Hoods," there are two primary classifications of exhaust hoods: constant volume and variable volume. Constant air volume is required for specific systems (perchloric acid fume hoods, radioisotope fume hoods, and biological safety cabinets) and is typically used in small installations with few hoods to simplify the installation and operation and to minimize the required capital investment. In the constant volume exhaust system, the volume of air exhausted from a laboratory room or hood is constant regardless of sash position or activity within the room. Typically, the constant airflow is an inherent characteristic of the exhaust hood. In some instances a flow controller is used to ensure that the exhaust from a hood or room is constant. The flow controller is identical to that used for VAV systems, except that it maintains a constant setpoint instead of varying the airflow with sash position or room

activity. Figure 6-11 details the key components of a generic flow controller.

Different manufacturers have different means of measuring the airflow (type and location of sensor) and for controlling the airflow (flat, airfoil, or plunger damper). However, the basic arrangement and principle of action shown in Figure 6-11 apply to all flow controllers.

In larger systems and where operating costs need to be minimized, variable volume systems are applied. In VAV systems, the air from a laboratory hood is varied to maintain a constant face velocity for all sash positions. In those systems, the flow controller uses an algorithm relating sash position to required airflow to achieve the constant face velocity. The controller maintains a minimum airflow to ensure airflow even when the sash is closed. Therefore, as the sash is closed, the airflow is reduced from its peak value to a minimum value. Since the power required to exhaust the air is proportional to the cube of the airflow (see Equation 6-1), it is often very economical to operate a VAV exhaust system.

$$P_{new} = P_{old}\left(\frac{Q_{new}}{Q_{old}}\right)^3 \qquad (6-1)$$

where
P = fan power, hp (kW)
Q = fan flow rate, cfm (L/s)
new = new operating point
old = old operating point

The primary difference between constant and variable air volume systems is that in the variable air volume system each hood and space must have a flow controller attached to it and the fan must have some form of flow control. Ideally, the fan flow control would be through the use of a variable speed drive that is controlled to maintain a constant negative pressure at some point in the exhaust ductwork (typically two-thirds of the way to the farthest run). However, fans have been installed to just "rind the fan curve." While there will be energy savings, the actual savings are approximately a third of what a variable speed drive system will have.

Determine Separation of Exhaust Systems

Exhaust systems within a building are separated for a number of reasons, including:
- Substances in exhaust air
- Type of exhaust (constant versus variable)
- Location of exhaust requirements

The key criterion for the separation of the exhaust airstream is the substances in the exhaust air. Several exhaust sources can be combined if they are from similar

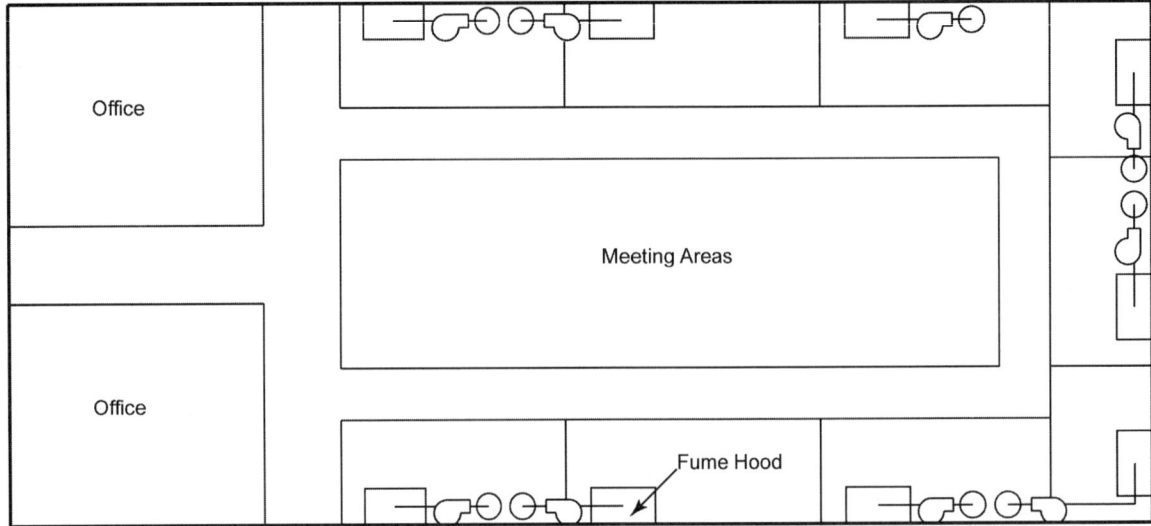

Figure 6-12 *Dedicated exhaust system—single-story.*

processes and classes of exhaust. For example, if all hoods within a facility are used for the purpose of teaching general chemistry with similar levels and types of chemicals, then the exhaust streams can be combined. However, the exhaust from Biosafety Level 1 and Level 3 laboratories, for example, should not be combined due to the different treatment and handling procedures of the exhausts. A further consideration on whether or not two streams can be mixed is if chemicals in the two streams will react with one another. One such example of when not to mix two exhaust streams is when one stream contains perchloric acid and another contains organic materials. When the two mix, an explosion may occur, creating a safety problem.

The second criterion, constant versus variable air volume, typically dictates the separation of systems. The only time when constant and variable air volume exhaust streams are mixed is when the constant volume source has a flow controller installed in it. It is also common that in larger systems the mixing of constant and variable airstreams is undesirable due to the use of one or both of the streams in heat recovery or treatment prior to leaving the building.

The final criterion in determining the separation of systems is the physical location of the exhaust points. As was determined during Chapter 3, "Laboratory Planning," the layout of service corridors, or shafts, and mechanical rooms dictates which spaces and hoods can be physically located together.

Accomplish Duct Design and Layout

Based on the level of separation of systems, there are two typical layouts of exhaust systems in laboratory facilities:

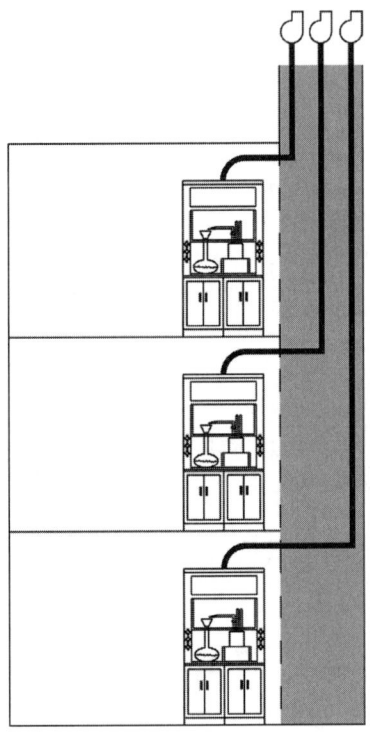

Figure 6-13 *Dedicated exhaust system—multi-story.*

- Dedicated
- Manifolded

Dedicated. In a dedicated exhaust system, each hood has its own exhaust fan and stack (see Chapter 9, "Exhaust Stack Design"). This type of system is commonly used due to its adaptability for various substance filtration requirements and simplicity of initial balancing.

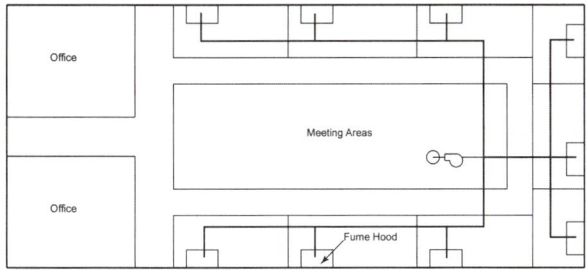

Figure 6-14 *Manifolded exhaust system—single-story.*

However, this arrangement does require numerous roof penetrations and stacks, which can result in higher initial and operating costs and can complicate design and space availability in multistory applications. Also, a dedicated system can make the addition of hoods and alteration of laboratories costly, both in initial design and later renovations.

It is less complicated to apply dedicated exhaust systems to single-story laboratory buildings. Exhaust fans dedicated to individual hoods are used to deliver exhaust gases directly to an exhaust stack on the roof of the laboratory building. Figure 6-12 shows an example of a single-story laboratory facility with a dedicated exhaust system.

In multi-story buildings, the use of dedicated exhaust systems is more complicated since each duct must end up at the roof. Therefore, mechanical chases must be used to channel the ductwork to the roof. Figure 6-13 details such a system.

Manifolded. The other option for the arrangement of hood exhaust is the use of manifolding. This type of system works by joining several hood exhausts into one main junction, which is then run to an individual stack on the roof. This type of system typically has several advantages, including:

- Lower capital and operational costs
- Fewer exhaust stacks
- Lower redundancy costs
- Greater adaptability of design
- Simpler effluent treatment
- Dilution and momentum

Using the information previously presented on which hoods can be combined, the layout of the exhaust system is simply a matter of convenience and location. Using the single-story and multi-story example from the dedicated system, Figure 6-14 and Figure 6-15 detail the layout of a manifolded system.

Duct Sizing. Laboratory duct velocities and static pressures are dependent upon numerous aspects of the exhaust materials they are designed to handle. For instance, the range for static pressure for exhaust ductwork is 0.25 to 0.50 in. w.g. (62.25 to 124.50 Pa). Higher levels of pressure should be applied to laboratories that contain hazardous materials to ensure containment, such

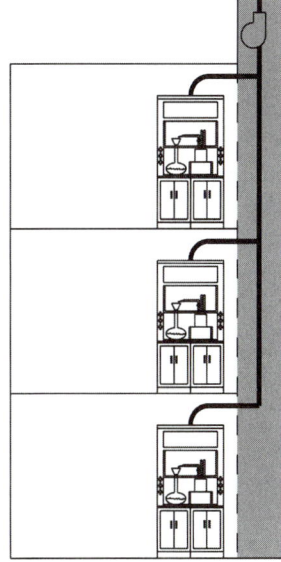

Figure 6-15 *Manifold exhaust system—multi-story.*

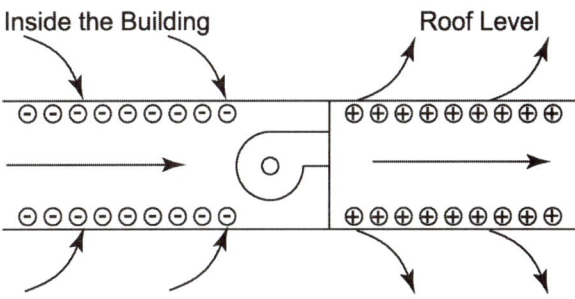

Figure 6-16 *Exhaust fan pressure relationships.*

as in Biosafety Level 2 and higher laboratories. Lower levels of pressurization thus can be used in such instances as teaching laboratories and low level chemical laboratories. In terms of velocities within ducting, these are typically in the range of 1,000 to 2,000 fpm (5.08 to 10.16 m/s) depending on the need to ensure that deposits do not form and sound control. This is particularly important in such applications as perchloric acid hood exhaust.

Accomplish Exhaust Stack Design

Detailed guidance on the design of exhaust stacks is in Chapter 9, "Exhaust Stack Design." The general characteristics of good exhaust stack design are:

- High discharge velocity (> 3000 fpm)
- Location away from air intakes and building entrances
- Consideration of prevailing wind direction
- Sufficient stack height

Select Exhaust Fan(s)

With the exhaust ductwork and stack designed, the exhaust airflow and static pressure losses can be calculated. However, the proper selection of fans for a specific

Table 6-2: Centrifugal Fan Type Application

Fan Type	Performance Characteristics	Applications
Airfoil	Peak efficiency at 50-60% maximum volume, with maximum volume at peak efficiency	Low-, medium-, and high-pressure applications with relatively clean exhaust air
Backward Curved	Similar, but lower efficiency than airfoil design	Good for corrosive environments where airfoil would corrode
Radial	Higher pressure characteristic but lower efficiencies	Material handling with high-pressure requirements – rugged wheel is sometimes coated and is simple to repair

(Source: Table 1, Chapter 18, *1996 ASHRAE Handbook – HVAC Systems and Equipment*.)

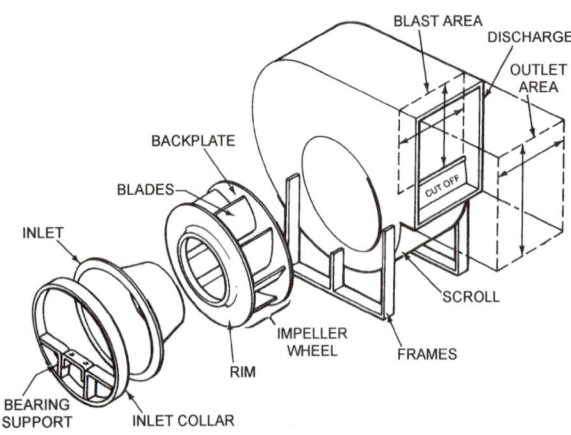

Figure 6-17 *Centrifugal fan isometric (courtesy of ASHRAE).*

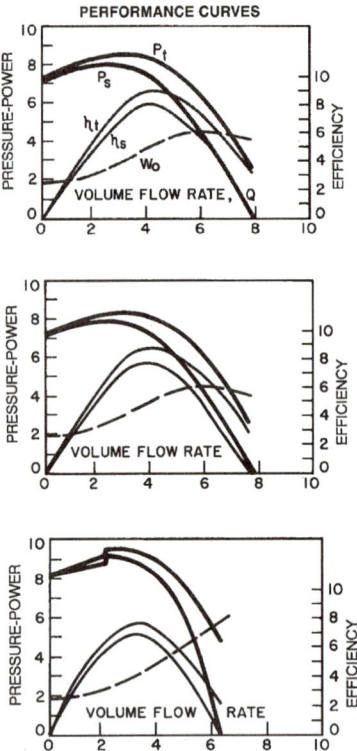

Figure 6-18 *Centrifugal fan performance characteristics (courstesy of ASHRAE).*

application goes far beyond just airflow and static pressure. Other items that must be considered include:

- Location
- Construction
- Controls
- Redundancy

Like the entire design process, the selection and application of a fan is an iterative process, with each of the above criteria affecting one another.

Location. Although hazard levels vary within different types of laboratories, it is generally good practice to locate the exhaust fan on the roof of the building. As is shown in Figure 6-16, the ductwork prior to the exhaust fan is negatively pressurized and after the exhaust fan, positively pressurized. Therefore, if the fan were located inside the building (below the roof level), the primary concern would be the potential of exhaust air leaking from the positively pressurized ductwork downstream of the exhaust fan. Locating exhaust fans on the roof, however, has a direct impact on the type of fan that can be used, as they must be able to withstand adverse weather conditions. Noise problems should be considered if the fan is to be located inside the building.

Construction. The construction of a fan is dependent upon several key criteria, including:

- Fan Type
- Pressure class
- Material composition
- Protective coating
- Special considerations

The three general types of exhaust fans used are centrifugal, vane-axial, and special application fans. Centrifugal fans (see Figure 6-17) operate by a centrifugal force that rotates the inlet air and moves it from inside to outside of an impeller wheel. There is also kinetic energy imparted to the air by the virtue of the air's velocity leaving the impeller.

Typically, centrifugal fans are belt-driven, with the fan inside or outside the airstream. For laboratory appli-

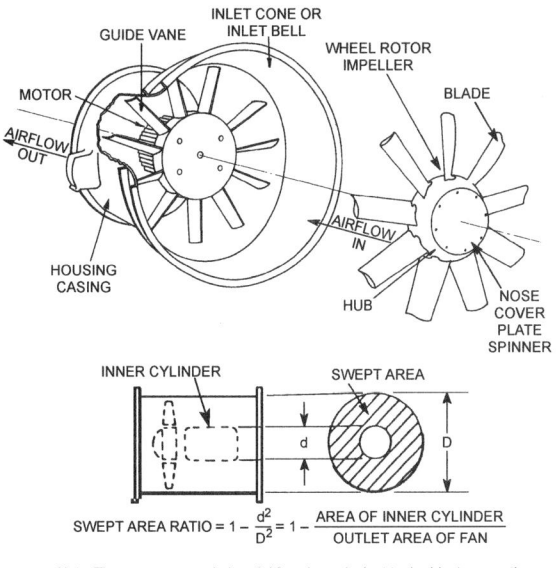

Figure 6-19 *Vane-axial fan isometric (courtesy of ASHRAE).*

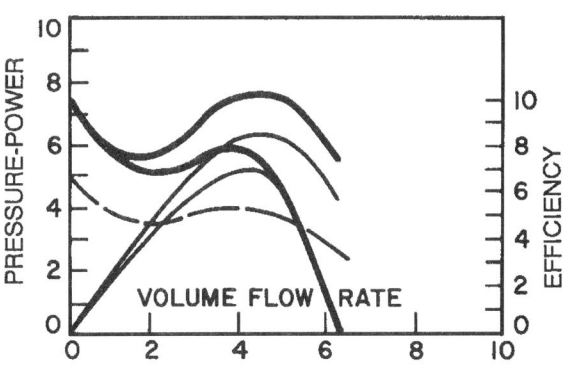

Figure 6-20 *Vane-axial fan performance characteristics (courtesy of ASHRAE).*

(Source: MK Plastics Corp.)

Figure 6-21 *Induced draft fan for perchloric acid fume hoods.*

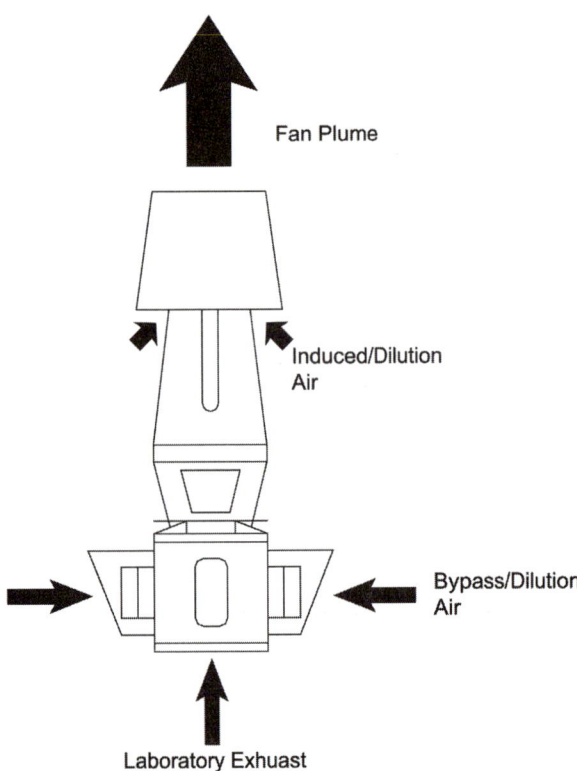

Figure 6-22 *Dilution fan schematic.*

cations, dual-belt models are used for redundancy purposes. The three centrifugal fan types applied to laboratory systems are airfoil, backward curved, and radial. Table 6-2 lists the application of each of these fans, and Figure 6-18 shows the general performance characteristics.

For laboratory applications, centrifugal fans are very reliable, easy to maintain, can be applied to a wide range of requirements, and have good pressure-volume characteristics. However, due to the belt drive, centrifugal fans are typically located within the building or mechanical room. Location outdoors typically requires special housing and motors.

For outdoor applications, vane-axial fans are more often used since the motor can be an integral part of the fan. Figure 6-19 shows an isometric view of a vane-axial fan. Due to the location of the motor, absence of belts, and inherent characteristics, vane-axial fans are more compact and have higher efficiencies than centrifugal fans. A primary benefit of the vane-axial fan is that it is located in-line with the airstream. This avoids the pressure losses of the 90° turn required for centrifugal fans. However, it should be noted that vane-axial fans could be belt-driven with the fan motor located outside the airstream. This is sometimes beneficial in highly corrosive applications that would damage the motor.

The performance characteristics for a vane-axial fan are shown in Figure 6-20.

Special application fans include laboratory induced draft fans and dilution fans. Induced draft fans are used in perchloric acid fume hoods and other highly corrosive or high temperature applications to avoid damage to the fan components and facilitate repair. A perchloric acid exhaust fan system (see Figure 6-21) should be separate from the laboratory's fume hood exhaust system.

Dilution fans (see Figure 6-22) are designed either as mix-flow or centrifugal. The fans have vertical discharge, provided with a windband, and may have a mixing intake plenum. Because the fan is designed for outdoor application, the ratio of outdoor air to indoor air is adjustable. These fans can be designed in multiple sets and provided with sound attenuation. Fan exit velocities can be as high as 7000-8000 fpm (35.6-40.6 m/s), creating an effective stack height as well as a physical stack height. Because the fan mixes outdoor air downstream, process concentrations can be reduced.

Regardless of the fan type used, the fan must meet the requirements of the Air Movement and Control Association (AMCA) through ASHRAE Standard 51-1985 (ANSI/AMCA Standard 210-1985) and those specified in NFPA Standard 70.

The fan types presented can be easily applied to low-, medium-, and high-pressure systems. It is fairly common to have negative pressures in a laboratory system ranging from 0.65 to 0.5 in. w.g. (62 to 124 Pa). As the negative pressure increases, sturdier fans and ductwork are required. The discharge velocity from the fan and exhaust stack should typically be at least 2,500 fpm (12.7 m/s) to avoid reentrainment of pollutants into the building and to ensure the exhaust airstream is properly diluted prior to reaching ground level or other occupant space.

The sturdiness of the exhaust fan and the corrosivity of the exhaust air determine the material composition of the fan. To avoid corrosion problems from the exhaust air or subsequent condensation, laboratory fans are often coated or constructed using special materials, such as stainless steel, fiber reinforced polyester (FRP), or polyvinyl chloride (PVC). Typically FRP is more popular than PVC due to its increased strength and resiliency. For general, noncorrosive exhaust airstreams, cast iron is used. Cast iron fans have low maintenance requirements and have a long life span. In less critical areas, steel fans can be used. However, these fans typically have a shorter life span and require additional maintenance relative to cast iron fans.

In addition to the fan material, protective coatings are often added to improve the fan's life span and to reduce maintenance and system downtime. In many instances, the coatings are used on lower cost fan materials (steel) to

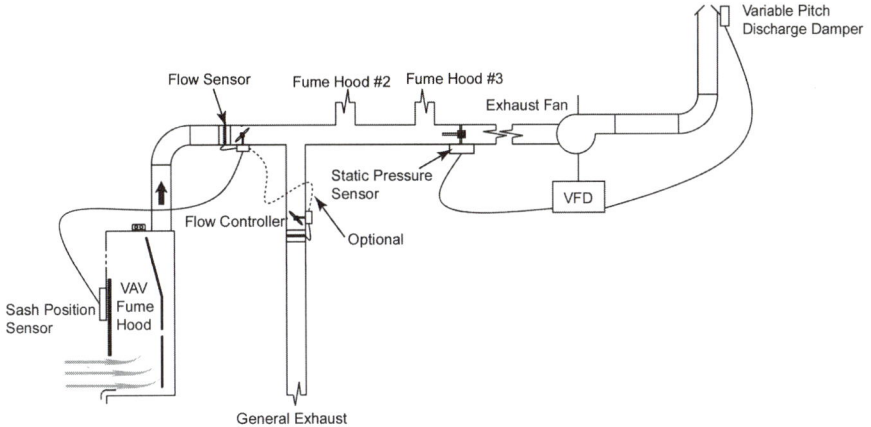

Figure 6-23 *Typical VAV exhaust control components.*

avoid the use of higher cost materials (stainless steel). Multi-layer epoxy can be used for chemical, salt, solvent, and corrosion resistance. Hot galvanization can be used for corrosion resistance and to retard tarnishing. Cadmium plating, vulcanized rubber, and zinc chromate paint are other protective coatings used.

The final three fan construction concerns are classified as special as they are dependent upon unique application. First is the use of a drainage connection for the fan. Since the use of caps on exhaust stacks is not recommended due to the adverse impact of the caps on dispersion of the exhaust air, the fan must either have an integral drain installed in it or one installed in the discharge/suction of the fan.

Second, in those cases where the lower explosion limit of the exhaust airstream is greater than 25%, nonsparking wheels and explosion-proof motors are required. The AMCA spark-resistant standards, which depend upon the level of explosion hazard, should be followed. They are:

- Type A Construction: Use of nonferrous material for all fan parts that are in contact with the exhaust air. This is to ensure that any mechanical failure or movement of the wheel will not cause a spark.
- Type B Construction: Use of nonferrous material for the wheel and the ring that the shaft of the fan passes through. Typically, this type of construction is suitable for most applications.
- Type C Construction: Construct fan in a manner so that two ferrous parts of the fan are unable to have contact. This type of construction is suggested for all exhaust fan construction for hoods.

Third, the fan, fan housing, and exhaust stack must be designed and constructed to withstand a wind load of 30 lb/ft^2 (146 kg/m^2). Good practice requires that the exhaust stack be self-supported to avoid excessive loading on the fan housing (Wunder 2000).

Controls. While the hoods, ductwork, fans, and exhaust stacks are designed for peak operating conditions, it is the controls that maintain the safety and consistency of operation at nonpeak conditions. For constant volume systems, the controls turn on and off equipment. There is no need for modulation of devices since the exhaust air volume is constant.

However, the need for proper controls is critical in VAV systems where the quantity and velocity of exhaust air changes based on sash position. For these systems, there are four key control loops that must be evaluated and integrated:

- Hood exhaust
- General exhaust
- Fan control
- Exhaust stack velocity

Figure 6-23 shows a general control system layout for a VAV laboratory exhaust system.

The hood exhaust is controlled based on sash position to maintain a constant hood face velocity. A minimum of 40 cfm/ft hood width (0.0155L/s-cm) is required in accordance with NFPA 45. The general exhaust is controlled either to maintain a constant exhaust airflow quantity or to track the hood exhaust flow. Typically, a constant volume is desired to maintain the indoor air quality.

The exhaust fan is controlled to maintain a constant negative pressure within the duct. This is required for the VAV controllers to function properly. While pressure-dependent VAV controllers are available, pressure-independent controllers are almost entirely used for their more consistent performance characteristics. The exhaust airflow rate through a pressure-dependent controller varies based upon the system pressure, whereas pressure-independent controllers adapt for the changes in system pressure.

The final control component in laboratory exhaust systems is the variable pitch discharge dampers. These dampers are used to maintain a constant discharge velocity of the exhaust airstream to avoid reentrainment or dilution problems. Typically, these dampers require no

Table 6-3: Exhaust and Supply Air Systems

Exhaust System	Supply System	Comments
Constant Volume	Constant Volume	Good combination – easily maintains pressure requirements but is energy intensive
	Variable Volume	Must be controlled to maintain pressure relationships, good for systems where exhausts are turned on and off
Variable Volume	Variable Volume	Excellent combination – supply controlled to maintain pressure relationship
	Constant Volume	Do not use – will result in loss of pressure control and serious problems

control interface as they operate off system pressure. However, it is possible to control the dampers based on the state of the VFD.

As with any VAV system, the sensors and controllers have a minimum airflow that is required for stable and safe operation. This minimum flow is often described using the turndown ratio defined in Equation 6-2:

$$Turndown\ Ratio = \frac{minimum\ airflow}{design\ airflow} 100 \quad (6\text{-}2)$$

A minimum turndown ratio of between 10% and 20% is recommended for most systems. For systems where precipitation of particles is a concern, the turndown ratio may be as high as 50% to 80%. Specifically, NFPA Standard 45-1996, Appendix A-6-4.5, recommends that, "a minimum exhaust volume of 40 cfm per lineal foot of interior hood width generally ensures that contaminants are exhausted from a hood; however, some activities require higher volumes to ensure that flammable atmospheres will not occur in the hood."

Redundancy. One of the more important issues in fan selection and design is the reliability of the fan. For the system to be 100% reliable, the system must be operational 24 hours a day, every day of the year. Therefore, many critical laboratory facilities have redundant (backup) exhaust fans. This ensures the safety and operability of the laboratory environment in case one or more of the fans experiences a malfunction or requires maintenance. In such cases fan downtime can create an intake of outdoor air from the roof through the affected hoods' exhaust ducts and cause great risks to personnel using those hoods. It is also for this reason that double belt-driven fans are common place in laboratories, due to their ease of belt replacement without requiring the shutdown of the hood or area it serves. In terms of cost, redundancy means higher capital costs, particularly in cases where dedicated exhaust has been chosen for the hoods.

In addition to the redundant fans and belts, many laboratory facilities have backup power generators to ensure the fans operate when electricity is not available from the main grid. In general, as the hazard level increases, the need for a reliable system increases. The higher reliability is achieved through better quality materials and components and the use of redundant components.

SUPPLY AIR SYSTEM

The primary role of the supply air system is to provide ventilation air to the laboratory space in sufficient quantities for the comfort and safety of the occupants. It is the type of exhaust air system (constant or VAV) and the specific space pressure relationship requirement that determines the type of supply air system (constant or VAV) and the volume of supply air required.

With the exhaust system type already determined and the exhaust air duct system designed, the design of the supply air system is simplified. It is common practice to match the type of supply system with the exhaust system. Table 6-3 lists the possible arrangements of exhaust and supply air systems that are commonly used.

Common practice is to have the type of supply system (constant or variable air volume) match the type of exhaust system. Therefore, determining the type of supply system is simple.

The layout of the supply system is also often a simple task as it usually parallels the exhaust system. As was determined during the planing phase (Chapter 3, "Laboratory Planning"), the distribution option for ductwork layout is integral to the layout of both spaces and the building. During the layout of ductwork, both supply and exhaust, the primary concern is accessibility for cleaning/decontamination and for maintenance of other systems (valves, piping, etc.).

With the type and layout determined, the design of the supply system focuses on the central system and controls. The specific steps to follow are:

- Verify supply air quantity
- Evaluate need for auxiliary air
- Select specific system type
- Select air-handling unit
- Determine control strategy

Verify Supply Air Quantity

Since the laboratory hood exhaust air quantity is based on capture velocity and the general exhaust air quantity is based on maintaining sufficient air exchange rates for safety, the determination of the supply air quantity simply focuses on maintaining the proper pressure (airflow direction) relationship. Due to a typical control error of ±5%, it is recommended that the supply air quan-

Table 6-4: Sample Exhaust and Supply Air Rate Verification Form

Room Data				Exhaust, cfm (L/s)			Pressure Relationship		Supply, cfm (L/s)	
No.	Use	Air Exchange Rates, ach	Room Volume, $ft^3(m^3)$	Required	Hood	General	±	%	Source*	Required

*Source is either clean (e.g., hallway or anteroom) or dirty (e.g., adjacent laboratory).

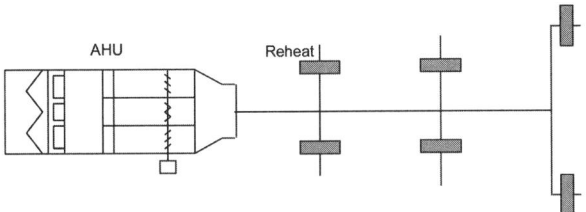

Figure 6-24 *Single-zone constant volume system.*

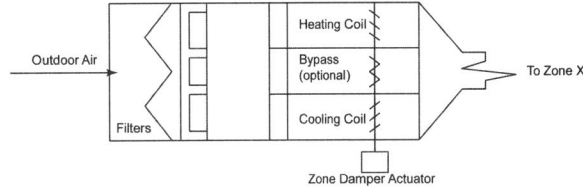

Figure 6-25 *Schematic of central multi-zone air-handling unit.*

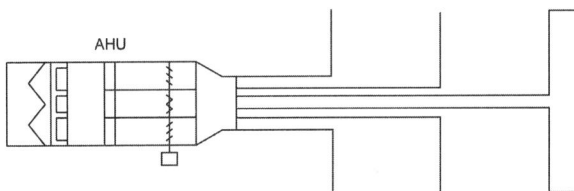

Figure 6-26 *Multi-zone constant volume system.*

tity be 10% greater than the exhaust air when the room must be positively pressurized and 10% less than the exhaust air when the room must be negatively pressurized. In situations with poor construction and high levels of infiltration/exfiltration, this differential may need to be increased to 15% or 20%. Thus, for pressure controlled spaces, better air conservation is achieved as the sealing of the building is improved.

Once the supply air quantity for a space has been determined, its value must be compared to the minimum air exchange rates for safety. This is especially critical for negatively pressurized spaces where the supply air volume is less than the exhaust air volume. In these instances, if the air being drawn into the space is not clean (from an adjacent laboratory) the supply and exhaust air volumes for the space may need to be increased to ensure the proper ventilation air exchange rate is maintained. Table 6-4 is an example form that can be used to document the proper exhaust and supply air rates.

Select Specific System Type

Depending upon the distribution layout being used, the supply air system can either be centrally located (central system) or can be located within the room (unitary system). Either of these systems can be constant or variable air volume.

Central Systems. The specific type of central system selected, as was previously determined, is either constant or variable volume. However, there are many options available in meeting the space requirements including:.

- Constant volume single zone
- Constant volume multi-zone
- Multi-speed
- Variable volume—single duct
- Variable volume—dual duct

The constant volume system is the simplest system in which a constant volume of air is supplied to the space. In a single-zone system, there is a central air-handling unit supplying air to the ductwork. The entire system is treated as a single zone as each space receives air of the same temperature. Space temperature is then controlled using some form of reheat. Figure 6-24 details a single-zone constant volume system.

A constant volume system provides a constant volume of air at varying temperatures to meet the space load.

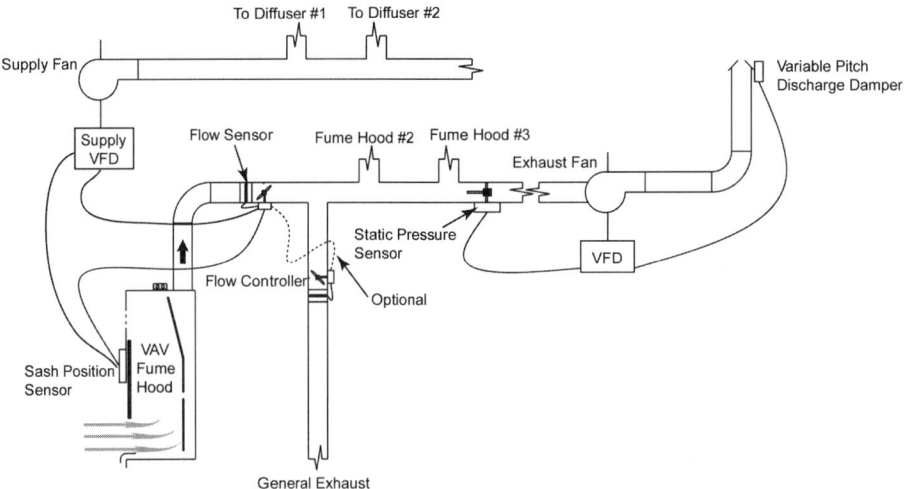

Figure 6-27 *Schematic of a VAV system.*

The warmest zone determines the supply air temperature. Zones that are overcooled then use some form of reheat to maintain comfort.

In an effort to reduce operating costs, while still maintaining the capital cost benefits of a central system, multi-zone constant volume systems are often used. These systems divide the space into two (often called a *dual-zone system*) or more (multi-zone system) zones. In a multi-zone system, the central air-handling unit has two or more "decks" where heating and cooling can be individually supplied to each zone. A schematic of the central multi-zone air-handling unit is shown in Figure 6-25 and a system layout in Figure 6-26.

Typically, a multi-zone system has a cooling and heating coil providing cool or heat. Since it is very inefficient to mix mechanically cooled or heated air, the cooling coil is turned off during the heating season and unconditioned air is bypassed through the cooling coil. The heating coil is then turned off during the cooling season. A way to be able to provide either cooling or heating to a space is to install a bypass section where either cooled air and bypass air or heated air and bypass air is supplied to the space. However, the main problem with multi-zone systems in laboratories is that any bypass air is unconditioned outdoor air. This results in poor comfort control in the occupied spaces due to lack of humidity control. Therefore, care must be taken in the design and control of multi-zone systems in laboratories. Additional information on the design and application of single- and multi-zone systems can be found in Chapter 2, *1996 ASHRAE Handbook—HVAC Systems and Equipment.*

A hybrid between a constant volume and a variable volume system is a multi-speed system. In this system, the ductwork is the same as a single-zone constant volume system, with the central fan having the ability to operate at multiple speeds/airflows. The fan is typically controlled based on the status of exhaust system, decreasing their speed as the exhaust systems shut down. With the cost of variable speed drives and system decreasing, the use of multi-speed system is infrequent.

Due to the potential energy savings and greater flexibility on control, most laboratory systems are designed as variable air volume (VAV). In VAV systems, the central system provides a constant temperature of air to the space. At each space is a terminal unit that varies the volume of air to the space to meet comfort and pressurization needs. When excess air must be provided to the space for pressurization requirements, the air is reheated to maintain comfort. Figure 6-27 is a schematic of a typical VAV system.

The two primary types of variable air volume systems are singe duct and dual duct. The single-duct VAV system is the traditional type system where constant temperature air is supplied to the space in varying quantities through a terminal unit. The terminal unit (VAV box) is simply a flow sensor, damper, and reheat coil with controls. Figure 6-28 shows a single-duct VAV system. The single-duct VAV system is simple to design, construct, and probably has the least expensive first cost of the supply systems available. Accurate room control with almost no instability is possible, and the system has the design advantage of being easier to modify in a remodel request. However, this kind of system uses a great quantity of energy.

The other VAV system uses two supply air ducts – a hot duct and a cold duct, hence, a dual-duct VAV system. Instead of having a reheat coil as in a single-duct system, which is energy intensive, the dual-duct system directly supplies heated or cooled air to the space as required to meet comfort conditions. Dual-duct VAV systems can be very efficient. However, special attention must be given in selecting and maintaining the two box dampers to minimize leakage of hot or cold air through a closed damper. Further, there must be sufficient space to lay out two par-

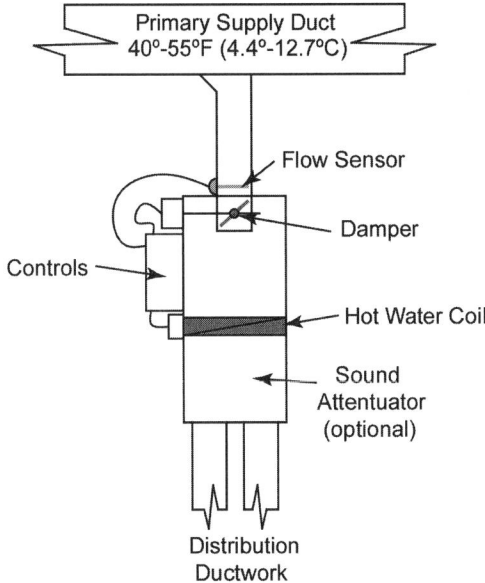

Figure 6-28 *Single-duct VAV system schematic.*

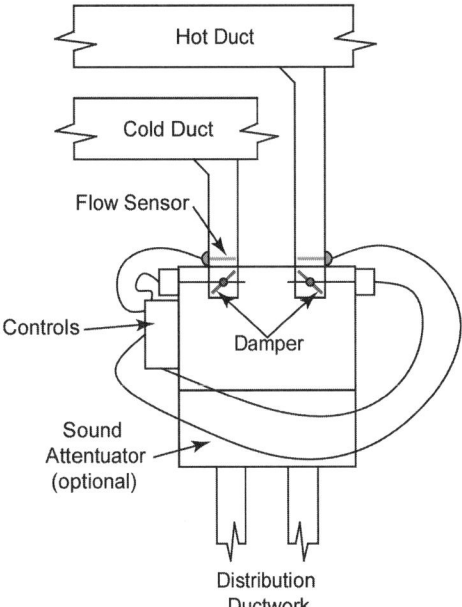

Figure 6-29 *Dual-duct VAV system schematic.*

allel duct runs for the hot and cold ducts. Figure 6-29 shows a dual-duct VAV system.

Unitary Systems. Converse to the central systems, unitary systems uses individual air-handling units to provide conditioning for each laboratory. Unitary systems are used for several reasons:

- Large laboratories
- Isolation of laboratories
- Limited distribution space

As the size of individual laboratories increases, there is a point when it is more economical and simpler to install separate air systems for each laboratory. A simple example is that of cleanrooms in industrial and pharmaceutical industries. The second need for unitary systems is when spaces need to be physically isolated from one another. This is the case when dealing with highly toxic and dangerous substances in biological laboratories. The final typical reason unitary systems are installed is a lack of distribution space. Since the exhaust air ductwork takes precedence, when there is limited space, the supply and outdoor air ductwork often must be contained within the laboratory space.

A typical unitary system is shown in Figure 6-30.

Similar to the central system in nearly all aspects, these unitary systems are sized to meet the specific requirements of each laboratory and can offer immense adaptability. They can also be operated using constant or variable air volume control methods to match the exhaust control. However, unitary systems do have a number of drawbacks. The primary drawback is that each unitary system must have access to outdoor air. This often requires a separate outdoor air intake louver (envelope penetration) for each system. This often eliminates the potential for central, economical heat reclaim and air treatment. Another drawback is chilled and hot water piping connections must be made to each system. This can be a substantial cost increase. Finally, in some situations, the unitary system cannot be physically located in the laboratory space but must be in a separate mechanical room.

Due to the need for accessibility for maintenance, a separate mechanical room for each unitary system will significantly increase the size of the building or significantly reduce the size of the laboratories. Biological, clinical, and animal research laboratories typically do not allow placement of the unitary system within the laboratory, due to concerns for possible system contamination from the experiments (system and ductwork upstream of supply fan under a negative pressure). Also, if the unitary system is located either above the space or adjacent to the space, significant sound pressure may be generated and must, therefore, be controlled.

Evaluate Need for Auxiliary Air Supply

Due to the high volume of air being exhausted through laboratory hoods relative to the volume of supply air required for comfort, methods have been developed to minimize the volume of conditioned air supplied to the laboratory. The primary method that is feasible is similar to a short circuit kitchen hood, where tempered outdoor air is supplied near the exhaust hood with the expectation that this tempered air will be directly exhausted and not impact the space conditions. For laboratory systems, this strategy is called auxiliary air.

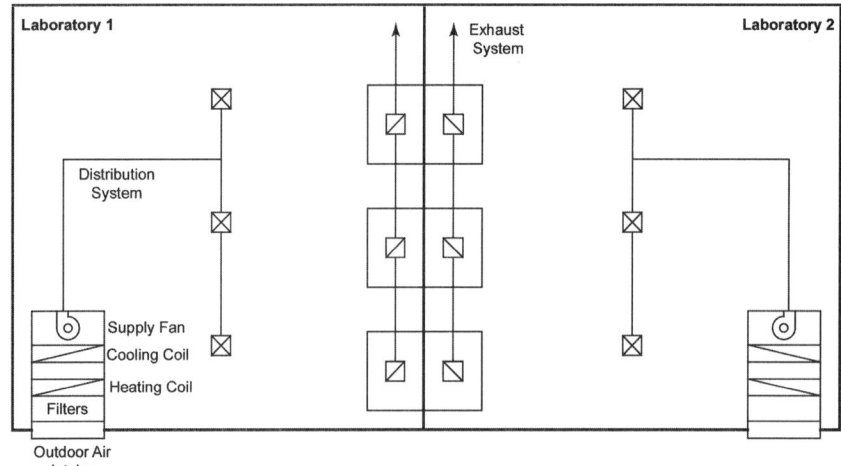

Figure 6-30 *Typical unitary system layout.*

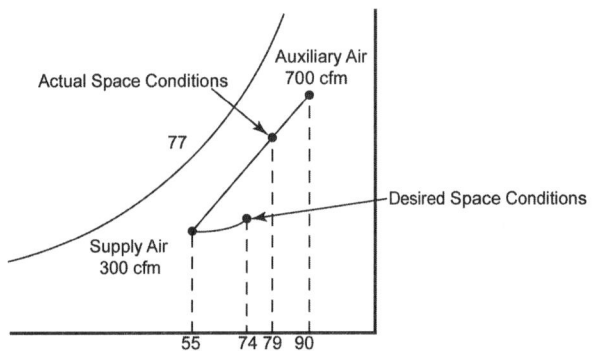

Figure 6-31 *Auxiliary air impact space conditions.*

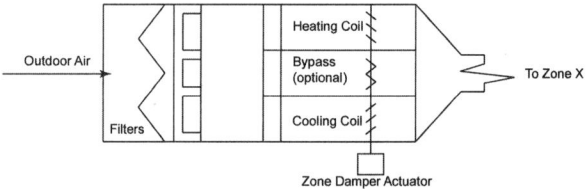

Figure 6-32 *Typical air-handling unit schematic.*

While each laboratory hood manufacturer accomplishes auxiliary air systems slightly differently, the basic principle is that tempered air is supplied near the face of the hood. This is typically above the hood opening or along the sides of the hood opening. Typically, the required supply volume of conditioned and ventilation air is reduced up to 70% in auxiliary air supply systems, where a constant speed blower is used to pull in a constant supply of outside air.

Tempering Only. Since the auxiliary air is directly exhausted, it is either unconditioned, filtered outdoor air or tempered, filtered outdoor air. Tempering is used to avoid cold drafts on the user of the laboratory hood. Therefore, if the outdoor air temperature falls below about 60°F (15.6°C), tempering is required.

The heating of the outdoor air is typically accomplished on a central system using steam, hot water, direct-fired natural gas, or some combination of the three. Often multiple direct-fired natural gas make-up air units are used for their simplicity and low capital cost. A summary of the auxiliary air heating options is given in Table 6-5.

Effects on Humidity and Latent Loads. The primary concern with auxiliary air systems is the impact that unconditioned or tempered air has on the comfort of the space. When designed properly, the auxiliary air should not mix with the occupied space. However, in improperly designed or operated systems, introduction of the auxiliary air into the occupied space will increase the space humidity through an increase of the space latent load. The result will be an increase in occupant discomfort due to a clammy, sticky perception of the skin.

The effect of introducing the auxiliary air into the space can be determined by a psychrometric analysis of the system. This is accomplished by using a psychrometric chart and plotting the mixing of the supply airstream and auxiliary airstream with the room air. Figure 6-31 is an example of such an analysis.

Therefore, to avoid adverse comfort effects, the auxiliary air must be supplied directly to the laboratory hood and not have a chance to mix with the occupied space air. However, about a 50% reduction in conditioning costs is often worth evaluating the application of auxiliary air systems, which in some cases could eliminate the savings from the reduced conditioning of supply air. Included in this evaluation should be considerations of some of the drawbacks of implementing such a system. These are discussed in Chapter 5, "Exhaust Hoods." VAV or reduced

Table 6-5: Auxiliary Air Heating Options

Heat Sources	Benefits	Drawbacks
Steam	Simple and economical	Must properly size coil and stream trap to avoid freezing at part loads
Hot water	Good part-load control characteristics	Typically not used when outdoor air temperature is below 32°F (0°C)
Direct-fired natural gas	Can be easily added to systems	Limited size availability

Table 6-6: Laboratory Filtration Options

Filter	Application	Notes
30% Efficient*	Prefilter	Used to extend life of higher efficiency filters
60% Efficient	Office, nonlaboratory spaces	Prefilter typically not required
85% Efficient	General laboratory	Minimum efficiency
95% Efficient	Biological and clinical laboratory	Sometimes used as final filters
99%+ Efficient	Animal and biomedical research and cleanrooms	HEPA (high efficiency particulate air) filters used as final filters
30% Efficient	Gas phase filtration	Used when outdoor air is polluted

*Efficiency defined using ASHRAE Standard 52.1 - 1992 (ASHRAE 1992).

air-flow systems are better designs used nowadays in lieu of the older auxiliary air design.

Select Air-Handling Unit

The selection of an air-handling unit, whether central, unitary, or auxiliary, is the same as for a typical nonlaboratory application. The uniqueness with the selection for laboratories is that 100% outdoor air is required. Therefore, special consideration must be given to the selection of the heating and cooling coils, filtration, and fans. A typical layout for an air-handling unit is shown in Figure 6-32.

Heat recovery systems are discussed in detail in Chapter 10, "Energy Recovery." Once through the heat recovery system, if present, the air is filtered, heated or cooled, humidified, and then supplied to the individual spaces. The role of the air-handling unit is to take the supply air at varying conditions and condition it to a consistent point. Inlet air can range from very dry and cold to very hot and humid. In this section, selection and sizing of each of the main components for laboratory applications of the air-handing unit are discussed.

Filtration. The treatment and filtration of the outdoor air is discussed in detail in Chapter 8, "Air Treatment." In general, unless there is reentrainment of exhaust air or the outdoor air is highly polluted, treatment of the outdoor air is limited to filtration of particles.

The filtration of the outdoor air is meant to protect the air-conditioning systems (clean coils, ductwork, etc.) and to protect the processes/experiments in the space. Therefore, the level of filtration required for a laboratory is typically determined by the use of the laboratory. Table 6-6 summarizes the possible filtration levels typically used in laboratories.

While higher filtration is often desirable, there is a tradeoff between better filtration and higher filter and operating costs. The operating costs increase with filter efficiency due to higher filter pressure drops and subsequent fan energy increase. In addition, when selecting and designing the filtration system, several criteria must be evaluated:

- *Eliminate moisture carryover* – Since filters collect and hold dust, dirt, and pollen, the introduction of moisture into filters can result in the growth of fungi and mold. To avoid this, any visible moisture from the outdoor air must be eliminated. The primary concern is carryover of rainwater through the outdoor air intake.
- *Loading pressure effect* – As a filter is used, it becomes loaded with materials. This results in an increase in the pressure drop across the filters. The effect of the increase in pressure drop from new to fully loaded on the system airflow must be addressed. This is of primary concern in constant volume systems but can also be deleterious in VAV systems if not addressed.
- *Filter bypass* – The typical construction of the rack holding the filters allows for some bypass. Consideration of the rack design must be taken to avoid any bypassing of the filters.
- *Maintenance* – To ensure laboratory operations or system integrity are not compromised, servicing of the filters must be addressed during the design.

Heating and Cooling Coils and Humidification. The purpose of the heating and cooling coils and humidification is to provide a consistent temperature/humidity supply air to the space. The actual sizing of the components is no different than any other application. However, due to the higher than normal capacities (temperature/humidity differentials) in laboratory application, there are a few key selection guidelines that should be followed:

- *Design conditions* – Design the heating coil and system for the absolute lowest temperature expected to ensure that the coil does not freeze up. If hot water is used as the heating media, add glycol to the desired concentration to avoid freezing. Use the ASHRAE 0.4% condition (ASHRAE 1997) for cooling design.
- *Cooling coil cleaning* – Due to the high latent loads, the cooling coil depth in a laboratory is typically greater than eight rows. To simplify maintenance and condensate management, it is recommended that the cooling coil be split into two. This can be two six-row coils or similar combination to make cleaning the coils simple. An access space must be left between all coils of at least 2 feet and preferably 3 feet.
- *Cooling coil face velocity* – To avoid carryover of condensate from the cooling coil to the supply air, the cooling coil face velocity should not exceed 500 fpm (2.54m/s). In situations with high condensate levels, a maximum face velocity of 400 fpm (2.03 m/s) is recommended.
- *Humidification* – Sufficient space upstream and downstream is required to obtain even flow distribution and avoid condensation of moisture on ductwork or supply fans. In some applications (cleanrooms, hospitals), the humidification must be upstream of the final filters.

Fan Types. Types of fans used for supply are not very different from exhaust fans. There is no need for special protective coating of these fans as the supplied air is not as contaminated as the exhaust air. However, in certain laboratory settings like cleanrooms, where the supply air is HEPA filtered, there will be a need for more powerful fans to pull the air through these filters. There also may be the need to provide multiple fans for redundancy.

Determine Control Strategy. Details on the selection and application of control strategies for air-handling systems in contained in Chapter 11, "Controls."

Air Quality. The air supply system has a key role in determining the air quality and comfort of the space. The key aspects where the air-handling unit directly affects the indoor air quality are:

- *Filters* – The filters primarily determines the cleanliness of the supply air from a particle viewpoint. However, the filters can also be a site of fungi and mold growth if moisture is present.
- *Cooling coil* – If dirt is present on the coils or the condensate pan is not draining properly, mold and fungi can grow in these locations, thus compromising the indoor air quality.
- *Humidifier* – Improper placement or maintenance of the humidifier can result in standing water and subsequent fungi and mold growth.
- *System integrity* – Poor integrity of the system (air leakage) between the filters and the supply fan can result in unfiltered or conditioned air moving into the system. Poor integrity downstream of the supply fan can result in inadequate capacity at the room or condensation and subsequent fungi and mold growth on ductwork or building structure.

DUCT CONSTRUCTION

Duct construction is an important part of the HVAC system in any building but is especially true in laboratories. This is due to the hazardous nature of the materials contained in the exhaust airstreams and the high-energy use of laboratory HVAC systems to condition and move large volumes of nonrecirculated air. Duct construction for laboratories includes general parameters that are common to all types of buildings and duct material selection and application.

General Parameters

In the construction of ductwork for laboratories, several parameters are universal. These parameters primarily deal with specifications established by SMACNA that pertain to almost all ductwork of any building type. They include such things as:

- Duct construction
- Duct system components
- Performance testing
- Duct materials

Duct Construction (in Accordance with SMACNA Standards). As with duct construction for all types of buildings, laboratories require dimensional stability in the ductwork to prevent shape deformation and maintain the strength of the ductwork. Dimensional stability for laboratories is especially important, as laboratories may require higher than normal velocities in the exhaust ductwork to keep contaminants suspended in the airstream. A ductwork shape that deforms can create additional pressure drops and cause the air velocity to decrease and increase the fan energy needed. Also, frequent and repeated shape deformation can cause fatigue and reduce the strength of the ductwork.

Containment of the airstream (minimize ductwork leakage) is needed in laboratories, as in other types of buildings, to minimize energy losses from conditioning and moving of the air that leaks and to provide the desired room conditions. Additionally, laboratories need to minimize ductwork leakage to control the spread of hazardous contaminants that are contained in the exhaust airstream, as this could result in the re-introduction of hazardous materials into the rooms and mechanical spaces in the laboratory building. Also, ductwork leakage needs to be minimized for the supply ductwork, as considerable energy (due to larger fans needed to overcome the pressure drop of high-efficiency filters) can be lost when conditioned 100% outside air or highly filtered supply air is not delivered to the desired spaces.

As with all ductwork, laboratory ductwork exposure to damage needs to be minimized. Types of damage include: exposure to weather, wind, temperature extremes, corrosive atmospheres, and biological contamination. As laboratory exhaust systems may frequently be exposed to corrosive chemicals, biological materials, and temperature extremes, the materials used to construct the ductwork need to be selected to withstand exposure to the conditions and materials that will be present in the laboratory.

Ductwork support includes both hangers for horizontal sections and risers for vertical sections. There are various methods available for hanging ductwork in buildings, including laboratory buildings. Hangers consist of three components: the upper attachment to the building structure, the hanger itself, and the lower attachment to the ductwork. Hangers with a maximum spacing of 8 to 10 feet (2.44 to 3.05 meters) typically can support ductwork, although the weight limits for the individual hangers must be considered. Wide ducts typically require closer spacing to reduce the load on the individual hangers. Ductwork risers should to be supported by angles or channels fastened to the sides of the duct with welds, bolts, sheet metal screws, or rivets. When fastening supports to the ductwork, caution must be used for ducts greater than 30 inches (76.2 cm) wide, as the internal pressure of the duct may cause expansion, which could tear fasteners out of the duct material. Riser supports should be at one- or two-story intervals (e.g., every 12 to 24 feet [3.66 to 7.32 meters]).

As with other building components, ductwork construction, hangers, and risers should follow appropriate seismic restraint requirements as applicable for the location of the laboratory. Typically, seismic restraint for ductwork will involve utilizing some form of vibration isolation or increased strength materials for the hanger and riser supports of the ductwork. For additional information on seismic restraint design, see Chapter 53, *1999 ASHRAE Handbook—HVAC Applications*.

The thermal conductivity of ductwork should always be considered, both for supply and exhaust air systems. During cooling, heat gain needs to be minimized for supply air ductwork so that sufficient cooling capacity can be delivered to the desired spaces. Similarly, heat loss from supply air ductwork needs to be minimized during heating to provide sufficient heating capacity to the desired spaces. During cooling, condensation can form on supply air ductwork that has a high thermal conductivity, e.g., steel ductwork that is not insulated. For exhaust systems, protecting against heat loss and condensation on the interior of ductwork are important considerations for laboratories. Heat loss through exhaust ductwork can heat up the surrounding spaces and increase the cooling load for exhaust systems, which primarily are used to remove heat from equipment and heat-producing chemical reactions. Condensation can form on the inside of exhaust ductwork if the exhaust air has a high amount of water/chemical vapors in it and the exhaust air is cooled significantly by the surrounding air due to a high thermal conductivity.

SMACNA defines seven static pressure classes for ductwork: 0.5, 1, 2, 3, 4, 6, and 10 in. W. g. (0.12, 0.25, 0.50, 0.75, 1.0, 1.5, and 2.5 kPa). Each class of ductwork is for static pressures above the previous class and up to the number designating the class. For example, 2 in. w.g. (0.50 kPa) ductwork would be for ductwork with a static pressure greater than 1 in. w.g. (0.25 kPa) up to 2 in. w.g. (0.50 kPa). For static pressures up to 0.5 in. w.g. (0.12 kPa), 0.5 in. w.g. (0.12 kPa) ductwork would be used. For each size of ductwork within a pressurization class, a minimum duct wall thickness, joint specification, and reinforcement specification are provided by SMACNA standards. HVAC drawings should specify each class of ductwork being used.

Duct systems used for perchloric acid fume hoods are seam welded stainless steel and generally vertical. Duct systems used in high containment biological laboratories (Biosafety Level 3 and Level 4) should be completely sealed joint construction to facilitate gas decontamination (Wunder 2000).

Duct System Components. For new construction (except low-rise residential), all air-handling ducts and plenums installed as part of an HVAC air distribution system should be thermally insulated in accordance with ASHRAE Standard 90.1, while existing buildings should meet the requirements of ASHRAE Standard 100. Due to the large amount of nonrecirculated air used in laboratories, it may be economically feasible to use more insulation for laboratory ductwork than the minimum values required by the applicable standards. Additional insulation and vapor retarders may be needed to limit vapor transmission and condensation. The selection of insulation determines the heat gain or loss from the ductwork,

Table 6-7: SMACNA Duct Sealing Classes

Seal Class	Sealing Required	Static Pressure Construction Class
A	All transverse joints, longitudinal seams, and duct wall penetrations	4 in. w. g. and up (1.0 kPa and up)
B	All transverse joints and longitudinal seams	3 in. w. g. (0.75 kPa)
C	All transverse joints	2 in. w. g. (0.50 kPa)

In addition to the above, any variable air volume system duct of 0.5 or 1 in. w. g. construction class that is upstream of the VAV boxes shall also meet Seal Class C.

Source: SMACNA 1985

Table 6-8: Applicable Leakage Rates

Duct Class	0.5, 1, 2 in w. g. (0.12, 0.25, 0.50 kPa)	3 in. w. g. (0.75 kPa)	4, 6, 10 in. w. g. (1.0, 1.5, 2.5 kPa)
Seal Class	C	B	A
Leakage Class			
Rectangular Metal Duct	24	12	6
Round Metal Duct	12	6	3

which impacts the supply air quantities, supply air temperatures, and coil loads. Therefore, in order to ensure that the required conditions for the laboratory are met, it is important to verify that the insulation requirements selected during the design phase are correctly implemented during construction.

Various sound control devices are available to minimize the noise of a ductwork system. First, a vibration isolator is used to support the supply or exhaust fan that serves the section of ductwork. Fan flex sleeves should be used when the fan is connected to the ductwork. Vibration isolators are also used to support terminal units, to connect the outlets of air-handling units and terminal units to ductwork, and sometimes used to support the ductwork itself, especially for ductwork located near a fan. Other sound control methods are also available, such as sound attenuators and duct linings. Care should be taken when using duct linings so they are not exposed to moist airstreams, and they should typically not be used in exhaust airstreams due to the their potential to trap harmful vapors.

One very important difference between laboratory and typical commercial building ductwork systems is how fire, smoke, and volume control is performed. Laboratory exhaust systems cannot use automatic fire dampers (NFPA Standard 45-1996, 6-10.3), as a malfunction or false alarm would cause the hazardous exhaust air to back up into the laboratory and compromise personnel safety. Similarly, fire detection and alarm systems should not automatically shut down exhaust fans. For these reasons, multiple story laboratories usually require a separate fire-rated mechanical shaft for the ductwork and other utilities servicing each floor.

In addition to national codes and standards from ASHRAE, SMACNA, NFPA, ANSI, and others, many laboratories may also have state and local codes that pertain to the construction of ductwork systems. Common items included in state and local codes include insulation requirements for smoke and flame spread, smoke evacuation procedures, and fire containment controls. High containment biological labs may have requirements for gas tight dampers and bag in/out filter housings to control the release of gas sterilants and pathogens.

Performance Testing. SMACNA defines three different classes for duct sealing requirements, which are defined in Table 6-7. Sealant types include liquids, mastics, gaskets, and tapes. Welding can seal some types of laboratory systems ductwork, such as the exhaust ductwork for perchloric acid fume hoods.

The permissible leakage rate and static pressure of the ductwork according to the following equation define the SMACNA duct leakage classification:

$$C_L = \frac{Q}{\Delta P^{0.65}} \text{ (IP Units)} \qquad C_L = \frac{710 \cdot Q}{\Delta P^{0.65}} \text{ (SI Units)}$$

(6-3)

where

Q = leakage rate in cfm per 100 ft² (L/s per m²) of duct surface

ΔP = static pressure in inches w. g. (Pa)

C_L = leakage class number

Table 6-8 lists applicable leakage classes based on pressure and sealant classifications. It is up to the laboratory designer, in consultation with laboratory staff and based on the materials being handled in the ductwork, to determine the appropriate leakage values for each ductwork system in a laboratory. The values contained in Table 6-8 are suggested minimum values, which may need to be exceeded for exhaust system ductwork that handles certain types of hazardous materials. These leak-

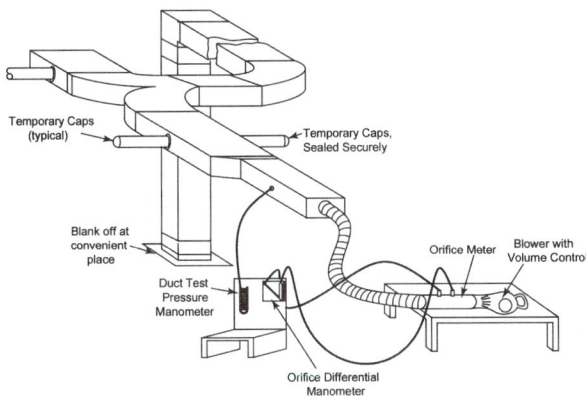

Figure 6-33 *Duct leakage testing apparatus. (Diagram modified with permission from SMACNA 1987.)*

approximately 15 to 30 minutes to verify that the flow rate and pressure can be sustained for an extended period of time. Calculate the leakage rate of the ductwork, and compare to the leakage requirement. If the leakage exceeds the requirement, inspect the ductwork for audible leaks and use a smoke test or soap solution to locate the leaks in the ductwork. Once the leaks have been located, depressurize the ductwork, repair the leaks, allow sealant to cure, and retest the ductwork.

Duct Materials. Numerous types of materials are available for the construction of ductwork for both the supply and exhaust systems. The choice of duct material depends upon which system (supply or exhaust) it will be used for, the types of substances to which the ductwork will be exposed, and applicable codes and standards. For instance, supply air ductwork for laboratories can generally be made from galvanized steel, as with typical commercial buildings. Exhaust ductwork requires materials that will not react or degrade when exposed to the hazardous materials in the exhaust air. For example, exhaust ductwork for perchloric acid fume hoods is typically made from stainless steel, and fume hoods that use large amounts of corrosive acids may require asphaltum-coated steel or a plastic surface (e.g., polyvinyl coating on steel) for exhaust ductwork. The following are commonly used materials for the construction of laboratory ductwork.

age classifications are shown as a corresponding leakage based on percentage of airflow in Table 6-9.

Duct leakage pressure tests are highly recommended for laboratory HVAC systems in order to verify that the leakage requirements of the ductwork were met and safety and energy use will not be negatively impacted. The procedure for testing laboratory ductwork leakage is the same as for typical buildings. Ductwork leakage testing can be performed on either a whole ductwork system or multiple portions of the ductwork system, and other equipment, such as coils, fans, and VAV boxes, should be isolated from the section of ductwork being tested. Leakage testing of ductwork is performed by using a flow meter, a differential pressure sensor, and a variable flow fan to positively pressurize the ductwork as shown in Figure 6-33.

Before performing the leakage test, calculate the volume of air that is permissible and make sure that the fan has enough capacity to supply that amount of air. Make sure to not overpressurize the ductwork, as the sealant and connections may develop leaks if exposed to excessive pressures. Starting the variable flow fan at a low speed or with the inlet damper closed and using an appropriately sized fan will help to avoid this problem. The speed of the fan should be increased until the pressure in the ductwork being tested reaches the desired value for the test. When this occurs, begin recording the flow rate from the flow meter at regular intervals (every 2 to 5 minutes) for

- Aluminum
- Asphalt-coated steel
- Epoxy-coated steel
- Galvanized steel
- Epoxy glass fiber reinforced
- Polyester glass fiber reinforced
- Polyethylene fluorocarbon
- Polyvinyl chloride (PVC): the NFPA code requires addition of a building sprinkler system to use this material for exhaust ductwork
- Polypropylene
- Stainless steel (304, 316)

Each of these materials should be chosen based on their reactivity to chemicals and flammability. Table 6-10 presents the chemical resistant properties and flame ratings for these materials.

Table 6-9: Leakage as Percentage of Airflow

Leakage Class	System cfm per ft² (L/s per m²) Duct Surface	Static Pressure, in. of water (kPa)					
		0.5 (0.12)	1 (0.25)	2 (0.50)	3 (0.75)	4 (1.0)	6 (1.5)
48	2	15	24	38	49	59	77
	2.5	12	19	30	39	47	62
	3	10	16	25	33	39	51
	4	7.7	12	19	25	30	38
	5	6.1	9.6	15	20	24	31
24	2	7.7	12	19	25	30	38
	2.5	6.1	9.6	15	20	24	31
	3	5.1	8.0	13	16	20	26
	4	3.8	6.0	9.4	12	15	19
	5	3.1	4.8	7.5	9.8	12	15
12	2	3.8	6	9.4	12	15	19
	2.5	3.1	4.8	7.5	9.8	12	15
	3	2.6	4.0	6.3	8.2	9.8	13
	4	1.9	3.0	4.7	6.1	7.4	9.6
	5	1.5	2.4	3.8	4.9	5.9	7.7
6	2	1.9	3	4.7	6.1	7.4	9.6
	2.5	1.5	2.4	3.8	4.9	5.9	7.7
	3	1.3	2.0	3.1	4.1	4.9	6.4
	4	1.0	1.5	2.4	3.1	3.7	4.8
	5	0.8	1.2	1.9	2.4	3.0	3.8
3	2	1	1.5	2.4	3.1	3.7	4.8
	2.5	0.8	1.2	1.9	2.4	3.0	3.8
	3	0.6	1.0	1.6	2.0	2.5	3.2
	4	0.5	0.8	1.3	1.6	2.0	2.6
	5	0.4	0.6	0.9	1.2	1.5	1.9
2	2	0.5	1.0	1.4	1.7	2.0	2.4
	2.5	0.4	0.8	1.1	1.4	1.6	2.0
	3	0.3	0.7	0.9	1.2	1.3	1.6
	4	0.3	0.5	0.7	0.9	1.0	1.2
	5	0.2	0.4	0.6	0.7	0.8	1.0
1.5	2	0.5	0.8	1.1	1.3	1.5	1.8
	2.5	0.4	0.6	0.8	1.0	1.2	1.5
	3	0.4	0.5	0.7	0.9	1.0	1.2
	4	0.3	0.4	0.5	0.6	0.8	0.9
	5	0.2	0.3	0.4	0.5	0.6	0.7
1	2	0.4	0.5	0.7	0.9	1.0	1.2
	2.5	0.3	0.4	0.6	0.7	0.8	1.0
	3	0.2	0.3	0.5	0.6	0.7	0.8
	4	0.2	0.3	0.4	0.4	0.5	0.6
	5	0.1	0.2	0.3	0.3	0.4	0.5

(Source: *1999 ASHRAE Handbook—Fundamentals*, p. 32.17.)

Table 6-10: Chemical Resistant Properties and Flame Ratings

Material	Acids[1] Weak	Acids[1] Strong	Alkalies[1] Weak	Alkalies[1] Strong	Organic Solvents[1]	Flammability[2]
Aluminum[3]	N	N	N	N	N	G
Asphalt-coated steel[4]	Y	Y	Y	Y	N	G
Epoxy-coated steel	Y	Y	Y	Y	Y	G
Galvanized steel[5]	N	N	N	N	Y	G
Epoxy glass fiber reinforced[6]	Y	Y	Y	N	Y	SL
Polyester glass fiber reinforced[7]	Y	Y	Y	N	Y	SL
Polethylene fluorocarbon[8]	Y	Y	Y	Y	Y	SE
Polyvinyl chloride[9]	Y	Y	Y	Y	N	SE
Polypropylene[10]	Y	N	Y	N	N	SE
316 Stainless steel[11]	Y	Y	Y	Y	Y	G
304 Stainless steel[11]	Y	N	Y	N	Y	G

1. N – attacked severely; Y – no attack or insignificant
2. G – good fire resistant; SE – self-extinguishing; SL – slow burning
3. Aluminum is not generally used due to its subjectivity to attacks by acids and alkalies
4. Asphalt-coated steel is resistant to acids, subject to solvent and oil attacks
5. Galvanized steel is subject to acid and alkaline attacks under wet conditions
6. Epoxy glass fiber reinforced is resistant to weak acids and weak alkalies and is slow burning
7. Polyester glass fiber reinforced can be used for all acids and weak alkalies but is attacked severely by strong alkalies and is slow burning
8. Polyethylene fluorocarbon is an excellent material for all chemicals
9. Polyvinyl chloride is an excellent material for most chemicals and is self-extinguishing but is attacked by some organic solvents
10. Polypropylene is resistant to most chemicals and is self-extinguishing but is subject to attack by strong acids, alkalies, gases, anhydrides, and ketones
11. Stainless steel 316 and 304 are subject to acid and chloride attacks varying with the chromium and nickel content

Duct Material Selection and Application

The selection of ductwork materials and applications for the laboratory involves the selection of materials and applications for the supply and exhaust systems. Generally, the selection does not require much additional guidance for laboratories, and general design considerations for typical commercial buildings can be used. However, the selection of exhaust ductwork for laboratories requires a specific set of criteria for material selection and application, which differs significantly from that of typical commercial buildings, as laboratory exhaust may contain a wide variety of hazardous materials. Selection criteria for both supply and exhaust ductwork are described in detail in the following sections.

Supply Ducts. Typically, general practices for all buildings can be used in the selection of supply duct materials for laboratories. The reasoning for this is that most supply ductwork will not come in contact with sufficient concentrations of hazardous laboratory materials to warrant special criteria such as that used for exhaust systems. Therefore, the following three items are the general considerations to make when designing supply ducts:

- *The ambient temperature* of the space surrounding the supply duct should be considered when selecting supply duct materials. Higher ambient temperatures relative to the supply air temperature can create condensation on the outside of the ductwork. Condensation on ductwork can eventually cause corrosion of the ductwork, leading to small leaks developing or premature failure of the ductwork. Also, condensation can lead to increased formation of mold and fungal organisms which can cause unacceptable odors or health impacts on occupants. To counter the effects of condensation, sufficient insulation, proper insulation sealing, and nonferrous duct materials can be used.

- *Duct velocities and pressures* also need to be considered, as they affect the choice of materials and overall energy use of the supply air system. For instance, low-pressure ductwork is often made of thinner material than higher-pressure systems, but it

Table 6-11: Exhaust Duct Velocities

Contaminant	Examples	Desired Velocity, fpm (m/s)
Vapors, gases, smoke	All vapors, gases, and smokes	1,400-2,000 (7.1-10.2)
Fumes	Zinc and aluminum oxide fumes	1,400-2,000 (7.1-10.2)
Very fine light dust	Cotton lint, wood flour, litho powder	2,000-2,500 (10.2-12.7)
Dry dust and powders	Cotton dust, light shavings	2,500-3,500 (12.7-17.8)
Average industrial dust	Sawdust, grinding dust	3,500-4,000 (17.8-20.3)
Heavy dusts	Metal turnings, lead dust	4,000-4,500 (20.3-22.9)
Heavy or moist dusts	Buffing lint (sticky), lead dust with small chips	4,500+ (22.9+)

(Source: SMACNA HVAC Systems Applications.)

requires a larger duct to move the same volume of air as a high-pressure system. High-pressure systems require engineering analysis for sound control. The selection of duct velocities and pressures will also affect the type, size, and energy use of the supply air fan.

- *The length and arrangement of supply ductwork* needs to be determined. Lengthy runs of ductwork result in larger supply air heat gains or losses and require additional cooling and heating capacity or additional insulation.

Exhaust Ducts. Due to the wide variation in types of hazardous materials that may be present in the exhaust airstream, exhaust ducts have more design considerations related to the selection of materials and applications to be made than supply ducts. The following criteria should be considered when selecting the materials and applications for exhaust ducts:

- The *nature of hood effluents* plays a significant role in determining the type of duct material to use and determines the *sealant type* and application procedure as well as the possible need for *protective coatings*. Therefore, knowledge of the current and future effluents and their possible concentrations is necessary to assess the materials required to handle the exhaust. The resistance of various materials to effluent attacks is in Table 6-10.

- The *ambient temperature* of the space surrounding the exhaust duct should be considered when selecting exhaust duct materials. As with supply ducts, the ambient air surrounding the exhaust ductwork can create condensation. However, since the exhaust air is generally warmer and more moist than the ambient air, the condensation will form on the inside of the exhaust ductwork. This may create problems for some exhaust systems, as the hazardous materials in the exhaust air may react with the duct after it condenses.

- *Effluent temperature* is also a concern for condensation. High-temperature effluents can cool off significantly in a lengthy section of duct. For selection purposes, an estimate of the lowest possible dew point of the effluent should be determined.

- *Duct velocities and pressures* affect the choice of duct materials and overall energy use of the exhaust air system, as was the case for supply air systems. However, the duct velocities for exhaust air systems are also determined by the type of materials in the exhaust airstream. Table 6-11 contains recommended design duct velocities for different types of exhaust air contaminants. Higher velocities may be required at the exhaust stack than are listed here for the ductwork.

- *The length and arrangement of ducts* is very important in the selection of exhaust duct materials and applications. Some materials that are exhausted, such as perchloric acid, need to have minimal or no horizontal runs to prevent condensation from forming. When using manifolded exhaust ductwork systems, the arrangement of the ductwork needs to ensure that the exhaust air from all of the sources is compatible with each other and the exhaust duct material.

- *Variance in exhaust air volume* can present a problem in maintaining needed duct velocities. In instances where variable air volume control is used for exhaust fans, the minimum acceptable duct velocity needs to be determined and verified that it will be met during operation from minimum to maximum flow of the VAV exhaust fan.

- *Flame spread and smoke development ratings* for duct materials need to be considered. Depending upon the national, local, and state regulations, some materials may not be acceptable for use as a duct material in some applications. Therefore, knowledge of the rating for the materials considered for ductwork is required and is included in Table 6-10.

ENERGY EFFICIENCIES

In the laboratory, energy consumption is typically high due to the use of 100% outdoor air. With the numerous pieces of equipment and the need to have constant exhaust and operation of hoods, energy efficient methods and equipment become all that more necessary. For this reason, a couple of possible methods for lowering or using energy more efficiently are discussed below.

Air Recirculation

Air recirculation is one method available to laboratory designers that can lower the operation costs of central systems. The idea behind this method is to recirculate or reintroduce exhaust from clean spaces into the supply air and laboratory spaces, thus requiring less conditioned air volumes. However, within the laboratory a few standards as to amounts and types of exhaust air that can be reentrained into supply air must be followed.

Chemical and Fume Exhaust. Chemical and hood exhaust can never be recirculated into the supply stream for laboratory spaces. This practice conflicts with the actual purpose of hoods and poses a great danger to personnel. Also, NFPA standards specifically state that this method is prohibited. However, some exhaust hoods do allow recirculation of exhaust air within the hood and laboratory depending upon filtration and hazard levels; they are described in Chapter 5, "Exhaust Hoods."

From Administrative Spaces. One method of air recirculation that is commonly allowed and practiced in laboratories is the use of exhaust air from administrative spaces and nonlaboratory zones. Except for laboratories that deal with highly toxic or hazardous materials, a constant leakage of air from administrative zones is allowed to enter laboratory areas. This is due to the negative pressurization of most laboratories and helps to maintain containment of pollutants. It is possible to increase this leakage amount by recirculating exhaust from administrative spaces into the laboratory. Caution in using this method must be taken to minimize cross-contamination. Also, if exhausts from the laboratory or hoods are combined into the general building exhaust, recirculation in this manner should not be used.

Within the Laboratory. Recirculation of laboratory general exhaust is prohibited. The reasoning for this is that not all effluents are captured by the hoods in the laboratory. Thus, introducing exhaust from the laboratory space would serve only to increase concentrations of the trace chemicals and contaminates to unsafe levels. For this reason it is typical that laboratory exhaust is 100%.

However, one instance where laboratory exhaust recirculation is acceptable and often used is in the cleanroom laboratory since high degrees of air change rates (600 to 900 per hour) are required and pose substantial costs for conditioning of these volumes.

Heat Recovery

Within the laboratory, a large portion of heat that is generated by equipment in the laboratory is just exhausted into the atmosphere. This is a giant energy loss for laboratories and typically is part of the reason that laboratories generally have extremely high operating costs. Therefore, the advent of some types of heat recovery equipment is an area of interest for the design engineer to lower a laboratory's annual cost. Yet to use this technology, careful planning and understanding of the operation and limitations of the equipment as it pertains to the laboratory are necessary. For this reason, a thorough discussion of possible heat recovery options for laboratories is presented in later chapters.

REFERENCES

American Society of Heating, Refrigerating and Air-Conditioning Engineers (ASHRAE). 1985. ANSI/ASHRAE Standard 51, *Laboratory Methods of Testing Fans for Rating*. Atlanta: ASHRAE.

American Society of Heating, Refrigerating and Air-Conditioning Engineers (ASHRAE). 1992. ANSI/ASHRAE 52.1, *Gravimetric and Dust Spot Procedures for Testing Air Cleaning Devices Used in General Ventilation for Removing Particulate Matter*. Atlanta: ASHRAE.

American Society of Heating, Refrigerating and Air-Conditioning Engineers (ASHRAE). 1996. *1996 ASHRAE Handbook—HVAC Systems and Equipment*. Atlanta: ASHRAE.

American Society of Heating, Refrigerating and Air-Conditioning Engineers (ASHRAE). 1997. *1997 ASHRAE Handbook—Fundamentals*. Atlanta: ASHRAE.

American Society of Heating, Refrigerating and Air-Conditioning Engineers (ASHRAE). 1995. ANSI/ASHRAE/IESNA Standard 100-1995, *Energy Conservation in Existing Buildings*. Atlanta: ASHRAE.

DiBerardinis, L.J., J.S. Baum, M.W. First, G.T. Gatwood, E. Groden, and A.K. Seth. 1993. *Guidelines for Laboratory Design: Health and Safety Considerations*, 2nd ed. New York: John Wiley & Sons, Inc.

National Fire Protection Association (NFPA). 2000. *NFPA 45, Fire Protection for Laboratories Using Chemicals*. Quincy, Mass.: NFPA Publications.

National Fire Protection Association (NFPA). 1999. *NFPA 70, National Electrical Code*. Quincy, Mass: NFPA Publications.

Sheet Metal and Air Conditioning Contractors National Association (SMACNA). 1987. *HVAC Systems Applications*. Vienna, Va.: SMACNA, Inc.

Sheet Metal and Air Conditioning Contractors National Association (SMACNA). 1985. *HVAC Duct Construction Standards—Metal and Flexible*. Vienna, Va.: SMACNA, Inc.

Wunder, J.S. 2000. Personal communication from operating experiences with laboratory equipment. University of Wisconsin–Madison.

Chapter 7
Process Cooling

OVERVIEW

Within the laboratory, particular pieces of equipment may require cooling that cannot be efficiently or sufficiently supplied by air cooling using mechanisms discussed in Chapter 6, "Primary Air Systems." In these cases the use of process cooling becomes necessary to ensure the safety of personnel, equipment, and experiments.

Process cooling is the supply of water or other fluid to various types of equipment for cooling purposes. The fluid is pumped through a series of pipes attached to equipment and heat is transferred from the equipment to the fluid.

For process cooling it is imperative to understand the types of water-cooled loads, water treatment and quality requirements, temperature and pressure requirements, and system pumping configurations.

TYPES OF WATER-COOLED LOADS

Several types of equipment require process cooling. Usually, this equipment generates high amounts of heat in very short periods that cannot be adequately handled solely by the primary air system. Typical types of equipment in a laboratory that require process cooling includes lasers, centrifuges, vacuum and diffusion pumps, and various others.

Lasers

Lasers are equipment that generally require process cooling. Typically used in physics laboratories, they can rapidly create large amounts of heat and thus need to be cooled. Here process cooling is used for personnel, experiment, and equipment safety since excessive heat within laser equipment can cause equipment failure and potential fire hazards. To further maintain safety, lasers are often equipped with interlocks that disable the heat production in the event of insufficient cooling. Figure 7-1 depicts a water-cooled laser arrangement.

Lasers are available in a range of sizes. Smaller lasers, 1 to 2 inches and with capacities in the milliwatt range, are mostly used for telecommunication. These lasers are typically cooled with thermoelectric coolers. Larger lasers, 2 to 6 feet in length with capacities ranging from 2 W to 20 W, need to be water cooled with their own dedicated cooling system.

The source of the heat produced in lasers may originate from its amplifying medium (solid, liquid, or gas) or from a particular target on which the laser is focused for cutting or boring. In the case of the latter, cooling may be needed to prevent the target from annealing or hardening.

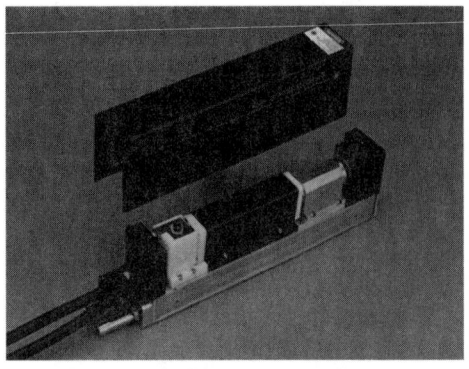

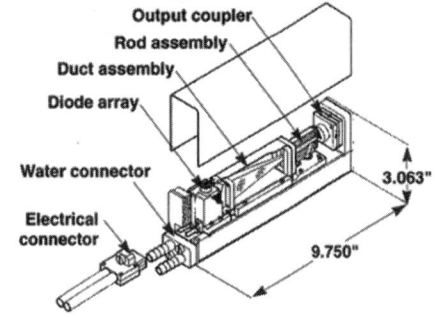

(Courtesy of Lawrence Livermore National Laboratory.)

Figure 7-1 *Water-cooled laser.*

Centrifuges

A centrifuge is a piece of equipment used for liquid/solid separation purposes. It does this by rotating samples at varying speeds using a rotor and an attached motor. Although the centrifuges generally found in clinical and biological laboratories are typically air-cooled by an integral fan system, those found in more industrial laboratories often have the option of using process cooling to maintain acceptable internal temperatures during their operation. Figure 7-2 illustrates a type of centrifuge used for vegetable oil separation.

Vacuum and Diffusion Pumps

Vacuum and diffusion pumps are used in processes where distillation or concentration of substances is required. Smaller versions of this equipment typically do not require any supplementary conditioning. However, larger operations, and thus larger pumps, usually do require the introduction of water cooling to ensure the safe operation and longevity of the pump. Figure 7-3 illustrates a diffusion pump.

Other

Other types of equipment, such as large blast ovens, tube furnaces, incubators, and autoclaves, may require the use of process cooling. Generally, the deciding factor for this type of cooling depends on manufacturers'

(Courtesy of Separators, Inc.)

Figure 7-2 *Centrifuge.*

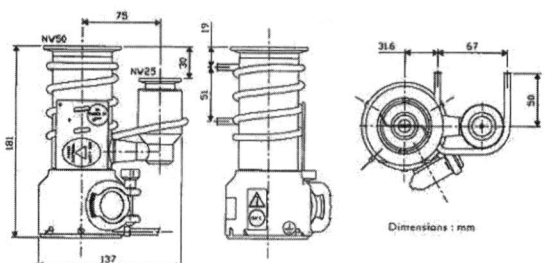

(Courtesy of Boc Edwards, Inc.)

Figure 7-3 *Diffusion pumps.*

requirements and power consumption, since larger consumption can be translated into greater heat gains from the equipment.

WATER TREATMENT AND QUALITY REQUIREMENTS

As in many processes, where water is used there is a need for its proper treatment to obtain and maintain quality. Treatment is used to combat various water problems, which include:

- *Corrosion* – This develops when metal is destroyed by chemical or electrochemical reactions within the process water. Corrosion inhibitors and pH control chemicals may be added to the water to minimize damage by reducing the amount of oxygen present.
- *Biological growth* – This is more prevalent in open systems and can cause blockages in the water distribution system. Excessive slime buildup can be expelled with the use of microbiocidal materials.
- *Scale formation* – This typically results from the precipitation of calcium carbonate and can be controlled by regulating the pH of the process water.
- *Solid solutes* – When present in systems, these reduce heat transfer and interfere with process water flow. After new piping is installed and before system operation, thoroughly flushing the system can eliminate these suspended materials. In addition, the use of strainers will remove larger solids.

When treating the process cooling water to reduce the unwanted problems outlined above, care must be taken when administering chemicals. Too high or too low a concentration can cause adverse effects. For example, at too high concentrations of chlorine, corrosion can actually speed up. However, at too low concentrations, pitting can develop.

Addressing the following quality requirements is helpful to maintaining effective water treatment:

- Power capacity should be available and adequate for chemical injection pumps
- Domestic water supply should be available for mixing chemicals
- A drain is necessary for bleed-off waste
- Turbulent flow through pipes is needed for water treatment chemicals to contact the metal surfaces to be protected
- Sufficient space must be provided for chemical equipment and storage

Consult Chapter 47, *1999 ASHRAE Handbook—Applications* (ASHRAE 1999) for further details on water treatment.

TEMPERATURE AND PRESSURE REQUIREMENTS

For process cooling the temperature and pressure requirements of the water used are dependent upon the cooling needs of the equipment or the desired effect the cooling is to accomplish. The typical temperature range for most chilled water systems is 40°F to 45°F (4.4°C to 7.2°C) at a pressure of 120 psi (827.4 kPa), which is suitable for most applications. However, some equipment may have greater cooling needs. Designers of process cooling systems should consult with the original equipment manufacturer (OEM) or distributor on the needs and requirements of the specific equipment.

Since additional cooling of the water for individual equipment can be costly, higher flow rates are often used. Higher flow rates are synonymous with higher costs due to this increased pumping power, but these costs are usually much less than the costs associated with an increase in compressor power for the refrigerant subsystem used to cool the water. There are some instances where the cooling needs of a piece of equipment warrant the use a dedicated system or the use of domestic water. However, this may require additional piping and can also be costly. Any proposed design should include the manufacturer's specifications for the pressure and temperature thresholds of the equipment.

PUMPING SYSTEM CONFIGURATIONS

This section presents the basic architecture of the overall chilled water system used for process cooling. Various pumping subsystems with their alternative pumping configurations are emphasized, and discussions include general practices for good design. More detailed information on overall system design is available in Chapter 12, *2000 ASHRAE Handbook—Systems and Equipment* (ASHRAE 2000).

System Basics

Most chilled water systems, including the ones used for process cooling, are closed instead of open. The main difference between closed and open water systems is the number of interfaces that exist between the water and a compressible gas (e.g., air). Whereas an open system has two or more such interfaces, closed systems have only one. The basic closed chilled water system comprises the following components/subsystems:

- *Source Subsystem* – The point where heat is removed from the process cooling system, e.g., chillers, heat pump evaporators, and heat exchangers.
- *Load* – The point where heat flows into the cooling system from the process thus rendering it cooler (e.g., lasers and associated heat-producing equipment).
- *Pump Subsystem* – The mechanical devices used and configured to provide sufficient energy to circulate the water in the process cooling system.
- *Distribution Subsystem* – The piping system that connects the various components of the entire cooling system and the main conduit through which all the water flows.
- *Expansion Chamber* – Also known as an expansion or compression tank, this provides a space into which the noncompressible water can expand or from which it can contract, as changes in temperature influences a volumetric change in the water.

Figure 7-4 shows the basic components of a closed hydronic system that can be used for process cooling. Note that this representation is only of the fundamental principles that are necessary for every process cooling

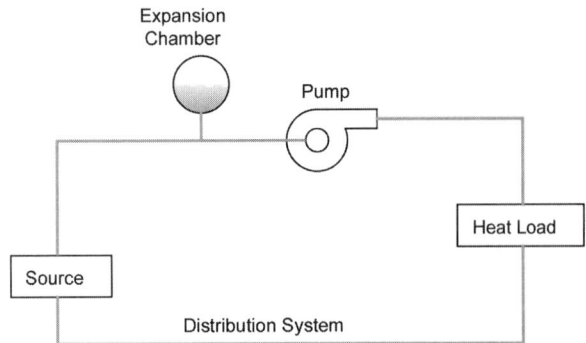

Figure 7-4 *Fundamental components of close hydronic system.*

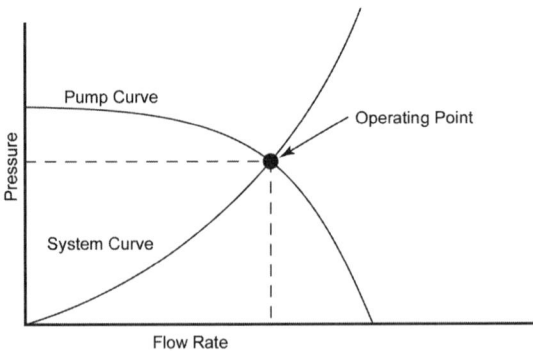

Figure 7-5 *Pump curve and system curve.*

system and excludes specific components such as valves, regulators, vents, etc.

Pumping Subsystems

The pumping system utilizes circulating pumps that may vary in size. This includes small in-line pumps that can deliver 5 gpm (0.32 L/s) at 6 or 7 ft (17.9 or 20.9 kPa) head to base-mounted or vertical, pipe mounted pumps that can handle much larger flows with head pressures restricted only by the overall system characteristics.

To effectively size a pump for a given process cooling system, the system operating requirements must be prudently matched to the pump operating characteristics. Figure 7-5 illustrates the operating point of a given pump, which is the point of intersection of the pump curve (representing pump operating characteristics) and the system curve (representing system operating requirements).

The operating point determined in this way can be quite variable depending on load conditions, piping connections and heat transfer elements, and control valves used. Therefore, when selecting pumps, the following steps are generally good practice:

- Select for design flow rates and use pressure drop charts that illustrate the actual closed loop system piping pressure drops.
- Select slightly to the left of the maximum efficiency point of the pump curve to avoid undesirable pump operation or overloading.
- Select a pump with a flat curve to make allowance for unbalanced circuitry and to provide a minimum differential pressure increase across two-way control valves.

Pumping systems are arranged using the following common configurations:

- Parallel pumping

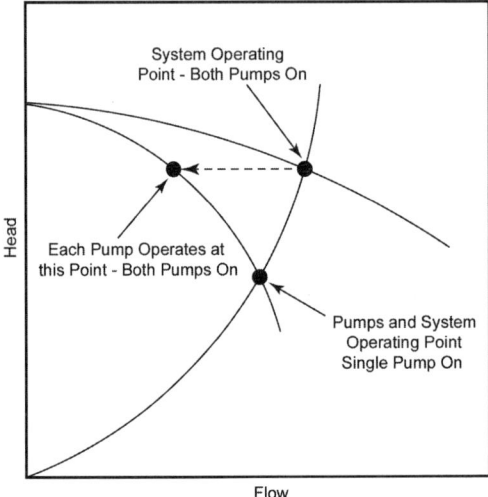

Figure 7-6 *Operating conditions for parallel pump installation.*

- Series pumping
- Compound pumping
- Multi-speed/variable speed pumping

Parallel Pumping. Pumps configured in parallel each operate at the same head pressure while providing their own share of the system flow at that pressure. Typically, the pumps used should be of the same size, and thus the characteristic curves for parallel pumps are attained by doubling the flow of the single pump curve as illustrated in Figure 7-6. Figure 7-6 also shows that the operating point when both parallel pumps are on is considerably to the right of the operating point when a single pump is on. This indicates that for a process cooled system using parallel pumps:

1. Overloading can be prevented during single-pump operation provided that the pumps in a parallel configuration are constantly powered.

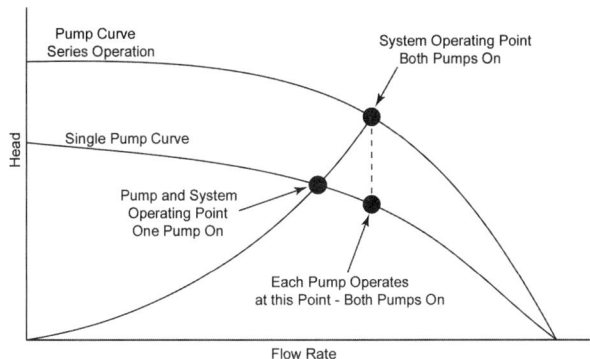

Figure 7-7 *Operating conditions for series pump installation.*

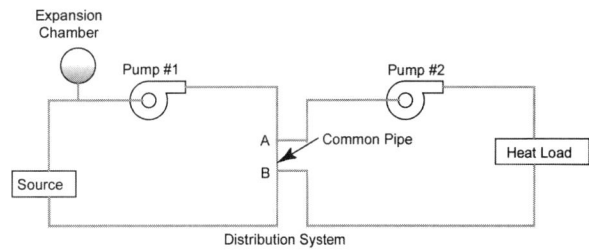

Figure 7-8 *Compound pumping schematic.*

2. A single pump can be used for standby service of up to 80% of design flow, depending on a given system curve and pump curve.

Series Pumping. Pumps configured in series each operate at the same flow rate while providing their own share of the total head pressure at that flow. Figure 7-7 illustrates the operating points for both single and series pump operation. Like the parallel pump configuration, a single pump in series configuration can provide up to approximately 80% flow for standby service, which is often adequate.

Compound Pumping. Compound pumping, also known as primary-secondary pumping, is typically used for larger systems. Figure 7-8 illustrates a basic compound pumping configuration. Pump No. 1 is the primary pump or source pump and Pump No. 2 is the secondary pump or load pump. The main advantages of using this type of configuration are:

1. A designer is able to achieve differing water temperatures and temperature ranges in different elements of the process cooling system.

2. The control, operation, and analysis of large systems are less complex because individual circuits are hydraulically isolated (i.e., they function dynamically independent of each other).

3. The primary and secondary circuits can be designed for different flow characteristics. For instance, the secondary circuit can incorporate two-way valves for improved control and energy efficiency, and the primary circuit can be operated at a constant flow to prevent freezing of chillers.

Two-speed/Variable Speed Pumping. Standard two-speed motors used in pumps are quite common. They are available in 1750/1150 rpm, 1750/850 rpm, 1150/850 rpm, and 3500/1750 rpm speeds. At a decrease in system flow, these pump motors will reduce overpressuring – the difference between the pump pressure and the system pressure and a condition that occurs in controlled flow systems.

In contrast to two-speed pumping, variable speed pumping uses variable frequency drives that are normally available with an infinite speed range. The pump with the correct controls will be able to follow the system curve and, therefore, like the two-speed pump, avoid any overpressure effects.

Systems Design Procedure

Provided the cooling loads, temperature and pressure requirements, and limitations of the equipment to be cooled are understood, engineers and designers may proceed with designing the process cooling system. For a given application, the following basic steps should be followed:

1. Preliminary equipment layout

    a. *Determine water flow rates in mains and laterals* – Starting from the most remote terminal and progressing toward the pump, sequentially list the cumulative flow in each of the mains and branch circuits for the entire distribution system.

    b. *Determine preliminary pipe sizes* – For each segment of the distribution system, select a pipe size from the unified flow chart (see Figure 1, Chapter 33, *1997 ASHRAE Handbook—Fundamentals* (ASHRAE 1997)).

    c. *Determine preliminary pressure drop* – For each pipe size determined in step 1b, determine the pressure drop in each segment of the distribution system.

    d. *Select preliminary pump* – A pump should be selected based on its ability to meet the determined capacity requirements.

2. Final pipe sizing and pressure drop determination

   a. *Determine final piping layout* – Make the necessary adjustments of the preliminary layout to optimize the design based on economics and functionality.

   b. *Determine final pressure drop* – Repeat step 1c based on the final piping layout.

   c. *Select final pump* – Based on the final pressure drop calculation in step 2b, plot a pump curve and a system curve and select the pump that operates closest to the design point (intersection of the two curves).

REFERENCES

American Society of Heating, Refrigerating and Air-Conditioning Engineers (ASHRAE). 1999. *1999 ASHRAE Handbook—HVAC Applications*, Chapter 47, Water Treatment. Atlanta: ASHRAE.

American Society of Heating, Refrigerating and Air-Conditioning Engineers (ASHRAE). 1997. *1997 ASHRAE Handbook—Fundamentals*, Chapter 33. Atlanta: ASHRAE.

American Society of Heating, Refrigerating and Air-Conditioning Engineers (ASHRAE). 2000. *2000 ASHRAE Handbook—HVAC Systems and Equipment*, Chapter 12, Hydronic Heating and Cooling System Design. Atlanta: ASHRAE.

Chapter 8
Air Treatment

OVERVIEW

Treatment of air, either before it is supplied to a space or before it is exhausted to the environment, is often required for the safety and health of occupants or people near the building or for processes that require special conditions. Air treatment technologies are applied in the exhaust or the supply based on what contaminants need to be reduced to an acceptable level.

This chapter reviews the requirements for concentration limits of typical gases and particulates set by codes and standards (acute, chronic, and carcinogenic) for short-term exposure, as well as acceptable concentrations to avoid health problems caused by long-term exposure and maximum concentration levels to prevent unwanted odors. This is followed by a discussion of the technologies available to achieve acceptable levels, including filtration, scrubbing, condensing, oxidizing, and fan-powered dilution.

REQUIREMENTS

The requirements for acceptable and safe levels of pollutants are based on various codes and standards, typically developed to protect industrial workers who are exposed to the chemicals. In this section, the allowable concentrations for chemicals are reviewed, followed by the consequences of exposure to excessive concentrations. Requirements for industrial processes (electronic manufacturing, etc.) are not discussed, as the laboratory designer will not determine them but rather be given the requirements as a prerequisite.

Allowable Concentrations

The allowable concentration limit of a pollutant depends on the type of substance and the length of exposure. Occupational health limits are for exposures that are relatively constant for an extended period of time (e.g., eight hours), whereas state and local ambient concentration limits are maximum limits for short-term exposure. Therefore, the state and local ambient code limits are typically significantly lower than occupational health limits.

Occupational Health Limit. Occupational health limits are typically expressed as an 8-hour time-weighted average (TWA) concentration to which healthy individuals can be exposed during a 40-hour work week with no significant health effects (Petersen and Ratcliff 1991). The Occupational Safety and Health Administration (OSHA) and the American Conference of Governmental Industrial Hygienists (ACGIH) have respectively published lists of permissible exposure limits (PEL) and threshold limit values (TLV) for various chemicals. The National Institute for Occupational Safety and Health (NIOSH) publish limits known as recommended exposure limits (REL), which are limits based on human and animal studies and recommended to OSHA. Whereas, RELs and TLVs are recommendations and guidelines, PELs are legally binding and OSHA has the authority to warn, cite, or fine violators if workers' exposure exceed PELs.

**Table 8-1:
OSHA and ACGIH
Thresholds for Selected Chemicals**

Chemicals	PEL (OSHA) ppm	TLV-TWA (ACGIH) ppm*
Acetaldehyde	200	100
Acetone	1000	750
Acetonitrile	40	40
Acrolein	0.1	0.1
Acrylonitrile	2	2
Ammonia	35	25
Benzene	1	10
Bromine	0.1	0.1
Carbon disulfide	4	10
Carbon monoxide	50	25
Carbon tetrachloride	2	5
Chlorine	1	0.5
Chloroform	50	10
Dichloromethane	500	50
Diethyl ether	400	400
Dioxane	100	25
Ethyl acetate	400	400
Ethylene oxide	1	1
Flourine	0.1	1
Formaldehyde	1	0.3
Hydrobromic acid and hydrogen bromide	3	3
Hydrogen cyanide	10	10
Hydrogen fluoride	3	3
Hydrogen sulfide	20	10
Methanol	200	200
Nitrogen dioxide	5	3
Ozone	0.1	0.1
Phenol	5	5
Phosgene	0.1	0.1
Sulfur dioxide	5	2
Toluene	200	50

* ppm = parts by volume of gas or vapor per million parts by volume of contaminated air

Detailed rationale for the determination of the TLV can be found in ACGIH (1999). Basically, the exposure is the average exposure per hour divided by the length of the workday (eight hours). The calculated exposure must not exceed the given limit. Some substances are also given a ceiling value, which cannot be exceeded during any 15-minute period. Other chemicals have an 8-hour weighted maximum value, an acceptable ceiling value (ACV), and a maximum peak value (MPV) for a given duration. The concentration is allowed to exceed the ACV but not the MPV.

Table 8-1 contains a summary of the OSHA and ACGIH limits for some of the chemicals typically found in laboratories.

The general guidance for acceptable pollutant concentrations in commercial office buildings is one-tenth the TLV (ASHRAE 1999).

State/Local Ambient Concentration Limits. Most state and local ambient concentration limits are intended to account for instantaneous or short-term exposure. Since these limits account for the exposure of sensitive individuals, they have a safety factor of 40 to 300 (depending on the chemical and the state) compared to the occupational health limits, and they must not be exceeded (Petersen and Ratcliff 1991).

To properly account for actual exposures during worst-case releases, usually spills or accidents, it is necessary to calculate the evaporation rate of the liquid chemical. Typically, information on chemical emission rates from existing or proposed local laboratory sites are insufficient. Petersen and Ratcliff (1991) suggest that a detailed survey of laboratory use should be conducted, where useful information from safety officers and knowledgeable personnel can be obtained.

Once armed with preliminary information of the key chemicals to be used at a local site, Equation 8-1, developed by Kawamura and McKay (1987), can be used to calculate the evaporation (mass emission) rate, which is primarily dependent on the saturation vapor pressure of the liquid.

$$E = k M P(T_s)/RT \qquad (8\text{-}1)$$

where

E = evaporation rate per unit area (g/m^{2}h)
k = mass transfer coefficient (m/h)
M = molecular weight
$P(T_s)$ = vapor pressure of the chemical evaluated at the surface of the pool (Pa)
R = gas constant (8.314 Pa m^3/mol·K)
T = absolute temperature (K)

Using the calculated emission rates of various chemicals along with their corresponding health threshold concentrations, a ratio of emission rate to thresholds can be used to rank chemicals in terms of their relative potential to be harmful. That is, a chemical with a high emission rate and low odor or health threshold would cause the most adverse effects. Table 8-2 presents a list of estimated emission rates and health thresholds for a few selected chemicals.

Odors. Odor perception is a very complicated process with significant individual variation. Chemicals with molecular masses greater than 300 are generally odorless (ASHRAE 2001a). Humans can perceive

Table 8-2:
Estimated Emission Rates and Health Thresholds for Selected Chemicals

Chemical	Emission Rate (Q), g/s (lbm/s)	Health Limit (HL), $\mu g/m^3$ (lbm/ft^3)	Limit Type	HL/Q * 1 g/s [5], $\mu g/m^3$ (lbm/ft^3)
Isopropyl mercaptan[4]	0.00167 (3.68 × 10^{-6})	NA	NA	NA
Bromine – 100g spill[2]	0.4 (8.82 × 10^{-4})	2000 (1.25 × 10^{-7})	ACGIH STEL	5000 (3.12 × 10^{-7})
Hydrogen fluoride[4]	0.2 (8.82 × 10^{-4})	2600 (1.62 × 10^{-7})	ACGIH CEIL	13000 (8.11 × 10^{-7})
Ethylene oxide[4]	3.026 (6.67 × 10^{-3})	9000 (5.61 × 10^{-7})	NIOSH - 10 min	2974 (1.86 × 10^{-7})
Chloroform[1]	4.28 (9.44 × 10^{-3})	10000 (6.26 × 10^{-7})	NIOSH - 60 min	2336 (1.46 × 10^{-7})
Ammonia – 2 lb spill[3]	1.08 (2.38 × 10^{-3})	24000 (1.50 × 10^{-6})	ACGIH STEL	22222 (1.39 × 10^{-6})
Dichloromethane[1]	4.26 (9.39 × 10^{-3})	174000 (1.09 × 10^{-5})	ACGIH TWA	40845 (2.55 × 10^{-6})
Acetonitrile[1]	0.94 (2.07 × 10^{-3})	101000 (6.31 × 10^{-6})	ACGIH STEL	107447 (6.70 × 10^{-6})
Tetrahydrofuran[1]	2.56 (5.64 × 10^{-3})	738000 (4.61 × 10^{-5})	ACGIH STEL	288281 (1.80 × 10^{-5})
Acetone[1]	2.29 (5.05 × 10^{-3})	2380000 (1.49 × 10^{-4})	ACGIH STEL	1039301 (6.49 × 10^{-5})
Hexane[1]	3 (6.62 × 10^{-3})	1800000 (1.12 × 10^{-4})	ACGIH 15 min	600000 (3.74 × 10^{-5})
Ethyl acetate[1]	2.03 (4.48 × 10^{-3})	1400000 (8.75 × 10^{-5})	ACGIH TWA	689655 (4.30 × 10^{-5})

(Source: Peterson and Ratcliff 1991)

Notes:

[1] Emission rate estimated assuming a 4 L spill and 2 x 2 m pool; thermal effects included

[2] Emission rate estimated assuming a 310 mL spill and 0.5 x 0.5 m pool; thermal effects included

[3] Emission rate estimated assuming a 1 L spill and 1 x 1 m pool; no thermal effects included (little variation of vapor pressure with temperature)

[4] Emission rate estimated with supplies mass and release durations

[5] These numbers used for comparison with wind tunnel results when $Q = 1$ g/s

chemicals with molecular weights less than 300 if the concentration is above an odor threshold limit, which can be as low as 0.1 ppb. This is often too low to be detected by a direct reading instrument. Table 8-3 relates the odor level to threshold limit value (TLV) and calculates a safety factor. A safety factor of greater than one is required to ensure that the occupants can detect odor before the hazardous levels of the gases are present. Ruth (1986) suggests that both upper and lower limits for odors be documented to account for differences in individuals and experimental techniques.

It is important from an occupant satisfaction basis to recognize that only chemicals with an odor threshold limit below the occupational limit are important, with safety factors greater than one. The reason for this is that occupants will smell the odor before it is harmful, and their satisfaction with the environment decreases with increasing odors. Conversely, from a health basis, it is those chemicals with odor threshold limits above occupational limits that are of concern. The reason is that once the occupants smell the chemical, they are already at risk for health problems.

In sensitive areas, such as universities and mixed-use spaces, it is recommended to use the lower limits (i.e., concentrations where only a few individuals can detect the odor). An average between the lower and upper limits can be used in less sensitive areas such as chemical manufacturing plants (Petersen and Ratcliff 1991).

Discussion of Symptoms of Inadequate Performance

Inadequate performance of systems that results in exposure of users to contaminants above limits given in codes and standards can cause both reversible and irreversible health effects. While skin irritation and headaches are milder reactions that usually are reversible, cancer, chronic respiratory disease, and heart diseases are reactions to exposure to hazardous substances that are often irreversible. Unfortunately, the causes of symptoms are hard to distinguish from other causes, such as viruses and stress reaction. Therefore, identifying the substance causing the symptoms is difficult, as odors can cause milder symptoms such as headaches, nausea, and loss of appetite without exposures that exceed safe limits.

FAN-POWERED DILUTION

The use of fan-powered dilution systems is the most common method for handling laboratory exhaust effluent, and some laboratory experts suggest that they should be thoroughly evaluated before selecting air

Table 8-3: Odor Thresholds and Safety Factors of Selected Gaseous Air Pollutants

Pollutant	Odor Threshold, mg/m³ (lbm/ft³)	TLV/Odor Threshold
Acetaldehyde	1.2 (0.08)	150
Acetone	47 (2.93)	38
Acrolein	0.35 (0.02)	0.7
Allyl chloride	1.4 (0.09)	2
Ammonia	33 (2.06)	0.5
Benzene	15 (936.45)	2
Benzyl chloride	0.2 (0.01)	25
2-Butanone (MEK)	30 (1.87)	20
Carbon dioxide	Infinite	0.0
Carbon monoxide	Infinite	0.0
Carbon disulfide	0.6 (0.04)	50
Carbon tetrachloride	130 (8.12)	0.2
Chlorine	0.007 (4.4×10^4)	430
Chloroform	1.5 (0.09)	33
p-Cresol	0.056 (3.5×10^{-3})	390
Dichlorodifluoromethane	5400 (337.12)	0.9
Dioxane	304 (18.98)	0.3
Ethylene dichloride	25 (1.56)	1.6
Ethylene oxide	196 (12.24)	0.01
Formaldehyde	1.2 (0.07)	1.3
n-Heptane	2.4 (0.15)	670
Hydrogen chloride	12 (0.75)	0.6
Hydrogen cyanide	1 (0.06)	10
Hydrogen fluoride	2.7 (0.17)	0.9
Hydrogen sulfide	0.007 (4.4×10^4)	2000
Mercury	Infinite	0.0
Methanol	130 (8.12)	2
Methyl chloride	595 (37.15)	0.2
Methylene chloride	750 (46.82)	0.2
Nitric oxide	>0	
Nitrogen dioxide	51 (3.18)	0.1
Ozone	0.2 (0.01)	1
Phenol	0.18 (0.01)	106
Phosgene	4 (0.25)	0.1
Propane	1800 (11.24)	
Sulfur dioxide	1.2 (0.07)	4
Sulfuric acid	1 (0.06)	
Tetrachloroethane	24 (1.50)	0.3
Tetrachloroethylene	140 (8.74)	2
o-Toluidene	24 (1.50)	
Toluene	8 (0.50)	47
Toluene diisocyanate	15 (0.94)	0.003
1,1,1-Trichloroethane	1.1 (0.07)	1730
Trichloroethylene	120 (7.49)	2
Vinyl chloride monomer	1400 (8.74)	0.007
Xylene	2 (0.12)	220

(Source: ASHRAE Fundamentals, 1997, Ch. 13)

treatment processes (Wunder 2000). These dilution systems induce clean (outdoor) air into the contaminated airstream and thereby dilute the contaminant exit concentration. The dilution increases the air volume out of the stack and thereby increases the effective stack height, which also ensures good mixing. This strategy can be used with variable air volume systems to ensure adequate stack exit velocity under all operating conditions. Chapter 6, "Primary Air Systems" briefly discusses the use of dilution fans.

FILTRATION

Filtration is a treatment process that traps particles or gas in a filter that, when it fills up, can be removed or cleaned. Particulate filters that can be used for laboratories include pleated bag, HEPA, electrostatic, and gas phase filters (activated carbon). Often, different filtration media are used in series to extend the life of the more expensive, higher efficiency filter. For example, pleated filters are usually used as pre-filters to bag, electrostatic, or activated carbon filters. The terms commonly used to describe particles include (ASHRAE 2001b):

- *Dust* – Particles less than 100 μm.
- *Fume* – Solid particles formed by condensation of vapors on solid materials.
- *Smoke* – Small solid and liquid particles (and gas) produced by incomplete combustion.
- *Bioaerosol* – Virus, bacteria, fungi, and pollen.
- *Mist* – Airborne small droplets.
- *Fog* – Fine airborne droplets (smaller than mist) normally produced by condensation.
- *Smog* – Air pollution of particles (solid, liquid, and gas) that impairs visibility and is irritating or harmful.

Particulate Filters

Particulate filters are mechanical devices that utilize five mechanisms to collect particles:

- *Straining* – The membrane openings are smaller than the larger particles, which are blocked from passing through the filter.
- *Direct interception* – The particle follows a flow line that results in direct contact with a filter fiber and thus becomes attached to the fiber.
- *Inertial deposition* – The particle does not follow a flow line around a fiber because of sufficient size or inertia and thereby gets in contact with a fiber in the filter and remains attached to the fiber. A higher percentage of the particles will collide with the fiber at high velocities due to increased inertia, but because of increased high drag and bouncing forces at high velocities, the fibers should have an adhesive coating to ensure that the particles stay in the filter.
- *Diffusion* – Very small particles have a random movement about their streamline (Brownian motion) that can cause the particle to contact a fiber

Air Treatment

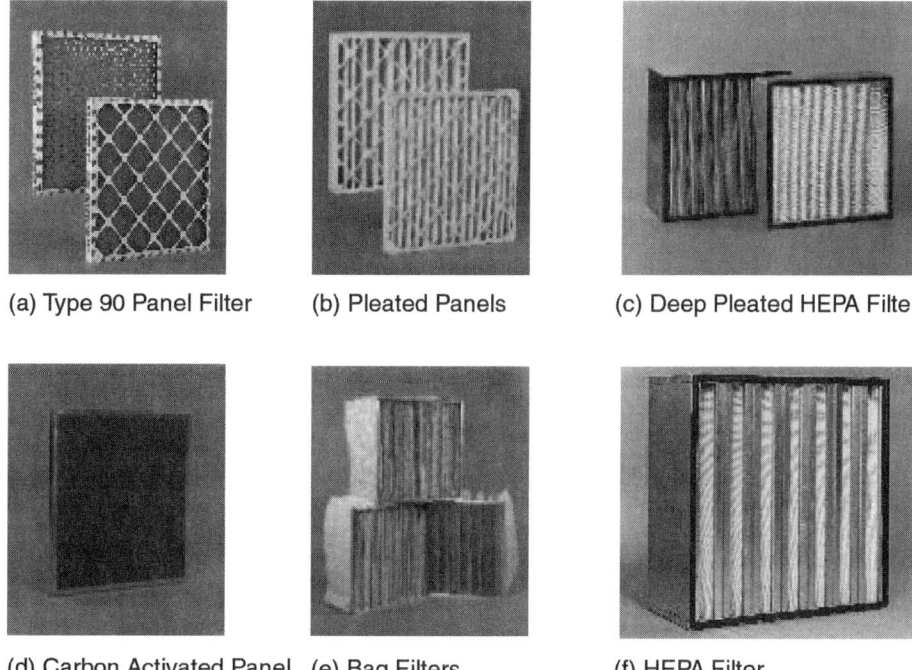

Figure 8-1 *Typical fibrous media unit filters (modified with permission from Jfilters.com).*

and thereby become attached. This effect is increased with decreasing velocity and decreasing particle size.

- *Electrostatic effect* – The particles are charged and pass by an oppositely charged surface to which they are attracted and attached.

The common types of particulate filters are:

- *Fibrous media unit filters* – Particles accumulate on the filter, causing the filtration efficiency and the filter, differential pressure to increase as the filter becomes loaded. However, at high particulate loading, the efficiency of some filters decreases due to offloading. These types of filters should be replaced when there is reduced filtration efficiency or the filters reach the upper limit for pressure drop. The efficiency of HEPA filters, however, increases with loading, but less air is moved due to an increase in pressure drop, thus leading to their replacement Fibrous media filters can range from coarse low-efficiency filters to ultra fine high-efficiency filters (with greater than 99.999% efficiency for 0.12 micron particles). Examples of typical fibrous media unit filters are shown in Figure 8-1.

- *Renewable media filters* – These filters remain at constant pressure and efficiency by periodically introducing new fibrous filter media into the filtration stream.

- *Electronic air cleaners* – This technology charges particles and collects them on an oppositely charged plate. Electronic air cleaners have relatively constant pressure loss and high filtration efficiency.

- *Combination air cleaners* – Any combination of the above techniques.

Gas Phase Filters

Gas phase filters adsorb gaseous contaminants by Van der Waals forces, which are created by physically broken or heated surfaces. Activated carbon is a commonly used adsorbent, as it has an enormous internal surface area compared to mass [1000 m^2 (10,800 ft^2) to 1 gram (0.035 ounces) of activated carbon]. Other common adsorbents include activated alumina silica gel and molecular sieves. Alumina silica gel will absorb water to the exclusion of other chemicals and is for that reason often used to dry airstreams. Molecular sieves are used for specialized pollution control where a specific contaminant is removed.

The adsorbent in gas filters can be impregnated with a chemically reactive substance to contain contaminates that are too loosely bound to the adsorbent alone to improve filtration efficiency. The temperature, concentration, and molecular mass of the contaminant determine the capacity of the adsorbent. The adsorbent process is reversible and high temperature or solution of contaminants in a liquid can reactivate the adsorbent.

There are three types of equipment for adsorption:

- *Fixed beds* – The absorbent is periodically regenerated or replaced.
- *Moving beds* – Moving the adsorbent as granular in cycles allows regeneration of the adsorbent and thereby better pollution control.
- *Fluidized beds* – Suspends fine particles of adsorbent in the gas stream.

Filter Retaining System

Particulate filters and gas phase filters used in laboratory applications may have to be secured in nuclear grade bag in-bag out housings to protect maintenance personnel from exposure to hazardous materials during a change-out procedure. Also, without an engineered filter clamping mechanism, filter bypass can take place contaminating downstream ductwork and associated equipment.

SCRUBBING

Scrubbers are air treatment systems that use a liquid (typically water) to dissolve or react with the gases to be removed. Scrubbers also remove particles. The liquid and air are in direct contact by spraying, dripping the liquid onto a packing through which the airstream is passed, or by bubbling the exhaust through the liquid. The by-product of scrubbing is a liquid solution or solid particles. The liquid can be disposed of or cleaned when the concentration of pollutant in the water is at a level where the efficiency of the scrubbing process is reduced. Particle filters can remove particles produced by the scrubbing process. The three main categories of scrubbers are dry, wet, and particle.

Dry scrubbing uses an alkaline water solution sprayed into an acid airstream. The acid is absorbed onto the water solution droplet and reacts with the alkaline to form a salt. Solid salt particles then form as the water evaporates and filters capture the particles. It is important that sufficient time be allotted between the introduction of the alkaline solution and the particle filter to ensure that the solution and acid reacted and that the water evaporated. Gases that typically are cleaned with this type of scrubber are hydrochloric acid (from biological waste incinerators), sulfuric acid, sulfur trioxide, sulfur oxides, and hydrogen fluoride.

Wet packed scrubbers remove gaseous contaminates by absorption on the water surface. Particles will be removed by impingement of the particles on the liquid droplets. While there is no limit to the particulate removal, the gaseous removal is limited by the partial vapor pressure of the gas. As long as the partial vapor pressure of the gas with respect to the liquid is above the partial pressure of the gas in the exhaust, the gas will be absorbed into the water. Packing material is typically used to distribute the water to provide a large contact area. The packing material should have a high void ratio to ensure a low pressure drop. Common wet packed scrubber configurations are:

- Horizontal cocurrent scrubber
- Vertical cocurrent scrubber
- Cross-flow scrubber
- Countercurrent scrubber

Particle scrubbers remove particles by impingement of the particles on the water droplets. Direct interception, inertial deposition, and diffusion are mechanisms in particle scrubbers that are similar to mechanisms in particle filters. An additional mechanism in particle scrubbers is condensation. Condensation occurs when the air or gas is cooled below its dew point. The vapor condenses on the dust particles, which serve as condensation nuclei. The dust particles become larger and the chance of removal is increased.

The advantages of particle scrubbers are:

- Constant operating pressure
- No secondary dust source
- Small parts requirements
- Ability to collect both gases and particles
- Low cost
- Ability to handle both high temperature and high humidity gas streams as well as to reduce the possibility of fire or explosions
- Reasonably small space requirements

The disadvantages for particle scrubbers are:

- High susceptibility to corrosion
- High humidity in discharge airstream can cause visible exhaust plume
- Large pressure drop, high power requirement
- Possibly difficult or high cost to dispose waste water
- Rapidly decreasing efficiency for particles less than 1 μm
- Freeze protection may be required in cold climates

For a given type of particle scrubber, the efficiency will increase as the power applied increases. The different types of particle scrubbers are:

- *Spray towers and impingement scrubbers.* The gas/air stream passes through a single or several sprays or a series of irrigated baffles. These scrubbers are low-energy scrubbers (up to 2 kJ/m^3, 0.25 - 1.5 kPa) and generally have a low degree of particle removal (50-99% efficient for particle sizes down to 2 μm).
- *Centrifugal type collectors.* The gas/air stream enters the scrubber tangentially. These scrubbers are medium-energy scrubbers (2-6 kJ/m^3, 1.5-4.5 kPa).

- *Orifice type collectors.* The gas/air stream passes through narrow openings where the gas/air and the liquid interact. These scrubbers are medium-energy scrubbers (2 – 6 kJ/m^3, 1.5 – 4.5 kPa).
- *Venturi scrubber.* The gas/air stream is accelerated up to 60 m/s through a venturi. The liquid is injected right before or at the throat. The rapid acceleration shears the water into a fine mist, which increases the chance of impact. These scrubbers are high-energy scrubbers (>6 kJ/m^3, >4.5 kPa).
- *Electrostatic augmented scrubber.* A combination of particle scrubber and electrostatic filter.

For a more detail discussion of scrubbers, refer to Chapter 25, *2000 ASHRAE Handbook—Systems and Equipment*.

CONDENSING

Condensing is a process where gas/vapor in an airstream condenses on a cold surface or on particles in the airstream. The airstream to be cleaned is cooled to a temperature lower than the dew point temperature of the gas/vapor in the airstream. The condensed gas/vapor can be removed as a liquid or by removing the larger particles containing the contaminant particle. The cooling capacity of the condenser must be larger than the latent and sensible heat that must be removed, including the latent heat capacity of other gases/vapors that condense.

Gases will be removed to a level equal to the partial vapor pressure for the contaminant at the leaving air temperature. When condensing is used to remove particles, the efficiency of the filtration process will depend on how much the particle sizes are increased by the condensing liquid and how efficient the filtration equipment is after the condensing has occurred.

OXIDIZATION (INCINERATION)

Oxidation of gas is a treatment that breaks down the contaminant into chemicals that are inert or less harmless. This process is, in essence, the same as incineration, but the contaminant is usually at a concentration so low that ignition is impossible. The treatment can be accomplished with the use of catalysts (catalytic incineration) by increasing the temperature to initialize the wanted reaction (thermal incineration) or by bleeding a strong oxidizer such as ozone into the airstream.

Adsorption and incineration can be combined to reduce the size of the incinerator and the energy consumption of the incineration process as much as 98%. The contaminants are accumulated in the adsorption material until it is saturated. The gas stream is then switched to another bed of adsorption material, while the contaminants in the saturated adsorption material are driven off by hot inert gas and incinerated. The volume of this contaminated gas stream is significantly lower, thereby reducing the energy needed for incineration.

REFERENCES

American Conference of Governmental Industrial Hygienist (ACGIH). 1999. *Documentation of the Threshold Limit Values and Biological Exposure Indices*, 6th ed. Cincinnati: ACGIH.

American Conference of Governmental Industrial Hygienist (ACGIH). 1998. *Threshold Limit Values for Chemical Substances and Physical Agents and Biological Exposure Indices*. Cincinnati: ACGIH.

American Society of Heating, Refrigerating and Air-Conditioning Engineers (ASHRAE). 2001a. *2001 ASHRAE Handbook—Fundamentals*, Chapter 13, Odors. Atlanta: ASHRAE.

American Society of Heating, Refrigerating and Air-Conditioning Engineers (ASHRAE). 2001b. *2001 ASHRAE Handbook—Fundamentals*, Chapter 12, Air contaminants. Atlanta: ASHRAE.

American Society of Heating, Refrigerating and Air-Conditioning Engineers (ASHRAE). 1996. *1996 ASHRAE Handbook—HVAC Systems and Equipment*, Chapter 24, Air Cleaners for Particulate Contaminants. Atlanta: ASHRAE.

American Society of Heating, Refrigerating and Air-Conditioning Engineers (ASHRAE). 2000. *2000 ASHRAE Handbook—HVAC Systems and Equipment Handbook*, Chapter 25, Industrial Gas Cleaning and Air Pollution Control. Atlanta: ASHRAE.

American Society of Heating, Refrigerating and Air-Conditioning Engineers (ASHRAE). 1999. *ANSI/ASHRAE Standard 62-1999, Ventilation for Acceptable Indoor Air Quality*. Atlanta: ASHRAE.

Kawamura, P.I., and D. MacKay. 1987. The evaporation of volatile liquids. Journal of Hazardous Materials, Vol. 15, pp. 343-364.

National Institute for Occupational Safety and Health (NIOSH). 1994. Pocket Guide to Chemical Hazards Department of Health and Human Services. Cincinnati, OH.

Petersen, R.L., and M.A. Ratcliff. 1991. An Objective Approach to Laboratory Stack Design. *ASHRAE Transactions* 97(2): 553-562.

Ruth, J.H. 1986. Odor Thresholds and Irritation Levels of Several Chemical Substances: A Review. *Journal of American Industrial Hygienists Association*, Vol. 47, p. A-142.

Wunder, J.S. 2000. Personal communication from operating experiences with laboratory equipment, University of Wisconsin–Madison.

Chapter 9
Exhaust Stack Design

OVERVIEW

Properly exhausting polluted air from the building to avoid concentrations of pollution in sensitive areas, such as air intakes, sidewalks, and building entrances that are higher than the limits given by codes and standards, is a critical function of a laboratory system. In this chapter, the required elements for good stack design and modeling techniques available to verify a design are presented in detail. The design issues that can accomplish this goal are then discussed. Finally, models that can help determine the effects of the design are explained. These models will document how the design of the exhaust performs.

ELEMENTS OF STACK DESIGN

The key elements of stack design are the parameters that influence the dispersion of the gas and the airflow around buildings. The latter is of particular interest as it has a significant impact on where the pollutants go and is a very complex issue. Those items that should be avoided, such as reentrainment into the building air intakes and contamination of building entrances, exits, and adjacent buildings, are presented. The effects of a building's shape and its neighboring buildings on airflow, as well as the effects of wind direction and static pressure around the building, are also discussed.

Once the design parameters have been presented, they are put into context in the "Design Issues" section. This includes aesthetics, energy use, noise and vibration, weather control and drainage, mixing of contaminated airstream, and maintaining stack velocity at all operating conditions. Situations that can cause failure of a stack design are also addressed.

Stack Design Parameters

The key parameters that affect stack design and location are:

- Stack height
- Exit velocity
- Volumetric flow rate
- Intake locations

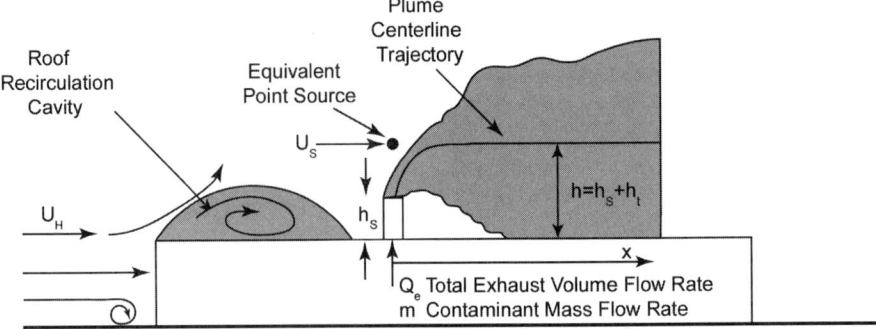

Figure 9-1 *Plume rise.*

The stack height must be sufficient to ensure that the exhaust plume is sufficiently diluted when it reaches sensitive areas or turbulent areas that include sensitive areas. The plume height should be calculated based on the stack height and not corrected for buoyancy or exit velocity, but reduced if downwash occurs. This is a conservative approach that includes safety margins. For design considerations, the effective stack height is calculated, which is the physical stack height plus the plume rise minus the downwash. Buoyancy effects are not included, to have a safety margin, as well as to account for local climatic effects, such as inversion, that can prevent buoyancy (see Equation 9-1).

$$h_e = h_s + h_r - h_d \qquad (9\text{-}1)$$

where

h_e = effective stack height, ft (m)
h_s = stack height, ft (m)
h_r = plume rise, ft (m)
h_d = stack downwash, ft (m)

For a plume with a negligible exit velocity, the lower edge of the plume will go downward one unit for every five units the plume goes forward. Since it is important that the plume be high enough to avoid contact with building objects and recirculation wake regions, the plume must rise (see Figure 9-1). This can be accomplished by increasing the exit velocity. However, if a raincap is added to the exhaust stack, the plume rise is zero. Therefore, raincaps are not recommended for laboratory exhaust stacks. The plume rise from a stack can be calculated using Equation 9-2:

$$h_r = 3.0 \cdot \beta \cdot \frac{V_e \cdot d}{U_H} \qquad (9\text{-}2)$$

where

h_r = plume rise, ft (m)
β = design parameter, 1.0 without cap, 0 with cap
V_e = stack exit velocity, fpm (m/s)
d = stack diameter, ft (m)
U_H = wind speed, fpm (m/s)

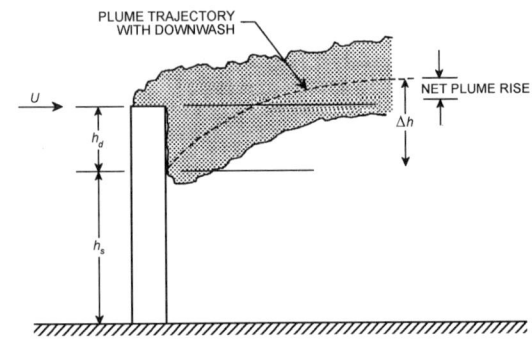

Figure 9-2 *Stack downwash (courtesy of ASHRAE).*

A counteractive force to the plume rise due to an increased exit velocity is a down-wash resulting from the exhaust air sticking to the leeward side of the stack relative to the wind direction. This occurs when the wind speed is high compared to the stack exit velocity. ASHRAE (1999) recommends that the stack velocity be at least 1.5 times higher than the design wind speed (exceeded 1% to 5% of the time) to minimize the effect of wind-induced downwash and provide good initial dilution. For situations where downwash will occur (V_e/U_H<1.5) the additional stack height needed (see Figure 9-2) can be calculated using Equation 9-3:

$$h_d = 2.0 \cdot d \cdot \left(1.5 - \frac{\beta \cdot V_e}{U_H}\right) \qquad (9\text{-}3)$$

where

h_d = additional stack height needed to compensate for downwash
d = stack diameter, ft (m)
β = design parameter (0.0 with cap)
V_e = stack exit velocity, fpm (m/s)
U_H = wind speed, fpm (m/s)

Equations 9-2 and 9-3 rely on an accurate estimation of the wind speed. Since the average wind speed is typically not measured at the actual building location, but at a meteorological station, the measured wind speed must be corrected for the actual location (see Figure 9-3).

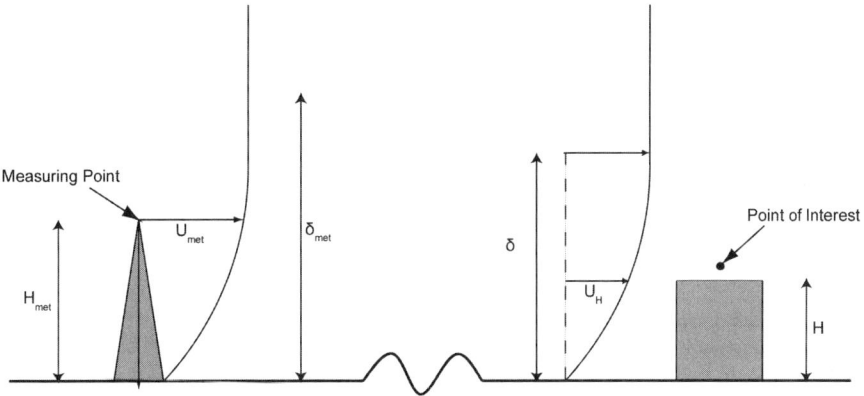

Figure 9-3 *Wind calculation from meteorological station to the point of interest.*

Table 9-1: Terrain Factors

Terrain	a	δ, ft (m)
Flat (airport)	0.10	640 (210)
Flat, obstacles less than 10 m (30 ft)	0.14	820 (270)
Suburban	0.22	1,130 (370)
Urban	0.33	1,400 (460)

This is accomplished using Equation 9-4:

$$U_H = U_{met} \cdot \left(\frac{\delta_{met}}{H_{met}}\right)^{a_{met}} \cdot \left(\frac{H}{\delta}\right)^{a} \qquad (9\text{-}4)$$

where

- U_H = free wind speed at height H approaching the building, fpm (m/s)
- U_{met} = wind speed at the meteorological station, fpm (m/s)
- δ_{met} = boundary layer thickness at the meteorological station, ft (m)
- H_{met} = height of the meteorological station, ft (m)
- a_{met} = roughness factor for the meteorological station, unitless
- H = upwind wall height, ft (m)
- δ = boundary layer thickness of airstream approaching the building
- a = exponent that varies with the roughness of the terrain for the local point, unitless

The boundary layer is defined as the transition from no velocity (ground) to uniform velocity. Height of the boundary layer (δ) and the exponent a, depends on the roughness factor of the local upwind terrain. Values for these variables have been experimentally defined for the selected terrain types shown in Table 9-1.

For the design of VAV systems, it is important to design the stack for all volumetric flow rates to maintain the exit velocity above a minimum level. This can be accomplished by

- sizing the stack for the minimum velocity at minimum exhaust flow,
- using a variable geometry exhaust stack that maintains constant velocity regardless of flow,
- introducing outdoor air into the exhaust stream prior to or after the exhaust fan to maintain a constant volumetric flow rate.

A final design parameter to consider in the location and sizing of exhaust stacks is the location of the outdoor air intakes. It is critical that the intake location be carefully evaluated to avoid contamination from the laboratory exhaust system. In addition, the air intake should be located so as to avoid other sources of contaminants, such as dust, fumes from traffic, kitchen exhaust, cooling towers, plumbing vents, loading docks, and leaf-shedding trees.

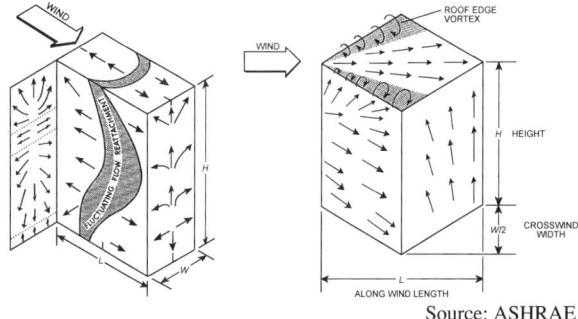

Source: ASHRAE

Figure 9-4 *Flow patterns around two buildings.*

Airflow Around Buildings

The local climate, in particular the wind, is an important parameter for designing stacks for laboratories. Knowledge of the airflow around and over buildings is necessary to avoid contaminating sensitive areas such as:

- Building air intakes
- Building entrances and exits

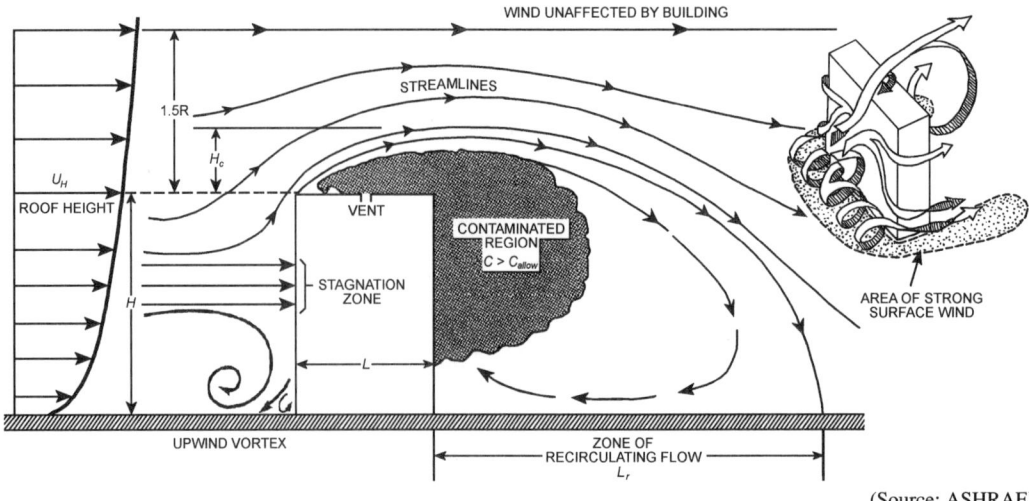

Figure 9-5 *Flow patterns around rectangular buildings.*

- Sidewalks
- Adjacent buildings' sensitive areas

Contaminating the air around building intakes and entrances results in reentrainment of the contamination exhausted from the laboratories back into the building. However, determining the airflow around buildings is complicated even for simple geometrical structures. The important parameters for understanding the effect wind has on pollution reentrainment are the airflow patterns, relative static pressures, and adjacent building effects. Wind data can be retrieved from the National Climatic Center in Asheville, North Carolina, and local meteorological stations.

Airflow Patterns. The upwind velocity profile, the upwind turbulence, the angle of the approach wind, and the shape of the building—all influence the airflow pattern around a building. The upwind terrain influences the upwind velocity profile. Rougher terrain has higher boundary layers and the free wind speed is reached at higher levels. The wind speed that influences the flow around the building (U_H) will be higher for a building where the terrain is flat compared to a building in an urban setting where the terrain is rougher when the free wind speed is the same. The upwind terrain also influences the turbulence of the wind hitting the building. This turbulence created by an upwind building can cause parts of the building to be in the wake or otherwise complicate the airflow pattern, which can be important when considering the stack height if large objects are close to the building. The angle of the approach wind significantly influences the pressure, surface flow pattern, and the size and shape of wake regions around the building (see flow patterns in Figure 9-4).

A stagnation point exists on the windward wall and separation occurs at sharp edges. Separation of the airflow creates turbulence and recirculation. The airflow can reattach to the building surface if the building extends far enough downwind. Figure 9-5 illustrates the flow pattern around a rectangular building.

As wind speeds increase, a larger stagnation pressure occurs at the top of the building, resulting in a downwash on the lower one-half or two-thirds of the upwind wall. The top one-quarter or one-third has a flow that is directed upward. However, a horizontal flow can exist in the intermediate zone if the building height is more than three times the width of the building.

The size of the recirculating zones created by separation of the flow at sharp edges can be estimated with the help of the scaling length, R (see Equation 9-7), which can be applied to the different locations of separation. While the size of these zones will vary over time due to the turbulent flow in the approaching wind and the wake region, the important fact to understand is where air is recirculating and where exhaust flows should not be located to avoid recirculation into the building.

Relative Static Pressures around Buildings. The difference between the static pressure outside buildings and inside buildings will influence where exfiltration and infiltration occur. The static pressure distribution on the outside of the building will, according to ASHRAE (1997), vary with the angle of attack of the wind. In general, the static pressure caused by wind will be positive on the upward wind surface until the angle of attack exceeds 45°. The static pressure will be negative on the upward wind surface when the angle of attack exceeds 75°, while both positive and negative pressures exist when the angle of attack is between 45° and 75°. On the leeward side, the static pressure will always be negative. The pressure on the roof caused by wind will, on average, be negative, but roof angles steeper than 20° will have parts of or the whole upwind side at a positive

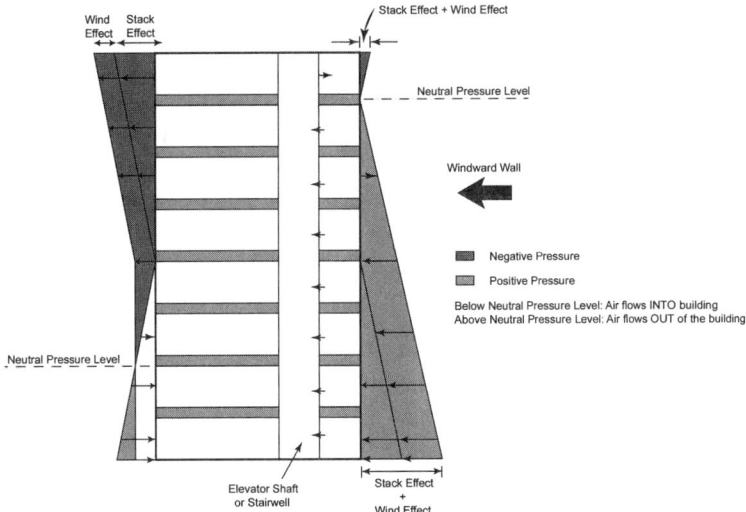

Figure 9-6 *Resulting building pressure difference due to wind and stack effect. (Diagram modified with permission from EPRI.)*

pressure. The static pressure can be calculated from Equation 9-5:

$$P_s = C_p \cdot P_v \qquad (9\text{-}5)$$

where

P_s = static pressure, in. H_2O (Pa)
C_p = static pressure coefficient, unitless
P_v = velocity pressure, in. H_2O (Pa)

C_p for different wind directions, building shapes, and location on the building can be found from figures and tables in ASHRAE (2001). The spatially average static pressure coefficient summarized around the building (all four walls) is approximately $C_{p,\,sum} = -0.2$.

The velocity pressure can be calculated from Equation 9-6:

$$P_v = \frac{\rho U_H^2}{2C_2} \qquad (9\text{-}6)$$

where

P_v = velocity pressure, in. H_2O (Pa)
C_2 = conversion coefficient, g_c (1)
g_c = $3.44 \times 10^{-8} \dfrac{\text{ft}^4}{\text{lbf}-\text{sec}^4}$
ρ = density of the air, lb/ft^3 (kg/m^3)
U_H = approach wind speed at upwind wall height H, ft/s (m/s)

Wind effects can disturb the fume hood operation by creating volume surges resulting in inadequate exhaust. This is unacceptable if highly toxic materials are handled. The exhaust hoods should be tested under both low and high wind conditions. If wind effects are suspected to cause inadequate exhaust, fume hoods should be provided with a flow sensing monitor that will create both visual and audio alarms.

The static pressure on the inside of the building is influenced by the stack effect and the pressurization caused by the air-handling system. The stack effect is caused by the movement of air in vertical enclosures (stairways, elevator shafts, etc.) and is induced by the density difference between the air in the enclosure and the ambient air (Dorgan and Dorgan 1996). Tall buildings in cold climates are significantly influenced by the stack effect in the heating season, and this will cause negative pressure in the bottom half of the building and positive pressure on the top half of the building.

The air-handling system can be used to pressurize the whole building, or parts of the building, to ensure exfiltration or infiltration or to overcome or reduce the pressure difference created by the stack effect or the wind pressure. Supplying more air than is exhausted and enough to obtain an interior pressure that is higher than the exterior pressure ensures exfiltration, while supplying less air than is exhausted and enough to obtain a pressure that is lower than the exterior pressure ensures infiltration. The difference in the amount of air supplied and exhausted depends on the pressure that is to be achieved and the leakage of the building structure. However, too high or too low a pressure inside the building can cause difficulties in opening or closing doors and windows and can create drafts, and most HVAC systems are designed to create a neutral pressure.

Adjacent Building Effects. Adjacent upstream buildings can change the upstream flow pattern around the building by creating wakes around parts of the building, or the whole building, by creating turbulent flow and by changing the wind profile. The exhaust gases can

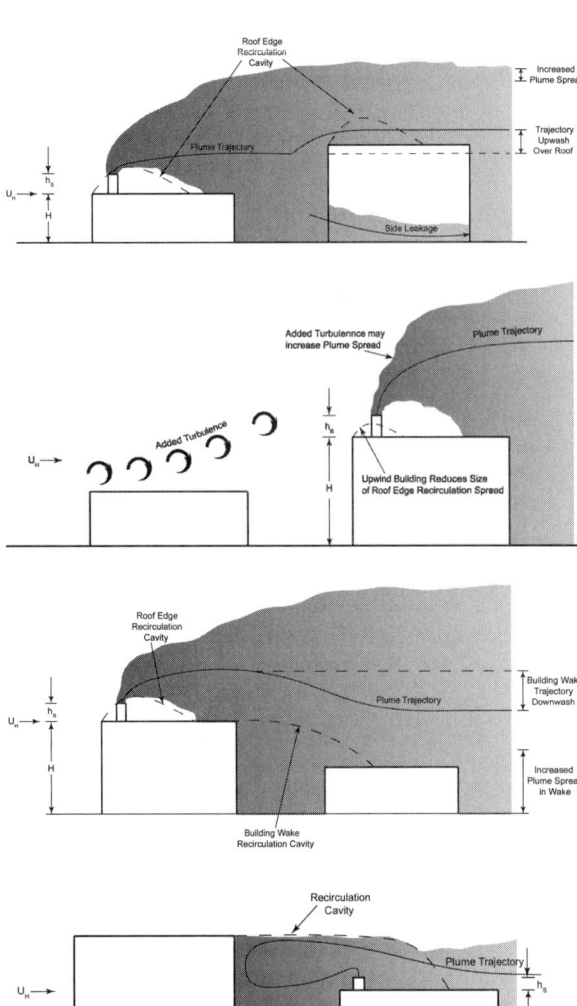

Figure 9-7 *Adjacent building effect on stack plume. (Diagrams modified with permission from Wilson et al. 1998.)*

contaminate downstream buildings if the exhaust is not diluted enough or does not pass over the downstream buildings. Figure 9-7 shows some examples of how an adjacent building can influence the plume from an exhaust stack.

The effect by or on adjacent buildings is often extremely complicated and may require wind model tests or computer simulations. However, some general guidelines are suggested by Wilson et al. (1998):

- Designers should avoid locating stacks near the edge of a roof, where the high wind speed can deflect the plume into the roof edge recirculation and reduce dilution by factors of 2 to 10.
- With the emitting building upwind, a lower, step down roof adjacent building will always have higher dilution on the step down roof than would

occur on a flat roof at the emitting building height. Ignoring the step down will produce conservative designs.

- If the lower adjacent building is upwind of the emitting building, it will block the flow approaching the emitting building, producing lower velocities and recirculation cavities on the emitting building roof and increasing dilution by factors of 2 to 10 on the emitting building
- Designers should use increased exhaust velocity to produce jet dilution when the plume will be trapped in the recirculating cavity from a high upwind adjacent building. This initial dilution is directly proportional to the exhaust velocity and will reduce the dilution everywhere on the emitting building roof. Using both a high stack and high exhaust velocities helps the exhaust jet escape from the upwind building recirculation cavity.
- When the adjacent building is higher than the emitting building, designers should try to avoid placing air intakes on the adjacent building at heights above the roof level of the emitting building. Then, use the highest stack and exit velocity possible on the emitting building.

Design Issues

In addition to design parameters, other design issues will influence the design of the stack:

- Stacks are usually highly visible and often aesthetically unappealing.
- Energy use is an issue that often will cause changed exhaust volumetric flow rates that will influence the stack design.
- Noise can cause problems both outdoors and indoors if not considered.

Aesthetics. The required stack height for proper expulsion and dilution of the contaminated air often makes the stack very visible. This visibility is normally unwanted. Therefore, many designers are forced to decide the placement of the stack not only on design parameters, such as location of air intake, recirculating zones, etc., but also on the desire of the owner or architect to make the stack less visible to the surroundings. Fans designed for laboratory exhaust dilution may be considered because they typically require less height than stacks and provide the necessary high exit velocities. Architectural screens and physical placement of the stack can reduce the visibility of the stack. However, care must be taken to ensure that the exhaust cannot enter the recirculating zones created by the architectural screens. The appearance of the stack can be altered to create a more pleasing appearance than plain metal stacks.

Energy Use. Laboratory buildings are often very high energy consumers due to the need for significant volumes of outdoor air to replace the air exhausted by the fume hoods. To reduce the amount of outdoor air needed, the fume hood exhaust system can be designed to be variable air volume (VAV). However, this also affects the stack design as the volumetric flow rate out of the stack will vary. The requirements for the exhaust outlet speed must be maintained at minimum exhaust air volumes to avoid downwash at the stack and ensure good initial dilution. The section "Maintaining Stack Velocity" in this chapter discusses options to accomplish this.

Noise and Vibration. High velocities out of exhaust stacks can cause excessive noise either from turbulence or from the exhaust fan. The noise to the surroundings should be evaluated and noise reduction equipment installed to protect noise-sensitive areas. The local code and noise level limits should be used as maximum values, with lower values desirable near residential areas. Noise traps and noise screens can effectively reduce noise to acceptable levels but may cause turbulence to effectively lower stack height.

Vibration from fans and motors can also be transferred to ductwork and cause noise problems from the stack that are not caused by the fan but rather the vibration of the duct against some other object. Further, wear and tear is increased due to vibrations, which can create increased maintenance and reduce the reliability and lifetime of the system. Vibrations can be reduced with flexible connections to the fan and other vibrating components. The connection must be of a material that is usable for the contaminants exhausted. Noise and vibration becomes important for exhaust velocities above 2,950-3,940 fpm (15-20 m/s).

Weather Control and Drainage. The stack exhaust velocity has to exceed 2,560 fpm (13 m/s) to prevent rain from entering the stack and to prevent condensed moisture from draining into the stack. If the exhaust speed is lower than this during operation, stack designs with internal drains have to be installed to prevent water from entering the fan or pooling at low spots in the exhaust ductwork. Caps and coverings or non-vertical exhaust exits will greatly reduce the performance of the stack to properly disperse the pollutants and should not be used.

Figure 9-8 shows both good and poor stack designs. It is important in cold climates to ensure the stack design prevents freezing by maintaining an internal stack temperature during the minimum design temperature condition. This is typically accomplished by insulating the stack or heating the stack. Freeze protection is especially important for exhaust systems with a washdown system (perchloric fume hoods), where water is used to clean the exhaust air of dangerous pollutants prior to their exhaust. In these systems, the exhaust air is saturated and will quickly freeze up the exhaust stack if allowed to cool and freeze on the stack walls.

Mixing of Contaminated Airstreams. It is possible that mixing of exhaust airstreams can result in an increased hazard—such as an explosive mixture, a more toxic chemical, or any other way that increases the potential hazards of the exhaust—that exceeds the design criteria for the stack and its components. Mixing of contaminated airstreams from other stacks containing the same substance will reduce dilution because the contam-

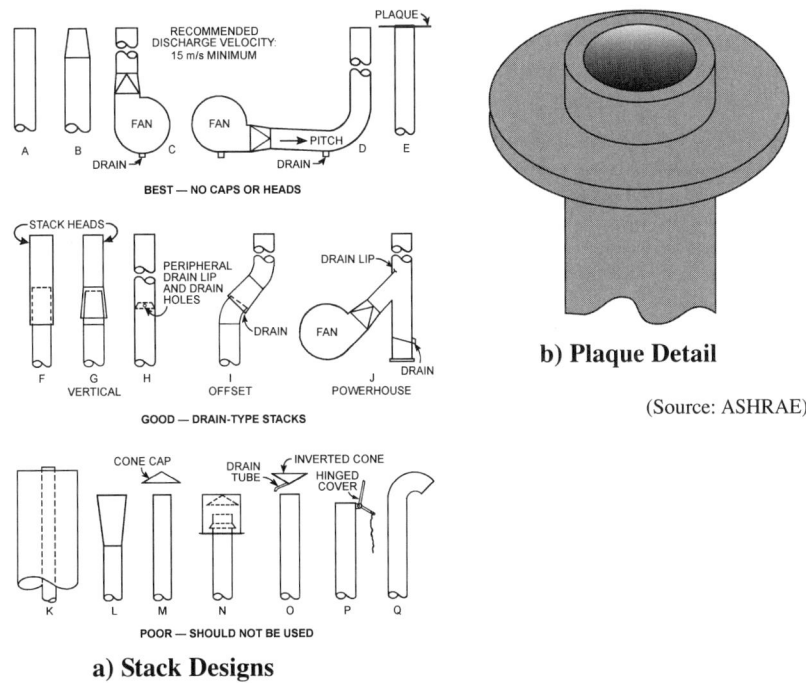

b) **Plaque Detail**

(Source: ASHRAE)

a) **Stack Designs**

Figure 9-8 *Good and poor exhaust stack design.*

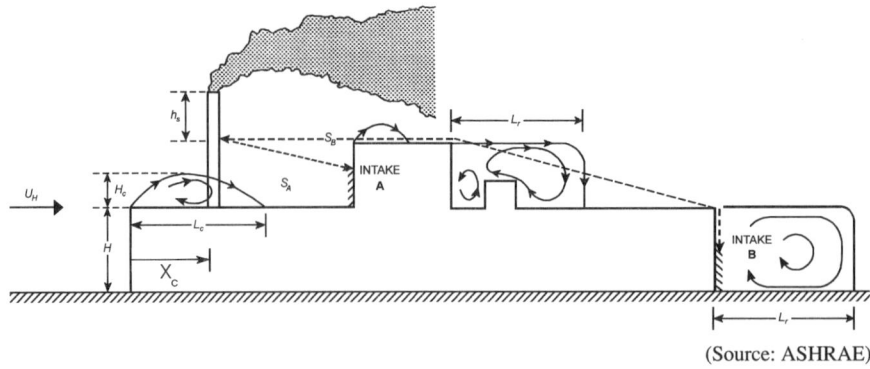

(Source: ASHRAE)

Figure 9-9 *Turbulent and recirculating zones on a building.*

inated flows will not mix with clean air but with already contaminated air.

Maintaining Stack Velocity. The problem of maintaining stack velocity at all operating conditions is related to VAV systems, which are primarily designed to save energy by reducing the need for outdoor air to replace the exhausted air from the fume hoods. These systems will be exhausting a varying amount of air, and the air speed out of the exhaust stack will subsequently vary. ASHRAE (1999) recommends designing the stack to meet the requirements at the minimum volume of air exhausted. The requirements at higher volumetric flow rates will thereby be met, but noise and vibrations should be considered if the exhaust velocity exceeds 3,000 to 4,000 fpm (15-20 m/s) at maximum volumetric flow rate.

There are other methods for maintaining the stack velocity:

- Variable flow geometry
- Induction of outdoor air
- Staging of multiple fans on a common inlet plenum

Causes of Problems

Many typical problems occur with exhaust stacks that result in significant performance degradation. These include:

- Release in turbulent airflow
- Insufficient plume rise
- Stack tip downwash
- Elevated receptors

Release in Turbulent Airflow. Release of the exhaust flow into a separated and highly turbulent airflow near the roof, caused by physical structure, can result in the accumulation of contaminants and prevent the exhaust plume from leaving the building. The turbulent zone may contain critical zones, such as air intakes and entrances, that can be polluted even if they are physically far away from the exhaust but are in the same turbulent and recirculating zone.

The structures that create turbulent and recirculating zones are architectural screens, roof-mounted air-handling units, penthouses, etc. Turbulent zones are also created behind all "leading edges" on a building, as shown in Figure 9-9 (Wilson 1982). Release in these turbulent and recirculating zones must be prevented. The stack must either be moved out of this area or the stack height increased to release the exhaust above the building. All wind directions should be considered when determining where turbulent and recirculating zones will be created.

Wilson and Winkel (1982) suggest that the dimensions of the turbulent zones be calculated using Equation 9-7:

$$R = B_s^{0.67} \cdot B_L^{0.33} \qquad (9\text{-}7)$$

where

R = scaling length, ft (m)

B_s = smaller of upwind building face dimensions, height or width, ft (m)

B_L = larger of upwind building face dimensions, height or width. If $B_L > 8B_s$ use $B_L = 8B_s$, ft (m)

For buildings with different roof levels, only the height under the roof and width of the roof in question should be used to calculate R, as long as the distance to other obstructions is at least B_s.

However, the shape and size of the turbulent zone is not constant, as both the height and length of the turbulent zone are smaller upwind on the same edge (see Figure 9-4). The turbulent zone at a downwind wall (L_r) extends approximately 1.0 R. For a flat-roofed building, the approximate extension of the turbulent zone is a maximum height, H_c, of 0.22 R, a position, X_c, of 0.5 R behind the leading edge, and to a length L_c, of 0.9 R behind the leading edge.

Insufficient Plume Rise. The plume has to rise high enough to avoid contact with any critical area, such as air intakes or entrances to the building. These critical areas can be contained in building recirculation wake (turbulent) regions. A poor design of the exhaust stack (rain cap, etc.) or a low exit velocity can reduce or eliminate the plume rise. A conservative design of the stack height does not take into account any plume rise above the height of the stack. This design approach provides a safety margin in case of unfavorable atmospheric conditions.

Stack Tip Downwash. The exhaust sticking to the leeward side of the stack due to high wind velocities causes downwash and thereby reduces the effective stack height. This occurs when the stack exit velocity is lower than 1.5 times the wind speed. Mounting a plaque on the tip of the stack can reduce the risk of stack downwash.

Elevated Receptors. The exhaust gas can contaminate elevated receptors if their height is within the exhaust plume. To avoid contaminating the receptor, the receptor height can be decreased, the stack height can be increased, or the stack or receptor can be moved to ensure that the receptor is not inside the plume or is far enough away to ensure enough dilution to be below the required level.

DISPERSION MODELING APPROACH

Dispersion modeling is used during design and system troubleshooting to determine the optimum stack design given local conditions. To understand how dispersion modeling works, the characterization of a pollution release must be accomplished including considerations of emergency and steady-state releases. Dispersion modeling approaches presented in this section include: the EPA model, dilution equations of Chapter 15, *ASHRAE Handbook—Fundamentals*, wind tunnel modeling, computer simulation, and computational fluid dynamics.

Emissions Characterization

Emissions can be characterized using Title III of the Superfund Amendments and Reauthorization Act (SARA) requirements. SARA Title III details how laboratories are required to plan for the accidental releases of hazardous chemicals. It stipulates how to respond and when to report the incident to the authorities. For planning purposes, the laboratory must maintain the Material Safety Data Sheets (MSDS) required under OSHA's "Hazardous Communication" standards. These MSDSs provide information on various materials such as physical data (boiling points, melting points, etc.), toxicity, health effects, first aid, reactivity, storage, disposal, protective equipment, and spill/leak procedures. Copies of MSDSs must be submitted to the authorities as well as an inventory of the hazardous chemicals present in the particular institution. The reason for these requirements is the community's "right to know" and for emergency response authorities to be able to efficiently handle emergency situations. Also, extremely hazardous chemicals have requirements that involve release notification and training of emergency personnel.

Accidental releases of liquid and compressed gas should be simulated to ensure that the contamination around the building does not exceed local or state limits for short-term exposure. The formulas for evaporation of volatile liquids are important to determine the concentration of the contaminant in the exhaust. See the discussion on ambient concentration limits in Chapter 8, "Air Treatment."

Steady-state processes are used to calculate the long-term exposure of downstream locations. This exposure must not exceed the 8-hour, 40-hour, and yearly limits to avoid health hazards for individuals that are exposed to the contaminant. The steady-state process is determined from the expected typical usage patterns.

By evaluating purchased amounts of chemicals, understanding how much is consumed in reactions, what portion is disposed of, and what portion is exhausted, it is possible to estimate the potential exposure of those downstream of the building. Further, some chemicals can only be purchased and stored in certain quantities for safety reasons.

Dispersion Modeling

The dispersion models are intended to help the designer investigate how the pollutants will be distributed in the atmosphere, around the building, and around adjacent buildings and areas. This will identify potential problems that could result in too high pollution concentrations by air intakes, entrances, or other sensitive areas that can fairly easily be corrected during design. These problems can then be corrected by changing design parameters such as exhaust exit velocity, location of stack, height of stack, etc. Identifying these problems during the design phase will allow for a less expensive, more efficient solution than trying to correct a real problem after the building is completed, when, for example, changing the location of the stack can be very costly.

Many models have been suggested and most of these are available as computer programs. ASHRAE presents models that are simple to use and can calculate the minimum dilution at a given critical location (such as an air intake). The EPA has developed models that are more detailed and give hourly results or monthly and annual distribution results. Many other models that are either more detailed, or focus on special applications, have been developed. Some of these models have been tested and compared (EPA 1995).

For more critical or especially complicated problems, the use of computer models that numerically solve simplified Navier-Stokes equations may be considered. These models can be both two- and three-dimensional and will produce very detailed information. However, some of these models have limitations (such as what geometry can be described or number of nodes that can be used), which should be considered, and a wind tunnel test may be needed in addition to, or instead of, computer simulations.

EPA Models. The dispersion model developed by EPA (1992) is called ICS2 (Industrial Source Complex regulatory plume model) and can be used to calculate short-term (hourly) exposure and long-term (monthly and annual) exposure. Both the short-term and long-term models are divided into three source classifications: (1) point source, (2) line source, and (3) area source. For exhaust stack design, the point source is the model of interest. The EPA guideline also describes a short- and a long-term dry deposition model.

The ICS2 model uses the Gaussian equation to calculate the concentration of the contaminant concentration downwind of the source. The models consider the wind speed profile, use plume rise formulas, calculate dispersions factors (which take into consideration different landscapes, building wakes and downwash, and buoyancy), calculate the vertical distribution, and consider decay of the contaminant.

ASHRAE Dilution Equations. ASHRAE dilution equations can be used to calculate the concentration of contaminants at air intakes. These models use a stretched string distance (S) to find the dilution factor (see Figure 9-9) and consider three special cases:

- Strong jets in a flow recirculation cavity
- Strong jets on multi-winged buildings
- Surface vents on flat-roofed buildings

In addition, ASHRAE presents a model for the normalized concentration coefficient when the exhaust source is in the recirculating (turbulent) zone. For surface vents on a flat roof, a critical wind speed can be calculated that gives the lowest dilution (highest concentration) for a given point when the stack is being evaluated. This critical wind speed can be used to calculate a critical dilution factor. The critical dilution factor does not change with increased exhaust velocity, but the wind speed where critical dilution occurs is increased. An increased exhaust velocity will usually reduce the amount of time when critical dilution occurs. A stack speed of at least 1.5 times the wind speed is needed to avoid downwash and provide good initial dilution.

Wind Tunnel Modeling. Wind tunnel modeling has shown very good results when compared to field observation, usually within a factor of 2. A wind tunnel model study should be accomplished when calculations show concentrations that are close to or exceed the acceptable limits. Scales of 1:50 down to 1:250 have been shown to have a good agreement with field observations (McQuaid and Roebuck 1984). However, the biggest scale possible should be used to get the best results from the wind tunnel model. Water channel models can also be used to determine the flow around a building.

A concern with wind tunnel modeling is that the boundary of the wind tunnel can interfere with the results from the model. To obtain good results, the model should obstruct less than 5% of the total cross-sectional area (ASHRAE 1997). The wind tunnel must also be long enough to allow the wind profile to develop before the flow reaches the building model. Certain conditions concerning flow rates, density rates, and flow parameters, such as Reynolds number and Froude number, should be equal in the wind tunnel test as it is for the actual case to ensure good results.

The concentrations measured in the wind tunnel must be converted to the full-scale concentration using Equation 9-8 (Petersen and Ratcliff 1991):

$$C_f = A \cdot \left(\frac{C \cdot U}{C_0 \cdot V}\right)_m \cdot \left(\frac{L_m}{L_f}\right)^2 \cdot \left(\frac{q}{U}\right)_f \cdot 10^6 \qquad (9\text{-}8)$$

where

- C = concentration of the gas of concern, lb_m/ft^3 ($\mu g/m^3$)
- U = wind speed, mph (m/s)
- C_0 = gas concentration at source (ppm)
- V = stack exit velocity, fpm (m/s)
- L = length scale, ft (m)
- Q = pollutant emission rate, lb_m/h (g/s)
- A = constant, SI=1, I-P=24.8
- m pertains to model
- f pertains to full-scale

Computer Simulations and Computational Fluid Dynamics (CFD). Computer simulations and computational fluid dynamics achieve the same results as wind tunnel testing but can often be quicker to implement and thereby less expensive. Computer models also offer the benefit of being less costly to change than wind tunnel models and can therefore be more viable for studies where many different models must be evaluated. Typically, a wind tunnel experiment is done in conjunction with computational models to ensure a higher degree of accuracy or to verify that the optimal model found by the computational model performs as simulated.

There are many programs capable of CFD analysis. CFD programs are also widely available via the Internet (CFDnet). These programs are becoming easier to use and have an increasingly improved output that visualizes the flow and concentrations calculated. However, care must be taken when choosing a CFD program to ensure that the program is capable of simulating the number of nodes required and that all necessary conditions can be modeled and considered (e.g., convection, turbulence, diffusion, boundary layers, heat transfer, different shapes, steady state, transient, etc.).

REFERENCES

American Society of Heating, Refrigerating and Air-Conditioning Engineers (ASHRAE). 2001. *2001 ASHRAE Handbook—Fundamentals*, Chapter 16, Airflow around Buildings. Atlanta: ASHRAE.

American Society of Heating, Refrigerating and Air-Conditioning Engineers (ASHRAE). 1999. *1999 ASHRAE Handbook—HVAC Applications*, Chapter 43, Air Intake and Exhaust Design. Atlanta: ASHRAE.

U.S. Environmental Protection Agency (EPA). 1995. *Testing of Meteorological and Dispersion Models for Use in Regional Air Quality Modeling*. Washington, D.C.: EPA.

U.S. Environmental Protection Agency (EPA). 1992. *User's Guide for the Industrial Source Complex (ISC2) Dispersion Models. Volume II – Description of Model Algorithms*. Washington, D.C.: EPA.

Dorgan, C.B., and C.E. Dorgan. 1996. *Ventilation Best Practices Guide*. Palo Alto: Electric Power Research Institute (EPRI).

Kawamura, P.I., and D. MacKay. 1987. The evaporation of volatile liquids. *Journal of Hazardous Materials*, Vol. 15, pp. 343-364.

McQuaid, J., and B. Roebuck. 1984. *Large Scale Field Trials on Dense Vapor Dispersion*. Final report to sponsors on the heavy gas dispersion trials at Thorney island, 1982-84, Health and Safety Executive Safety Engineering Laboratory, Sheffield UK.

Petersen, R.L., and M.A. Ratcliff. 1991. An Objective Approach to Laboratory Stack Design. *ASHRAE Transactions* 97(2), 553-562.

Wilson, D.J., I. Fabris, and M.Y. Ackerman. 1998. Measuring Adjacent Building Effects on Laboratory Exhaust Stack Design. *ASHRAE Transactions* 104(2): 1012-1028.

Wilson, D. J., and G. Winkel. 1982. The Effect of Varying Exhaust Stack Height on Contaminant Concentration at Roof Level. *ASHRAE Transactions* 88(1): 515-533.

Chapter 10
Energy Recovery

OVERVIEW

Energy recovery utilizes the temperature or temperature/humidity in the exhaust air or other source of heat/cool that would otherwise be wasted to pre-treat (heat or cool) the supply airstream, thereby saving energy and power capacity of the heating, cooling, dehumidifying, or humidifying equipment. These savings can be substantial, with lifetime savings and payback period of energy recovery systems being very favorable. Since laboratories often use 100% outdoor air, energy recovery is an important consideration for the design. In some projects, energy recovery has been the main contributor to savings from energy-efficient installations, with others involving VAV and diversification considerations (Streets and Setty 1983). Due to the strong influence of the climate on potential energy savings, each heat recovery system must be evaluated based on variations in the local climatic condition and energy costs.

Energy can be recovered from exhaust air, process water, and cooling water from air-conditioning equipment. Heat recovery from water only transfers sensible heat, while heat recovery from air offers the opportunity to transfer both sensible and latent loads. However, care must be taken to avoid contaminating the supply airstream. NFPA Standard 45, Appendix 6-4.2, states that "the use of devices for energy conservation purposes should consider the potential contamination of the fresh air supply by exhaust air containing vapors of flammable or toxic chemicals." Biological contamination is also a concern.

In this chapter, air-to-air, water-to-air heat, and hot gas-to-air recovery systems are discussed followed by a presentation of selection parameters for these systems. More details on the theory of heat recovery equipment are available in *2000 ASHRAE Handbook—HVAC Systems and Equipment*.

AIR-TO-AIR HEAT RECOVERY

The intent of air-to-air heat recovery systems is to transfer heat (sensible only) air energy (sensible or latent) between two airstreams, typically the exhaust air and supply air. Heat recovery equipment includes runaround loops, heat pipe, and fixed plate. Hygroscopic fixed plate heat exchangers, heat wheels, liquid desiccants, and evaporative cooling devices are energy recovery technologies.

There are several design issues that must be addressed when selecting the type of air-to-air device, including:

- *Relative airstream location* – Some air-to-air systems require the two airstreams to be located adjacent to each other, whereas other systems can have a reasonable distance between the airstreams.
- *Cross-contamination potential* – Unless there is a physical barrier between the two airstreams, contamination of the supply air could occur.

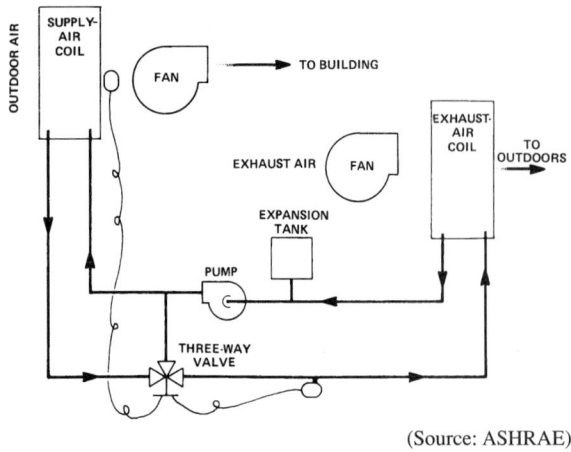
(Source: ASHRAE)

Figure 10-1 *Runaround loop recovery system.*

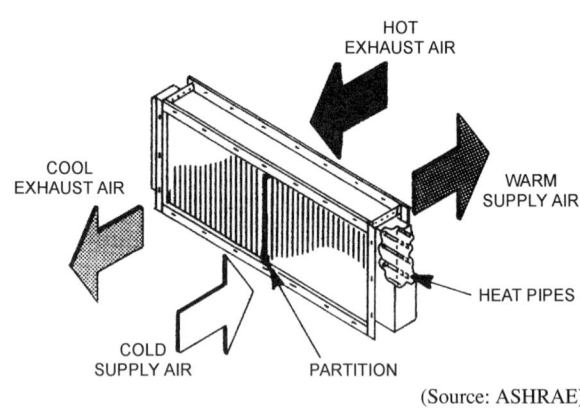
(Source: ASHRAE)

Figure 10-2 *Heat pipe recovery system.*

- *Corrosion* – The chemical in the exhaust air can corrode the heat recovery device, either directly or due to the formation of condensate when the warm moist exhaust air comes in contact with the cold heat exchanger surface. Corrosion can lead to cross-contamination by breaking down the barrier between the two airstreams. Perchloric acid fume exhausts should not be considered for energy recovery due to corrosion and explosion hazards.
- *Fouling* – Fouling of the heat exchanger reduces the supply or exhaust air volume, resulting in reduced heat exchanger and system performance, which is critical to maintaining system performance.
- *Freeze-ups* – Freeze-ups occur when the exhaust air contains sufficient moisture and the outdoor air is below freezing. The moisture in the exhaust air condenses and freezes. As frost builds up, airflow is decreased until there is no flow, which seriously compromises the occupant's health and safety.

When cross-contamination is a major concern, physically separating the two airstreams by using a second media is recommended. Runaround loops, heat pipes, or evaporative cooling should be used. In some instances, the exhaust air is cleaned using a scrubber or washer prior to the heat exchanger to reduce the potential for cross-contamination.

Runaround Loop

The runaround loop is a heat recovery system where the heat from the exhaust air is transferred to a water-based medium (glycol) using a typical coil heat exchanger. It is then circulated through pipes to the supply air where the water releases the energy recovered from the exhaust air using another coil heat exchanger. The water then returns to the exhaust air coil. Figure 10-1 depicts this heat recovery system, which is usually used in systems where it is uneconomical or not desired to have the exhaust duct and the supply duct next to each other. These systems have a relatively high initial cost and low performance compared to other heat recovery systems. Runaround loops have a sensible heat transfer effectiveness of 55% to 65% (ASHRAE 2000a).

Runaround loops are excellent heat recovery systems for laboratory applications where it is critical that no cross-contamination occur. To achieve increased effectiveness, the exhaust coil is typically a closely spaced fin coil that requires filtration upstream to keep dirt from accumulating inside the coil. The "ideal" runaround loop system would have exhaust HEPA filters upstream of the exhaust coils to eliminate cleaning entirely. These systems require periodic cleaning and thus coil static pressure drop should be monitored to ensure that the coil has acceptable pressure drop and that the laboratory exhaust has acceptable airflow. In order to clean the exhaust coil, high-pressure water has been used, which can push particles deeper inside the coil. This forces coil replacement to maintain airflow. Therefore, unless a long-term maintenance plan is developed to maintain the runaround loop, with effective pretreatment of the incoming exhaust air, this method should not be used.

Heat Pipe

Heat pipes rely on refrigerant migration within sealed pipes with half of the pipe in the exhaust airstream and the other half in the supply airstream. The heat pipe transfers heat from one airstream to the other by having the refrigerant evaporate (cools the airstream) at the hot side of the tube and condense (heats the airstream) at the cold side. The refrigerant is then returned to the warm side. Figure 10-2 shows the configuration of a heat pipe recovery system, and Figure 10-3 depicts a single heat pipe.

The heat pipe is usually made of copper pipes with aluminum fins. The heat transfer fluid must be selected

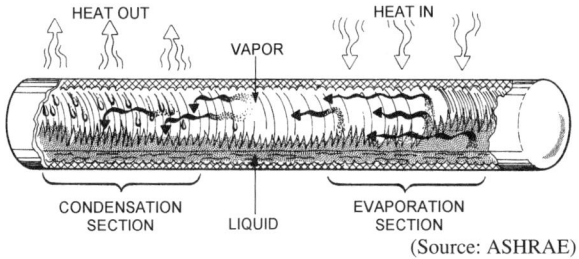

Figure 10-3 *Heat pipe.*

to work for the operating temperatures to avoid decomposition and deterioration of performance. Since heat pipes can have the supply and exhaust ducts physically separated, there is no cross-contamination. The effectiveness of sensible heat transfer for heat pipes is between 45% and 65% (ASHRAE 2000a).

Heat Wheel

A heat wheel consists of a cylinder with a large internal surface where one half is in the exhaust air while the other half is in the supply air. The cylinder rotates and heats up on the warm side and transfers this heat to the cool side. Similar to a heat wheel, an energy wheel also transfers humidity in addition to sensible heat using a desiccant coating or molecular science technology. However, the risk of transferring pollutants increases with increased humidity transfer. Figure 10-4 depicts a heat wheel recovery system.

There will always be some cross-contamination with heat wheels. The cross-contamination can happen by carryover or leakage. Carryover is air that is entrained in the heat recovery wheel as it moves from the exhaust to the supply. Leakage occurs because of a pressure difference between the exhaust and supply side. The amount of cross-contamination is dependent on the pressure of the exhaust relative to the supply airstream at the heat wheel and the design of the heat wheel. The normal recirculation rate is from 1% to 10%. With a purge section, the cross-contamination can be reduced to below 0.1%. The effectiveness of sensible heat transfer is between 50% and 80% and for sensible and latent heat transfer between 55% and 85% (ASHRAE 2000a).

Fixed Plate

Fixed plate heat exchangers (depicted in Figure 10-5) have alternating layers of exhaust and supply airstreams separated by a plate. The primary benefit of the fixed plate system is no moving parts. To maximize heat transfer, the airflow is typically arranged in a cross-flow pattern. Even though a plate separates the airstreams, latent energy can be transferred between the

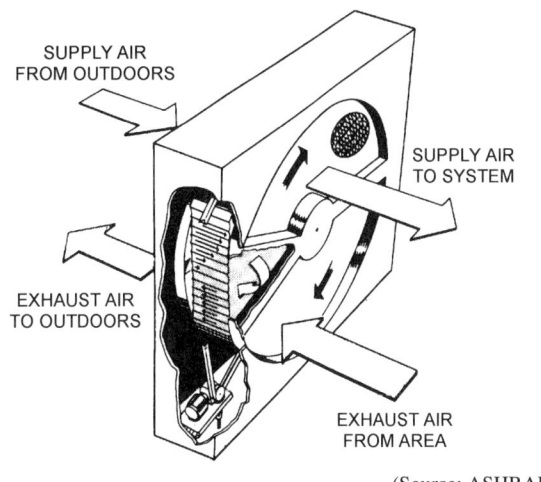

Figure 10-4 *Heat wheel recovery system.*

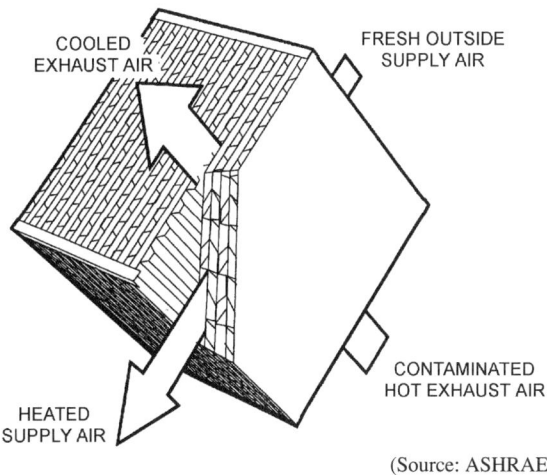

Figure 10-5 *Fixed plate recovery system.*

airstreams by using a hydroscopic material for the "plate" between the supply and exhaust streams.

Even though fixed plate heat exchangers are often considered to have no cross-contamination, the risk of contamination is always present, as the thin plates are all that separates the exhaust and supply. Therefore, fixed plate heat exchangers usually have between 0% to 5% cross-contamination due to leakage. Corrosion, freezing, and cleaning of the fixed plate heat exchanger can damage the plates and can increase the leakage rate. The effectiveness of sensible heat transfer is typically between 50% to 80% (ASHRAE 2000a).

Liquid Desiccant (Lithium Chloride)

A liquid desiccant heat recovery system consists of a liquid desiccant that is circulated between the supply and the exhaust airstreams. The desiccant transfers sensible and latent energy by being in direct contact with

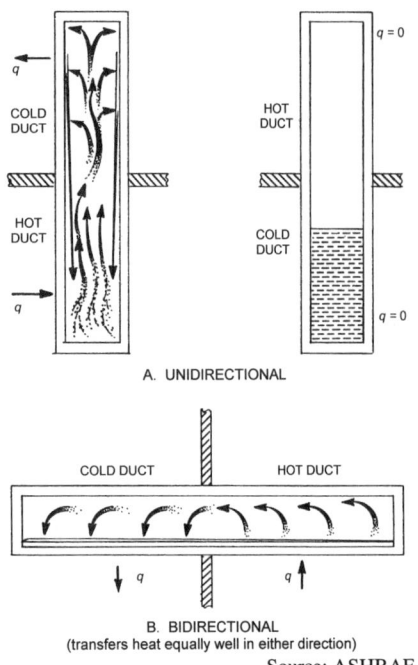

Figure 10-6 *Sealed tube thermosiphon recovery system.*

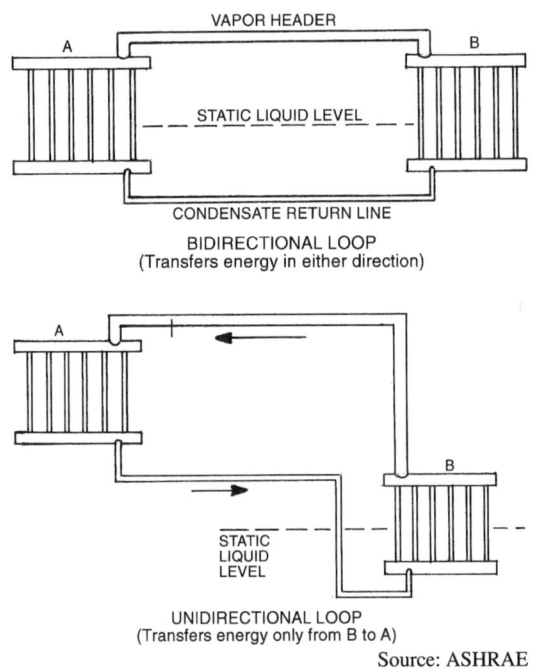

Figure 10-7 *Coil tube thermosiphon recovery system.*

the air, using contactor towers in both the supply and exhaust. The exhaust air must be filtered to remove particles such as animal hair, food particles, etc., to avoid contamination of the liquid desiccant.

However, any particulates captured in the exhaust air do not typically pose a contamination threat as the particle cannot be separated from the liquid intake supply airstream. The primary concern is the potential for limited gaseous cross-contamination, which is dependent upon the solubility of the gas in the sorbent solution. Tests with sulfur hexafluoride as a tracer gas have shown a typical gaseous cross-contamination rate of 0.025%. The cross-contamination rate should be investigated for a system with large amounts of gaseous contaminates, as well as the gaseous effect on the sorbent solution. The effectiveness of sensible heat transfer is typically between 40% and 60% (ASHRAE 2000a). The capital and operating costs of these systems are high.

Thermosiphon

A thermosiphon is used in two different heat recovery systems – a sealed tube design and a coil loop design. The sealed tube design is illustrated in Figure 10-6 and is essentially the same as heat pipes with the exception of the wicker material in the heat pipes. This thermosiphon is dependent on gravity to get the condensate back to the warm side. The coil tube design is shown in Figure 10-7 and is a thermosiphon coil that is essentially the same as the runaround loop heat recovery system. However, this thermosiphon system does not require any pumping. Thermosiphons typically require a significant temperature difference to initiate boiling. The effectiveness of sensible heat transfer for thermosiphons is between 40% and 60% (ASHRAE 2000a).

Evaporative Cooling

Evaporative cooling is not really a heat recovery method, but one where the evaporation of water is used to cool the air temperature. The cooling occurs when water is evaporated in air and thereby converts the sensible heat in the air to latent heat and reduces the temperature of the air.

Evaporative cooling may be accomplished directly or indirectly. Air is cooled via direct evaporative cooling by direct contact with the water. This is done either by a series of sprays or by an extended wet surface material. In very dry climates, the supply air can be humidified directly to reduce the air temperature without resulting in excessively humid supply air. In indirect evaporative cooling, either the exhaust air or outside air is humidified and the dry-bulb temperature decreased. The cooled air or water can then be used to cool the supply air with a sensible heat exchanger. A combination of both indirect and direct evaporative cooling can also be used.

Figure 10-8 is a psychrometric chart that illustrates what happens when air is passed through either a direct or indirect evaporative cooler. Chapter 19 of the *2000 ASHRAE Handbook—HVAC Systems and Equipment* and Chapter 50 of the *1999 ASHRAE Handbook—HVAC Applications* explain in detail how evaporative cooling

Energy Recovery

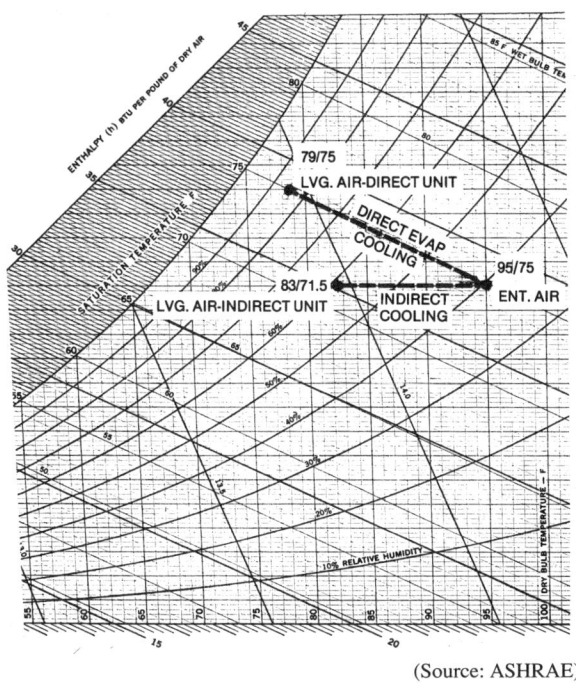

(Source: ASHRAE)

Figure 10-8 *Psychrometrics of evaporative cooling.*

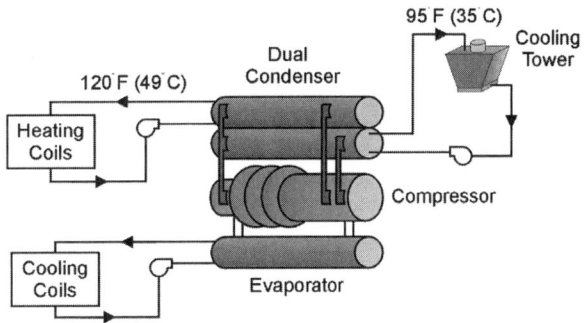

Figure 10-9 *Dual-condenser system.*

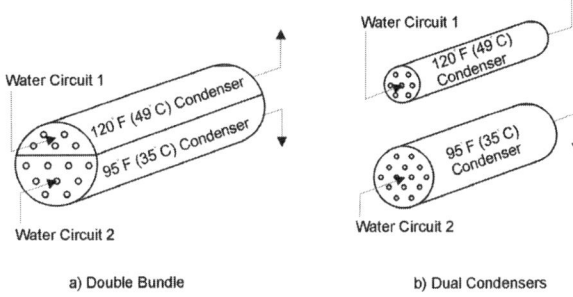

Figure 10-10 *Double-bundle and condenser system arrangement.*

is accomplished. Cross-contamination in indirect evaporative cooling will depend on what kind of heat exchanger is used. Direct evaporative cooling does not cause cross-contamination. However, growth of algae, slime, fungus, or bacteria must be prevented in direct evaporative cooling systems.

WATER-TO-AIR HEAT RECOVERY

In order to overcome the concerns with cross-contamination using air-to-air heat recovery systems, water-to-air heat recovery systems can be used. These systems require piping to the air supply duct and a heating coil that transfers the heat from the water to the air. Potential systems include:

- Refrigeration machine heat recovery
- Condenser water heat recovery
- Hot water waste heat recovery

Refrigeration Machine Heat Recovery

Refrigeration machine heat recovery utilizes a dual air-cooled condenser to heat an airstream. This type of system is used when there is a cooling load (interior zone) and heating load (perimeter zone) at the same time. Instead of rejecting the condenser heat to the outdoors, it is rejected to the exterior zone supply airstream. In some instances, a heat pump is placed across the exhaust and supply air to economically transfer heat from one to the other.

Condenser Water Heat Recovery

Condenser water heat recovery uses the heat typically rejected by a chiller to a cooling tower. A double-bundle or split-bundle condenser system has historically been the most common configuration used. It comprises a double-bundled condenser, which has a single condenser shell with two independent water/refrigeration loops. One of the water loops is connected to a cooling tower to reject excess heat and the other is connected to the heating load equipment.

However, double-bundle condensers are not as widely available since there have been recent changes in using newer refrigerants. Instead, dual-condenser chillers for heat recovery have been more readily offered by chiller manufacturers due to their greater ease of manufacture. Figure 10-9 illustrates this system configuration. A dual condenser has two distinct condenser shells. These have water loops that are, respectively, connected to the cooling tower and to the heating equipment (e.g., heating coils). Superheated refrigerant gases after the compression process flow into both condenser shells, and the heat is rejected to one or both of the condenser water loops.

The double-bundle and dual-condenser arrangements are constructed differently but are operated equivalently. Figure 10-10 illustrates these condenser arrangements. ASHRAE's *Chiller Heat Recovery Design Guide* contains details on identifying and sizing double-bundle arrangements.

Hot Water Waste Heat Recovery

Hot water waste heat is often available from industrial processes, including high-temperature process cooling water and combustion gases. The temperature is usually very good for preheating the supply air and often high enough and plentiful enough to heat the supply air to the desired temperature. The temperature of hot water heat recovery can be close to the boiling temperature of water. Physical configuration for this type of heat recovery is nearly identical to runaround systems.

SELECTION PARAMETERS

The selection of the type of heat recovery system to use is based on parameters such as laboratory requirements, climate, and exhaust and supply duct location. The economics are based on the initial cost, expected savings, and space implications.

Table 10-1 presents a summary of various kinds of energy recovery devices along with their physical features, advantages, and limitations.

Table 10-1: Comparison of Energy Recovery Devices

	Fixed Plate	Rotary Wheel	Heat Pipe	Runaround Coil Loop	Thermosiphon
Airflow arrangements	Counterflow Cross-flow Parallel flow	Counterflow Parallel flow	Counterflow Parallel flow	Counterflow Parallel flow	Counterflow Parallel flow
Equipment size range, L/s	25 and up	25 to 35000 and up	50 and up	50 and up	50 and up
Type of heat transfer (typical effectiveness)	Sensible (50% to 80%)	Sensible (50% to 80%) Total (55% to 85%)	Sensible (45% to 65%)	Sensible (55% to 65%)	Sensible (40% to 60%)
Face velocity, m/s (most common design velocity)	0.5 to 5 (1 to 5)	2.5 to 5	2 to 4 (2.2 to 2.7)	1.5 to 3	2 to 4 (2.2 to 2.7)
Pressure drop, Pa (most likely pressure)	4 to 450 (25 to 370)	(100 to 170)	(100 to 500)	(100 to 500)	(100 to 500)
Temperature range	−60°F to 800°F	−60°F to 800°F	−40°F to 35°F	−45°F to 500°F	−40°F to 40°F
Typical mode of purchase	Exchanger only Exchanger in case Exchanger and blowers Complete system	Exchanger only Exchanger in case Exchanger and blowers Complete system	Exchanger only Exchanger in case	Coil only Complete system	Exchanger only Exchanger in case
Unique advantages	No moving parts Low pressure drop Easily cleaned	Latent transfer Compact large sizes Low pressure drop	No moving parts except tilt Fan location not critical Allowable pressure differential up to 15 kPa	Exhaust airstream can be separated from supply air Fan location not critical	No moving parts Exhaust airstream can be separated from supply air Fan location not critical
Limitations	Latent available in hygroscopic units only	Cold climates may increase service Cross-air contamination possible	Effectiveness limited by pressure drop and cost Few suppliers	High effectiveness requires accurate simulation model	Effectiveness may be limited by pressure drop and cost Few suppliers
Cross-leakage	0 to 5%	1 to 10%	0%	0%	0%
Heat rate control (HRC) schemes	Bypass dampers and ducting	Wheel speed control over full range	Tilt angle down to 10% of maximum heat rate	Bypass valve or pump speed control over full range	Control valve over full range

Laboratory Type

The proposed energy recovery device must adequately meet the needs of the laboratory without compromising the safety of the occupants, functionality of equipment, or integrity of experiments. Some recovery devices are more appropriate for certain types of laboratories than for others. In general, most energy recovery applications are applicable to most types of laboratories. The key question to ask in order to eliminate a particular option is, what is the risk of cross-contamination that will threaten the safety of the laboratory occupants, products, or experiments? For example, it would not be recommended to use heat wheels or fixed plate recovery systems, which expel infectious substances, toxic chemicals, and animal dander, in a biological, chemical, or animal laboratory, respectively, where there is a possibility of cross-contamination.

Climate

The climate determines how much energy is necessary to heat or cool the supply air for a building. The potential energy savings will depend on the amount of energy spent. If the supply temperature is close to the ambient temperature for large parts of the year, then the energy savings are often small.

Exhaust and Supply Location

Most energy recovery systems need the exhaust and supply ducts next to each other. This can cause an increased amount of ductwork unless this is planned from the initial stages of the project. Also, the safety of adjoining contaminated exhaust air to supply air is a concern. Heat recovery systems that allow the exhaust air and supply air to be at different locations, such as a runaround system, can be a solution.

Economy

Heat recovery is usually aimed at reducing the amount of energy consumed. Most of the savings will come from this reduced energy usage. However, heat recovery systems that do not freeze (i.e., have a lower efficiency during subfreezing temperatures) or have to shut down during subfreezing temperatures can also reduce the component size (or power demand). This demand reduction can reduce or even eliminate the initial cost of the heat recovery system.

To evaluate the economic benefits of selected energy recovery systems, an economic analysis can be done using a simple payback (SP) analysis or a life-cycle cost (LCC) analysis. Chapter 15, "HVAC System Economics," gives details of these techniques.

REFERENCES

American Society of Heating, Refrigerating and Air-Conditioning Engineers (ASHRAE). 2000. *1999 ASHRAE Handbook—HVAC Applications,* Chapter 13, Laboratories. Atlanta: ASHRAE.

American Society of Heating, Refrigerating and Air-Conditioning Engineers (ASHRAE). 1999. *1999 ASHRAE Handbook—HVAC Applications*, Chapter 50, Evaporative Cooling Applications. Atlanta: ASHRAE.

American Society of Heating, Refrigerating and Air-Conditioning Engineers (ASHRAE). 2000a. *2000 ASHRAE Handbook—HVAC Systems and Equipment*, Chapter 44, Air-to-Air Energy Recovery. Atlanta: ASHRAE.

American Society of Heating, Refrigerating and Air-Conditioning Engineers (ASHRAE). 2000b. *2000 ASHRAE Handbook—HVAC Systems and Equipment*, Chapter 19, Evaporative Air Cooling. Atlanta: ASHRAE.

Dorgan, C.B., R.J. Linder, and C.E. Dorgan. 1999. *Application Guide Chiller Heat Recovery.* Atlanta: ASHRAE.

Streets, R.A., and B.S.V. Setty. 1983. Energy Conservation in Institutional Laboratory and Fume Hood Systems. *ASHRAE Transactions* 89(2B): 542-551.

BIBLIOGRAPHY

National Fire Protection Association (NFPA). 2000. *NFPA 45, Fire Protection for Laboratories Using Chemicals*. Quincy, Mass.: NFPA Publications.

Chapter 11
Controls

OVERVIEW

All buildings require a properly functioning control system to operate comfortably, safely, and energy efficiently. This is especially true in laboratories, where there are complex HVAC systems and dangerous pollutants that must be kept away from the occupants to maintain health and safety. Whereas the physical systems are designed for the peak operating condition, the control system must maintain proper temperature, humidity, air speed, air volume flow, and pressure. This is to ensure the safety of the occupants and the accuracy of experiments and research conducted during all nonpeak periods. In addition, the HVAC system for a laboratory is often subjected to rapid disturbances or changes, such as the opening of doors, opening sashes on numerous fume hoods at once, or turning on large pieces of heat-generating equipment. This requires a fast response from the control system to maintain the precise conditions required in the laboratory. Further, as the materials used in laboratories are commonly hazardous, corrosive, or flammable, the control system components need to withstand exposure to the materials that are used in the laboratory.

A final complication is that laboratories typically operate using 100% outside air, which requires considerable energy to condition before being supplied to the individual rooms. Therefore, it is very important that a laboratory HVAC control system be robustly designed in order to accurately provide a fast response to changing conditions outdoors and indoors.

In this chapter, equipment control, including constant volume fume hoods, variable volume fume hoods, and other exhaust equipment, will be discussed. Room control will then be discussed, including the theory of room control, minimum ventilation air changes per hour, control stability, variable and constant volume control strategies, temperature and humidity control of critical spaces, and building pressurization. Finally, emergency situations for the central system control will be discussed.

EQUIPMENT CONTROL

Due to the critical nature of the exhaust equipment to maintain containment, the development of a control system typically starts with the exhaust equipment. While fume hoods are the most common type of exhaust equipment specific to laboratories, there are several other types of exhaust, such as biological safety cabinets, snorkel exhausts, glove boxes, chemical storage cabinets, general exhaust, and direct equipment exhaust that must be considered when designing the control system for a laboratory. Background information on fume hoods and other types of exhaust equipment can be found in Chapter 5, "Exhaust Hoods."

Fume hoods are used for a wide variety of experimentation, research, and temporary material storage in most types of laboratories. Fume hoods can be classified into two types: constant volume, which exhausts a fixed rate of air

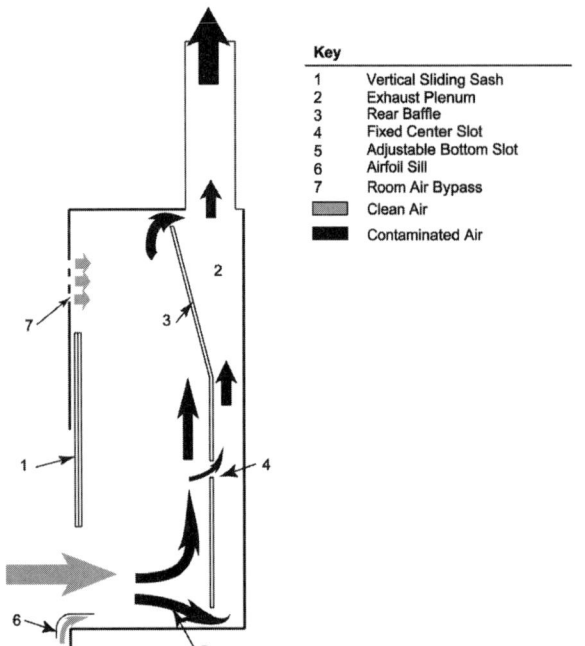

Figure 11-1 *Typical constant volume fume hood. (Diagram modified with permission from DiBerardinis et al. 1993.)*

regardless of the fume hood sash position, and variable volume, which exhausts only the air needed to maintain a specified face velocity for the current position of the sash.

Constant Volume Fume Hood Control

Constant volume fume hoods continuously exhaust a given airflow, regardless of sash position. Bypass constant volume hoods, which are the most common, are manually controlled by physically allowing additional air to enter the hood through a bypass opening, as the sash is closed, in order to maintain a relatively constant face velocity. Manual constant volume fume hoods are simple and reliable; they can maintain proper hood face velocity regardless of sash position or dependence on controllers and sensors to maintain safe face velocities. For this reason, constant volume fume hoods have been used extensively in laboratories in the past. Figure 11-1 is a schematic of a typical constant volume fume hood.

While high energy use for constant volume hoods has begun to reduce their usage, understanding control of these systems is important for proper operation of the installed systems and to understand their operation when considering renovations. For constant volume fume hoods, the control system should monitor the fume hood fan (when present) or main exhaust fan to note it is operating and provide an alarm or switch to a backup fan if it is not operating. The sash position or face velocity of these constant volume fume hoods should also be real-time monitored as recommended by ANSI Standard

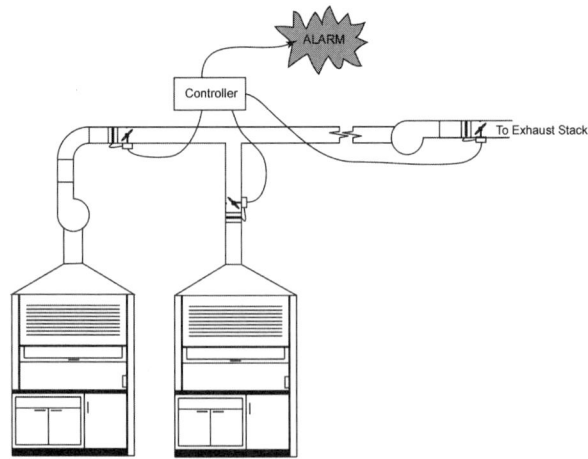

Figure 11-2 *Constant volume hood system.*

Z9.5 (ANSI 1992) for all new and remodeled fume hoods (e.g., using a static pressure monitoring device). This will give users a local indication of correct face velocity, which is especially important when the fume hood is designed to operate at a position less than full open.

To accomplish this in retrofit situations, where the exhaust system is being replaced while the constant volume hood is being reused due to cost issues, a variable air volume controller/monitor is sometimes installed if the hood is to be converted to variable volume. This controller is used as an alarm to signal when insufficient air is being exhausted (low-pressure differential), and once the hood is retrofitted, the controls operate as detailed in the next section. If there are no plans to convert to VAV, a face velocity monitor can be installed. Figure 11-2 details a typical control schematic for a manually operated constant volume hood system.

Most fume hoods are only used periodically and should have their sashes closed when not in use. However, when constant volume fume hoods are not in use and their sashes are closed, considerably more air is exhausted than is needed to maintain safe working conditions in the laboratory. As laboratories frequently use 100% outside air, this unnecessary exhausting of air requires significant energy to condition and move the air.

Variable Volume Fume Hood Control

To reduce the energy requirements of constant volume hood systems, variable volume fume hoods have been developed to vary the exhaust airflow from the hood and maintain a specified face velocity at the hood opening as the sash position is varied. Typically, using a variable frequency drive (VFD) on the hood exhaust motor or a damper in the exhaust duct, the airflow is regulated. Variable volume fume hoods use one of two methods to determine the desired airflow for the fume

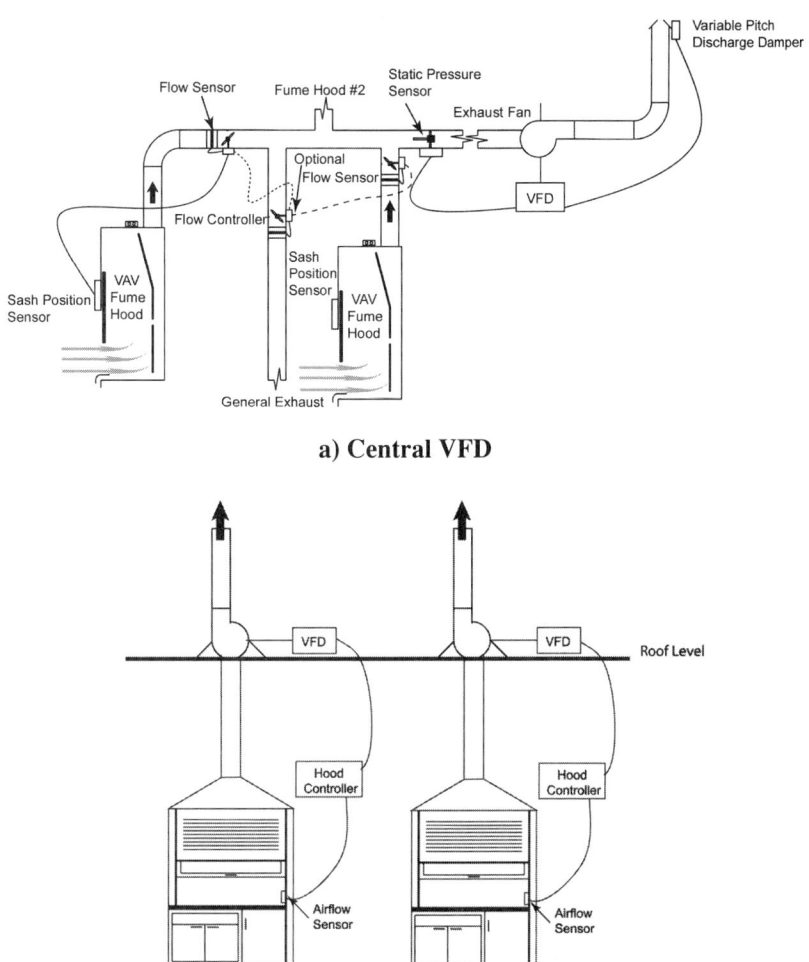

Figure 11-3 *Typical VAV fume hood control system.*

hood. One method is to use an air velocity sensor (anemometer) to directly measure the face velocity at the sash. The other method is to use a duct airflow rate sensor and a sash position sensor to calculate the average face velocity. When a damper is used to control the volume of air, a central VFD exhaust fan is often used to vary the total system volume. This fan is typically controlled to maintain a constant negative pressure in the exhaust ductwork. Figure 11-3 shows a typical control schematic for a variable volume hood control system.

Why Provide Variable Volume Control for a Fume Hood? As with any laboratory hood, the primary goal is to contain pollutants to maintain the health and safety of laboratory personnel. Variable volume fume hoods achieve this goal by ensuring that the proper hood face velocity is maintained at all times. The face velocity needed for safety can be actively and continually verified, since the face velocity must either be measured directly from an airflow sensor or calculated from the sash position and the exhaust airflow rate to control the exhaust airflow of a variable volume fume hood. Provided the sensor is calibrated on a regular basis and the sensor is not damaged or counteracted by laboratory operations (e.g., blocking the airflow sensor while using the hood or spilling chemicals on the sensor), this can provide valuable documentation to building owners. Specifically, it can document that the design airflow rate and face velocity are being met. It can also be used to track fume hood usage by measuring sash position and thereby identify fume hoods that are left open (and exhausting excessive air) during unoccupied hours of the laboratory.

A secondary goal that VAV hoods achieve is the minimization of system energy. By controlling the exhaust airflow needed to maintain safety in the laboratory and not exhausting additional air beyond the level needed for safety, variable volume fume hoods require less supply air to maintain the proper pressure relationship of the laboratory. Energy consumption for exhaust fans and supply fans, treating exhaust air, and condition-

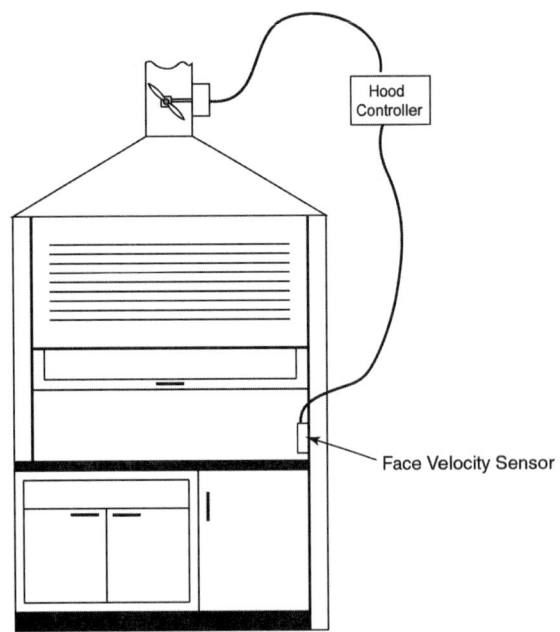

Figure 11-4 *Direct air velocity measurement control.*

As only the air velocity is known when using this type of control, and not the exhaust airflow rate, direct pressure control must be used (unless other sensing equipment is added), which modulates the supply airflow to maintain a fixed pressure differential between the laboratory and the adjacent rooms. However, if a sash position sensor or flow stations for fume hoods and general exhaust are added, the exhaust airflow rate can be determined, and flow tracking control may be used with the supply and exhaust airflows, which modulates the supply airflow to maintain a fixed exhaust to supply airflow differential. Direct pressure and flow tracking control are discussed later in this chapter in the variable air volume laboratory control strategies. Figure 11-4 contains a schematic of how direct air velocity measurement can be used to control a variable air volume fume hood.

The other method of determining face velocity is to measure both the sash position of the fume hood and the exhaust airflow rate. The sash position is either determined by a rheostat (variable resistor) attached to the sash or by a magnet on the sash that passes through a sensing strip. For a fume hood with a vertical sash, the sensor reads the vertical position of the fume hood sash, and by knowing the fixed width of the fume hood, the open area of the fume hood face can be calculated by the control system. For a fume hood with a horizontal sash, the sensor reads the horizontal position of the fume hood sash, and by knowing the fixed height of the fume hood, the open area of the fume hood face can be calculated by the control system. A flow rate sensor in a straight run of exhaust ductwork above the fume hood is then used to measure the flow rate of the exhaust air. The exhaust airflow rate is then controlled by dividing the exhaust airflow rate by the open area of the fume hood, giving the required face velocity.

The benefit of using this method of control is that the air velocity obtained is an averaged value over the entire face of the hood, and accurate control of the hood is not susceptible to the airflow sensor being blocked by the fume hood operator or other equipment. However, this method of control may make the fume hood more difficult to properly maintain, as the flow sensor would typically be located above a suspended ceiling or in a mechanical chase, which may make it difficult to access for maintenance and calibration. Further, the hood sash position sensor must be periodically calibrated to ensure proper system performance. Figure 11-5 is a schematic

ing of supply air is thus reduced. As laboratories often use 100% outside air, the energy reduction for conditioning of supply air associated with variable air volume control can be significantly more than that attainable in a typical building that uses a large portion of recirculated air.

Types of VAV Fume Hood Controls. There are two methods for controlling variable air volume fume hoods: directly measure the air velocity or indirectly measure air velocity by measuring sash position and exhaust airflow rate and calculating the fume hood face velocity.

Direct measurement of the fume hood face velocity is typically accomplished by using a small air velocity sensor (hot wire or vane anemometer), which is placed along the side wall of the fume hood near the sash opening. The exhaust airflow rate is then controlled using a variable speed hood fan or a damper by the air velocity sensor to the design value. The benefit of using this method of control is that it is simple (only one sensor) and provides a direct measurement of the air velocity. However, the reading at the sensor may not be representative of the air velocity over the entire fume hood area, as the sensor may be partially blocked by the fume hood operator or the equipment inside of the hood. Also, locations in the center of the fume hood opening (i.e., far from the sensor) may have substantially higher or lower velocities (SBT 1999) that are not detected or accounted for when using this method of control.

Controls

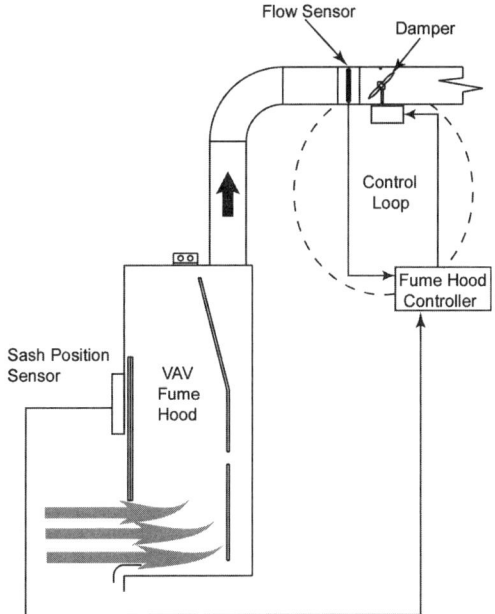

Figure 11-5 *VAV hood with sash position and airflow measurements.*

of how sash position and airflow rate measurements can be used to control a variable air volume fume hood.

System Response Time. System response time is an important factor for maintaining proper airflow velocity and direction to prevent spillage from fume hoods during disturbances, such as opening the sash from closed to fully open. Careful selection and programming of controllers are important to provide a fast and accurate response to minimize or eliminate fume hood spillage. ASHRAE Standard 110-1995 defines an acceptable response time for a VAV hood as the time needed for the face velocity to stabilize within 10% of the design value. This is determined while the sash is opened from a working height of 25% open to fully opened at both 1.0 ft/sec (0.30 m/s) and 1.5 ft/sec (0.46 m/s). The response time of the control system should be designed to maintain proper laboratory safety and pressurization differentials when the system is disturbed. Response times on the order of two seconds are typically acceptable (Monger 1994).

Control of Other Laboratory Equipment

In addition to fume hoods, the control of other equipment, including snorkel exhausts, glove boxes/biological safety cabinets, flammable and solvent storage cabinets, general laboratory exhaust, and direct equipment exhaust must be properly integrated into the control of the primary air systems for laboratories to maintain proper room conditions. Additional general information on these types of equipment can be found in Chapter 5, "Exhaust Hoods."

Snorkel Exhausts. Snorkel exhausts are used to provide small, movable exhaust intakes for benchtop laboratory equipment. While they remove heat and non-toxic particle emissions, they do not provide adequate protection against harmful chemicals. The typical control of these systems is the use of a switchable independent exhaust fan or the use of a manual damper to connect to a manifolded exhaust system.

Glove Boxes/Biological Safety Cabinets. Biological safety cabinets are used to protect laboratory workers, and occasionally the research work being performed, from various biological agents. There are three levels of safety cabinets, Class I, Class II, and Class III, which provide increasing protection. Some biological safety cabinets recirculate some or all of the exhaust air back into the laboratory once it has been HEPA filtered, while others exhaust all of the air to the atmosphere after being HEPA filtered. Exhaust from biological safety cabinets is either exhausted through individual exhaust stacks or a manifolded system, which only serves biological exhaust sources of similar hazard level.

Class I cabinets provide protection against low- and moderate-risk biological agents. It does not filter or treat the supply air that is introduced to the cabinet and should not be used for high-risk biological substances, as disruptions in the inward airflow to the cabinet can allow airborne particles to escape through the inlet. The exhaust air is HEPA filtered. Class I cabinets use constant volume control with a fixed opening (no sash).

Class II cabinets are used for moderate risk biological agents; they use HEPA filtered supply and exhaust air. There are currently four types of Class II cabinets, Types A, B1, B2, and B3, each of which uses a different combination of recirculated air and filtering. Each type of Class II cabinet has its benefits, is used for different types of work, and has a sash that can be opened for performing work. Class II cabinets use constant volume control with a movable sash and a filtered bypass opening.

Class III cabinets, or glove boxes, are used for highly infections materials. They maintain physical separation between the user and the substances inside the cabinet by using arm-length gloves, rather than a sash opening, to interact with the work. The cabinet is constantly maintained under a negative pressure by drawing supply air through a HEPA filter and removing exhaust air contaminants by either HEPA filtration or incineration. To control Class III cabinets, the cabinet is on at all times, operates at a constant volume, and may require redundant exhaust sources and controls. Figure 11-6 is a schematic that shows the differences between the three classes of biological safety cabinets.

Flammable and Solvent Storage Cabinets. Flammable and solvent storage cabinets protect laboratory personnel from the dangers of stored gases and chemicals, including possible explosions and volatile fumes. These cabinets may come with a variety of features, such as fire sprinklers, access openings, and safety controls. Flammable and solvent storage cabinets exhaust air from a high strength (explosion proof) cabinet, provide supply air (typically room air) through a filtered opening in the cabinet, and are maintained at a slightly negative pressure to prevent fumes from entering the laboratory.

These systems are typically controlled by the use of an independent exhaust fan or the use of a manual damper connected to a manifolded exhaust system. When a manual damper is used, its minimum value is set to provide the design airflow. Depending on the location of the damper, the damper can then be opened farther in case of an emergency, such as a spill inside of the storage cabinet. The fan for the main exhaust system is run continuously, as turning off the exhaust to the storage cabinet could allow flammable vapors to build up to dangerous levels. Figure 11-7 contains a schematic of a flammable and solvent storage cabinet.

General Laboratory Exhaust. General laboratory exhaust provides ventilation to the laboratory as a whole, where equipment exhausts such as fume hoods provide ventilation to a confined area within the laboratory. General laboratory exhaust serves two purposes; it provides ventilation of the room to remove odors and contaminants that were not captured by the equipment exhausts and it allows for higher supply airflows to pro-

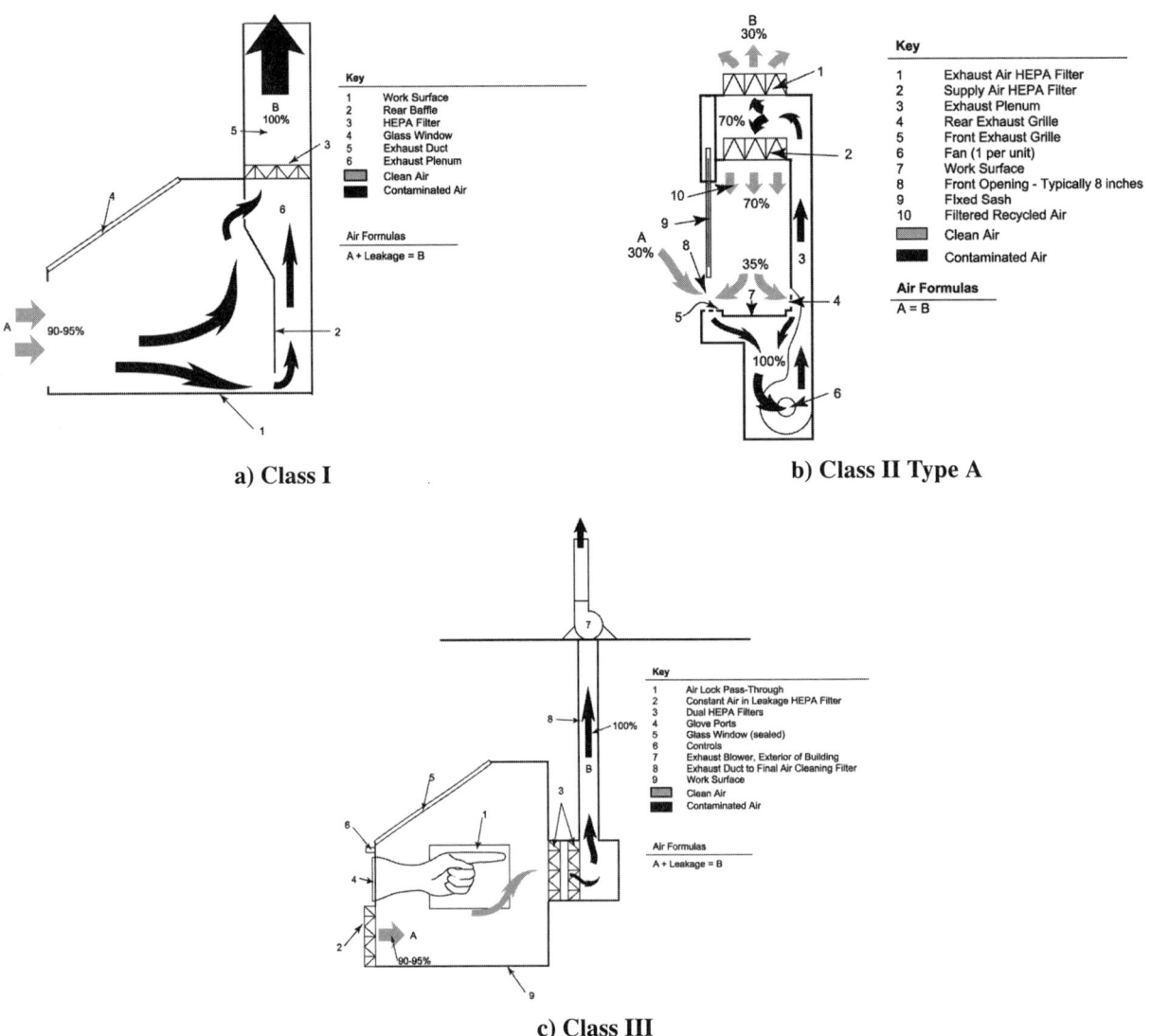

Figure 11-6 *Classes of biological safety cabinets. (Diagram modified with permission from Diberardinis et al. 1993.)*

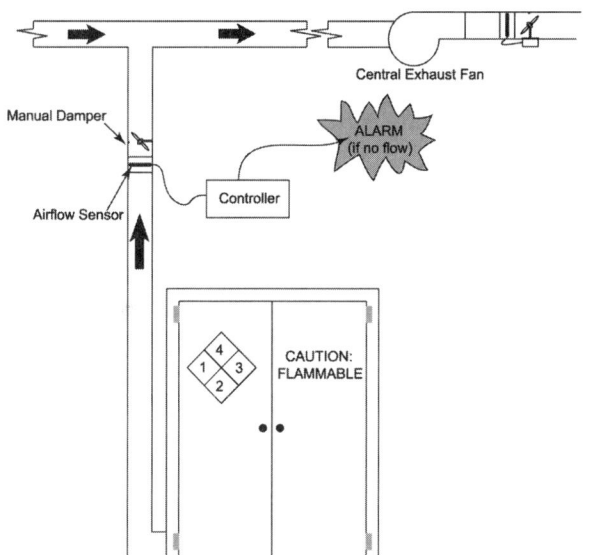

Figure 11-7 *Flammable and solvent storage cabinet.*

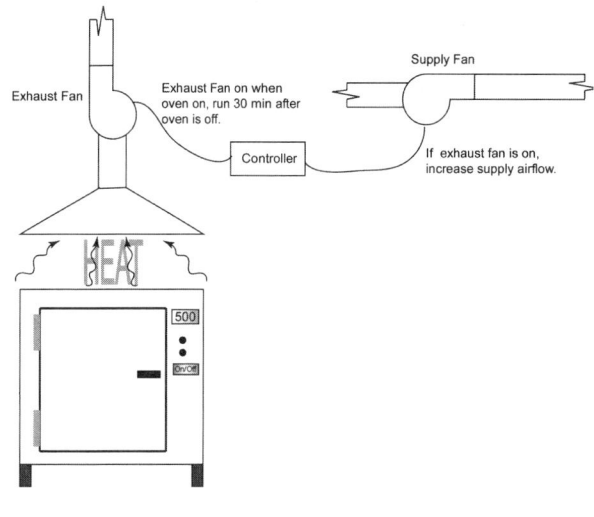

Figure 11-8 *Example of direct equipment exhaust using a canopy hood.*

vide the necessary cooling in cases where the cooling load, not the exhaust airflow, dominates the design of the laboratory systems.

Typical control of general laboratory exhaust is through the use of a manifolded exhaust system and balancing dampers. However, individual exhaust fans may be needed in some cases where the materials handled in the room are not compatible with those from other rooms. General laboratory exhaust can be either constant volume or variable volume, depending on the control strategy for the room.

Direct Equipment Exhaust. Direct equipment exhaust is used for pieces of equipment that generate considerable amounts of heat, humidity, or pollutants. Figure 11-8 illustrates a canopy hood used for direct exhausting. By providing adequate direct exhaust, the heat/humidity gain from the equipment does not contribute a load to the space. However, additional supply air is required to make up for the exhaust air. Equipment that may have direct exhaust includes large furnaces, ovens, centrifuges, autoclaves, and glassware wash machine.

The typical control of these systems is by an independent exhaust fan or by a damper connected to a manifolded exhaust system, which enables the exhaust whenever the equipment is on. Independent exhaust fans are typically wired to automatically turn on when the equipment is turned on, or they have a separate switch to enable the fan. When a damper is used to connect to a manifolded exhaust system, the damper is either manually opened by the operator or automatically opened by an actuator, which monitors the on-off status of the equipment. The fan for the main exhaust system is run continuously while the equipment requires direct exhaust, and an alarm or interlock may sometimes be used to make sure that while the equipment is running it has exhaust from the main system. Depending on the type of equipment, direct equipment exhaust may use a delay timer to continue providing exhaust for a certain length of time after the equipment is turned off, such as for a large furnace or autoclave, which would stay hot for a considerable time after being turned off.

ROOM CONTROL

Laboratories require accurate control of the temperature and humidity levels in individual rooms as well as pressure differentials between adjoining spaces. Temperature and humidity are accurately and consistently controlled in a laboratory to ensure the safety of the occupants and protect the research that is being conducted. Rapid increases in exhaust airflow from the room (e.g., opening a sash on a fume hood) will require a similarly large increase in supply air to the room. As room temperature sensors can have a slow response time relative to changes in exhaust and supply airflow, the control system may need to adjust the supply air temperature when large changes in airflow are made to maintain the temperature setpoint.

Pressure control is used to prevent transfer of pollutants to and from unwanted areas. Cleanrooms need to be positively pressurized to remain cleaner than surrounding areas, while most other laboratories need to be negatively pressurized to prevent airborne pollutants generated in the laboratories from migrating to other areas of the building. Typical values for room pressurization differentials are between 0.01 and 0.03 in. w.g. (2.5 and 7.5 Pa) (SBT 1999). Figure 11-9 shows an

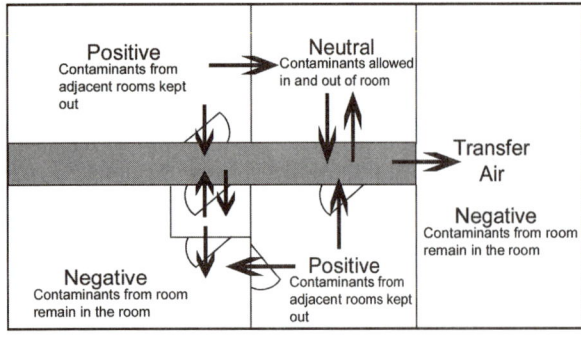

Figure 11-9 *Room pressurization to prevent contamination spread.*

example of how room pressurization is used to prevent contaminants from spreading.

Theory of Room Control

The general theory of room control is to maintain a differential between the supply and exhaust air, to either pressurize or depressurize the room relative to its surroundings. This is either an active or passive process.

In an active process, the actual relative room pressure is monitored using a differential pressure sensor, and the supply or exhaust air is controlled accordingly. The active process of maintaining room pressurization is called direct pressure control, and it is used most frequently in variable air volume systems.

In a passive system, the control system measures the exhaust and supply airflow rates, and either the supply or exhaust system is controlled to maintain a constant differential between the supply and exhaust flow rates, but the room pressure is not actively measured. Passive pressure control for variable air volume systems is referred to as "flow tracking," while for constant air volume systems it is called "fixed offset," both of which are discussed later in this chapter.

Maintaining Minimum Ventilation Air Changes per Hour

Many codes and standards for laboratories require that a minimum ventilation rate in air changes per hour be maintained. Therefore, the systems and controls must be designed and operated to maintain minimum outdoor airflow rates to the laboratory during all occupied periods and a reduced airflow rate during unoccupied periods. In constant volume systems, this is easily achieved, as 100% of the maximum airflow is always supplied to the laboratory. For VAV systems, the supply and exhaust airflow rates must have minimum values for occupied and unoccupied periods, including a minimum value per period for each fume hood.

Minimum recommended air changes for most laboratories are between 4 and 12 air changes per hour.

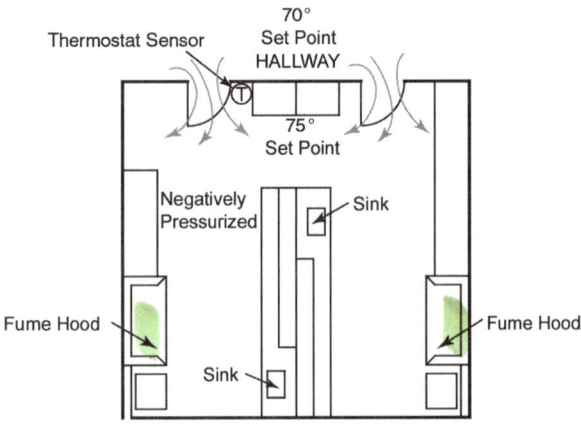

Figure 11-10 *Example of poor sensor location.*

However, certain laboratories such as cleanrooms may require higher ventilation rates. Additional details on recommended air changes can be found in Chapter 3, "Laboratory Planning."

Temperature Sensor Location

The temperature sensor location can affect the ability to accurately control the temperature in a laboratory. The sensor should be located where it can read the actual temperature of the laboratory space. The pressure differentials between spaces and the possibility of significantly different temperature requirements between adjacent spaces, which are associated with laboratories, can pose difficulties in getting the temperature sensor to read the actual room temperature.

For example, if a temperature sensor is located near the door of a negatively pressurized laboratory, air from the hallway will continually flow into the room near the temperature sensor. Depending on the exact location of the sensor with respect to the wall and door configuration, the temperature sensor can read the temperature of the hallway. This can be an increasing problem as the temperature differential between the hallway and laboratory increases. Figure 11-10 is a schematic, which shows an example of poor sensor location.

Another example is a temperature sensor mounted in or over the top of a hole in the wall (or loose electrical box). If the temperature sensor is loose around the hole in the wall, the negative pressure of the laboratory can cause the temperature sensor to read the temperature in the wall cavity.

Sensor Accuracy

The accuracy of sensors should be considered in the design of laboratory control systems, as the sum of the error in multiple sensors may create undesirable conditions. For example, consider a variable volume system with airflow sensors that have an accuracy of ±5% of the rated airflow, a supply airflow of 1,000 cfm, and exhaust airflow of 1,100 cfm (100 cfm differential). If one sensor was used for supply and one for exhaust, the total error would be ±105 cfm (5%*1,100 +1,000*5% = 55 + 50 = 105 cfm). Therefore, in this example it would be possible for the error in the sensor to either eliminate the airflow differential that was intended to keep the room at a negative pressure or to double the airflow differential in the room.

Control Stability in Interruptions

A laboratory control system needs to maintain stability during interruptions, such as the opening of doors, occupants moving around, and infiltration through the building envelope. Different control options, such as direct pressure control, are more sensitive to disturbances than others, such as flow tracking. Additional equipment and control sequences may be needed to maintain stability for control options that are sensitive to disturbances. For example, a direct pressure controlled laboratory that is maintained at a negative pressure may need a sensor to detect the position of the door and prevent the control system from trying to maintain the required pressure level when the door is open and then generate an alarm after a given length of time if the door does not close. Some laboratories that are highly sensitive to pressure fluctuations may require the use of an air lock to enter the room. When airlocks are used, evacuation procedures for fires and other emergencies should be considered. Other interruptions, such as those created by occupant movement, can be minimized by locating fume hoods away from high traffic areas in laboratories.

The key to stable operation of a control system during interruptions is a stable, accurate, and predictable response of the control system to changes in environmental conditions. This is more critical in VAV systems, as it is directly affected by the changes. Therefore, the control system for VAV systems must have a quick response that does not swing wildly trying to find the new control point, sometime referred to as hunting.

Variable Air Volume Laboratory Control Strategies

Since variable volume fume hoods vary the exhaust airflow in response to the sash position of the hood, the control system needs to make corresponding changes to the supply airflow to maintain the pressure differential between laboratory and nonlaboratory spaces. There are two different methods to ensure that the proper pressur-

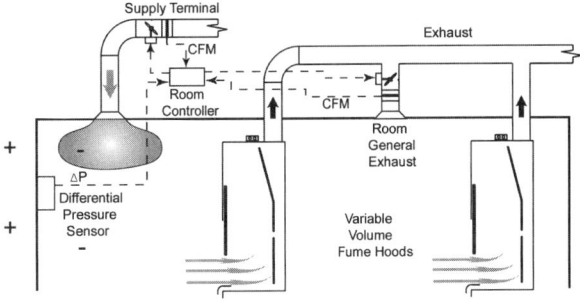

Figure 11-11 *Direct pressure control for variable volume laboratories.*

ization is maintained in a laboratory: direct pressure control and flow tracking.

Direct Pressure Control. Direct pressure control uses a pressure sensor across the envelope of the laboratory to directly measure the pressure difference. The supply airflow is then controlled to maintain a set differential pressure across the envelope. Tracking of all supply and exhaust airflows is not required, and direct pressure control can maintain directional airflow. Figure 11-11 is a schematic of direct pressure control.

However, direct pressure control is sensitive to disturbances, such as the opening of doors and infiltration. Tight envelope construction is required to both physically and economically maintain the proper pressure differential. In order to maintain system stability when a laboratory door is open, an air lock may be required. Also, a switch may be installed on the door to turn off the control response (provide more supply air) when the door is open. Direct pressure control sensors are not subject to minimum air velocity requirements to produce an accurate reading.

Flow Tracking Control. Flow tracking, or mass tracking, monitors the supply air and exhaust air to the room and controls the supply airflow to maintain a set differential between supply and exhaust volume. Tracking of all supply and exhaust airflows in the room is required to ensure flow-tracking works accurately. Figure 11-12 is a schematic of flow tracking control.

This method does not recognize disturbances in the laboratory, such as the opening of doors or infiltration. For this reason flow tracking does not guarantee directional airflow, which can be very important, depending on the uses of the laboratory. Also, airflow sensors are subject to minimum velocity (flow) requirements, which need to be considered when selecting minimum flow rates for the laboratory and laboratory equipment. Proper and regular calibration and maintenance of airflow sensors are required to ensure flow tracking remains working as originally intended.

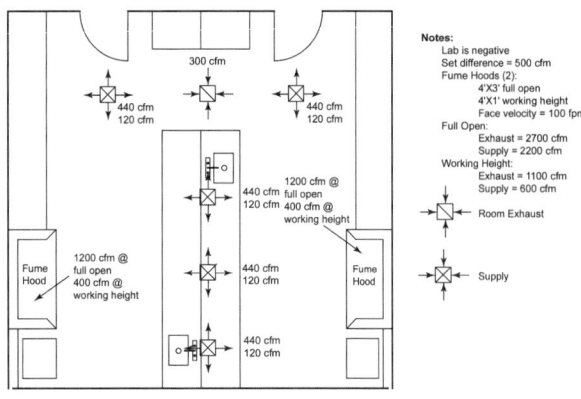

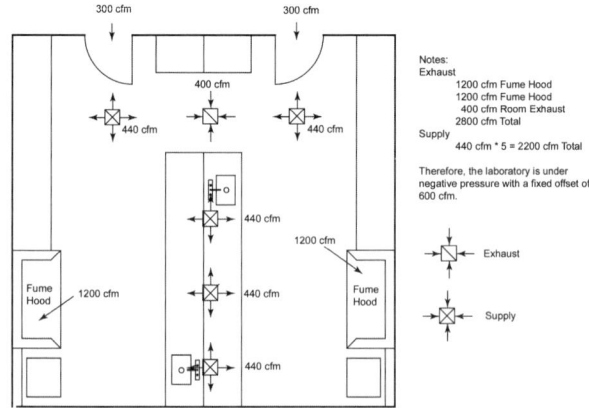

Figure 11-13 *Fixed offset control for constant volume laboratories.*

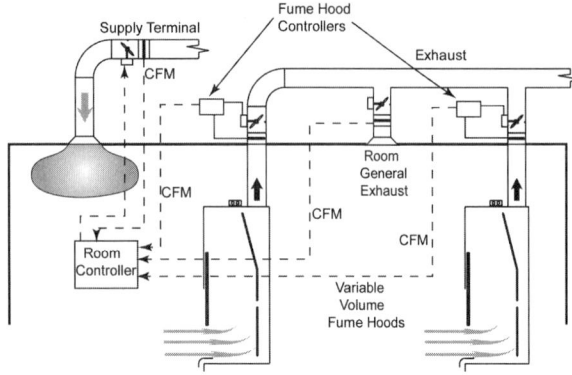

Figure 11-12 *Flow tracking control for variable volume laboratories.*

Constant Air Volume Laboratory Control Strategies

The control strategy for constant volume laboratory systems is typically not an active control strategy. Rather, the control strategy is inherent to constant volume equipment (i.e., is accomplished by having the bypass air physically linked to the sash), so as the sash is closed, the bypass area open to the room is increased. Typically, fixed offset control is used, which is the simpler constant volume control strategy. However, auxiliary air control is also available for use with constant volume systems but generally requires additional controls.

Fixed Offset. Fixed offset control is a passive constant volume control strategy, which during design specifies a fixed offset between the supply and exhaust to maintain flow direction. Unlike flow tracking for VAV systems, which constantly monitors supply and exhaust airflows, fixed offset control uses a value selected during design and is only periodically verified, usually during performance testing of fume hoods. As the actual difference between supply and exhaust airflow is determined based on construction methods and materials and testing and balancing, construction and TAB must be of high quality to ensure the required offset is met and the health and safety of the occupants will be maintained. Figure 11-13 is a schematic of how fixed offset control is used for constant volume laboratory systems.

Auxiliary Air. Auxiliary air fume hoods use a separate supply airstream, which is unconditioned or only partially conditioned at the hood opening to minimize the amount of room air that is exhausted. These hoods require careful coordination during the design of the exhaust, auxiliary, and supply air systems of the laboratory to ensure proper airflow direction is maintained for both the hood itself and the room. For example, the volume of auxiliary air must be selected carefully so that all, or nearly all, of it is drawn into the fume hood and exhausted; otherwise the control of the room temperature, humidity, and pressure could be negatively impacted.

There are several other requirements for auxiliary fume hoods, such as: the hood must have a vertical sash, the auxiliary air can't degrade the flow of air into the fume hood, and the auxiliary air must be conditioned to avoid creating cold drafts and additional heating loads. Also, the control system for laboratories with auxiliary air fume hoods is generally more complex, as each fume hood has its own supply air system that is separate from the supply air system for the room. Figure 11-14 is a schematic of a typical auxiliary air fume hood.

Temperature and Humidity Control of Critical Spaces

Laboratories typically require more accurate and consistent control of temperature and humidity than typical commercial buildings. For critical laboratory spaces, accurate and consistent temperature and humidity control are essential to ensure the accuracy of the work performed in the laboratory and the reliability of the information gained from the work. Laboratories for research and testing, as well as animal facilities, require the most precise control of temperature and humidity.

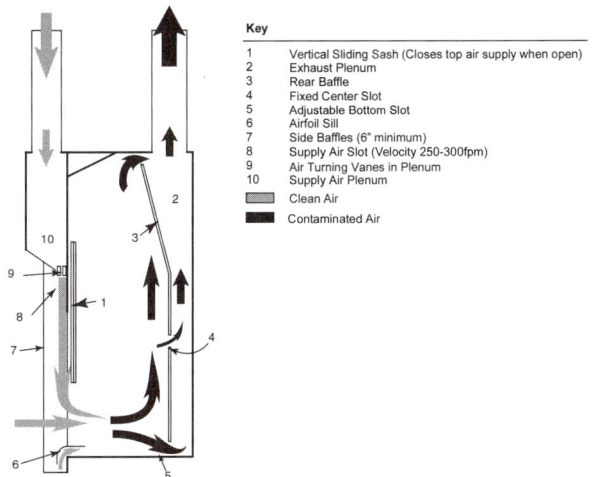

Figure 11-14 *Typical auxiliary air fume hood. (Diagram modified with permission, DiBerardinis et al. 1993.)*

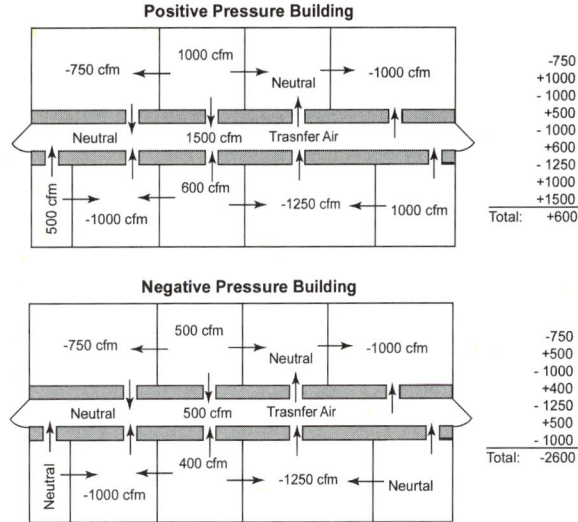

Figure 11-15 *Positive vs. negative pressurized buildings.*

Laboratories. Temperature and humidity can be controlled locally for experiments using heaters and coolers. However, the proper design, maintenance, and operation of the cooling and heating systems for general laboratory spaces are critical for many experiments investigating new processes or materials, to ensure that the results obtained were not affected by fluctuating temperature and humidity levels in the laboratory. Therefore, critical spaces require precise control and adjustability of laboratory conditions. This includes the use of high-accuracy sensors, fast acting actuators and control sequences, and HVAC systems that are designed and constructed well and tested thoroughly.

Animal Facilities. Often in animal laboratories, temperature and humidity need to be maintained at a specified control point as part of the research. Different laboratory temperatures may be needed for different types of animals, and ventilation and air-conditioning loads often vary with the type of animal and their activity level. Also, the types of animals housed in the laboratory may change several times throughout the life of the laboratory. Consultation with laboratory researchers is necessary to ensure that the HVAC system will be able to provide the conditions necessary for the type of work that will be performed in the laboratory during its lifetime. This includes sizing the heating and cooling systems to account for the maximum anticipated loads, allowing for a wide range of temperature and humidity setpoints, allowing for a range of ventilation rates, and allowing for a flexible ventilation system that can be easily reconfigured.

Effects of Laboratory Controls on Building Pressurization

Due to the mixed use of most laboratory buildings (offices, storage, cafeterias) the control of the laboratory airflows can have a significant impact on airflows in other areas and on the building pressurization. Unfortunately, many laboratory buildings are under a severely negative pressure due to the high exhaust rates. To counteract this problem requires a robust design and control sequence to properly manage air transfers in the building to maintain building pressurization. Robust design and control sequences includes analyzing all of the supply and exhaust airflows under various load conditions (minimum airflow, maximum airflow, direct exhaust equipment on, direct exhaust equipment off, when filters are heavily loaded, etc.) to ensure that the building as a whole will be maintained at a positive pressure. Figure 11-15 shows examples of negative and positive pressurized buildings.

CENTRAL SYSTEM EMERGENCY SITUATIONS

Laboratory control systems need to be able to automatically respond to emergency situations in order to maintain a safe working environment for the laboratory personnel. One of the most important emergency situations is a fan failure. This can result in exhaust air not being available for laboratory equipment, such as fume hoods, or supply air not being provided, which would create a large negative pressure in the laboratory and reduce the amount of air that is available to be exhausted. Fan failures require different control responses for one- and two-fan systems.

Failure in a One-Fan System

In a one-fan system, failure of a fan will result in no airflow through the system served by the fan. When this occurs, the desired room pressure differentials cannot be maintained unless action is taken to modify the operation of the remaining functioning system. In some cases, a backup fan may be available, in which case the backup fan would need to be started quickly in order to minimize the risk to the laboratory personnel. If there isn't a backup fan, then the other system that is still operating would need to be turned off or have its airflow greatly reduced. For example, if the exhaust fan failed, the supply fan would need to be turned off (or greatly reduced if the laboratory was intended to be positively pressurized) to prevent the laboratory from changing from negative to positive pressurization. In another example, if the supply fan failed, the exhaust fan would need to be turned off if the room was intended to be positively pressurized and the exhaust airflow would need to be reduced significantly if the room was intended to be negatively pressurized. In the event of a fan failure in a one-fan system, an alarm is needed to notify the laboratory personnel that a failure has occurred and that the room may need to be evacuated.

Failure in a Two-Fan System

A two-fan system for either supply or exhaust consists of two fans that operate continuously in parallel. This configuration is used so that if one fan fails, the other fan would be able to continue to provide some airflow after the failure. This differs from a primary/backup fan system, which would have a brief interruption in airflow after the failure of the primary fan while the backup fan was being started. Typically, a two-fan system will use two fans for supply and two for exhaust.

In the event of a failure in a two-fan system, the control system needs to be able to maintain the pressure relationships of the laboratory. To do this, the airflow from the system that has two fans still operating needs to be reduced. For example, if one of the two exhaust fans were to fail, the supply airflow would need to be reduced to prevent the laboratory from becoming very positively pressurized. For constant volume systems, this would typically be accomplished by turning off one of the supply fans. For variable volume systems this could be accomplished by either turning off one of the fans or reducing the output of the variable frequency drive by 50% (assuming both fans were sized the same). In the event of a failure in a two-fan system, an alarm should notify the laboratory personnel.

REFERENCES

American National Standards Institute (ANSI). 1992. *Standard Z9.5, Laboratory Ventilation, American National Standard Practices for Respiratory Protection.* New York: ANSI.

American Society of Heating, Refrigerating and Air-Conditioning Engineers (ASHRAE). 1995. *ASHRAE Standard 110-1995, Method for Testing Performance of Laboratory Fume Hoods.* Atlanta: ASHRAE.

DiBerardinis, L.J., J.S. Baum, M.W. First, G.T. Gatwood, E. Groden, and A.K. Seth. 1993. *Guidelines for Laboratory Design: Health and Safety Considerations,* 2nd ed. New York: John Wiley & Sons, Inc.

Monger, S.C. 1994. Designing or Renovating Fume Hood Laboratories. *TAB Journal,* pages 13-19. Associated Air Balance Council.

Siemens Building Technologies, Inc. 1999. *Laboratory Control and Safety Solutions Application Guide.* Buffalo Grove, Ill.: SBT.

Chapter 12
Airflow Patterns and Air Balance

OVERVIEW

Understanding how airflow patterns protect laboratory occupants and the procedures to test for proper airflow are important for the safe design and operation of laboratories. Airflow needs to be maintained in a specific direction and velocity in order to protect against harmful substances and is influenced by many aspects of a laboratory, such as equipment placement and traffic flows. Once the desired airflow patterns for laboratories are understood, testing procedures need to be developed and followed to ensure that these airflow patterns are maintained for the life of the laboratory.

AIRFLOW PATTERNS

Airflow patterns are used in a laboratory to provide separation between contaminants and occupants and to ensure the integrity and accuracy of work performed in the laboratory. The separation is accomplished by not using recirculated air in laboratory work areas and by maintaining airflow direction from clean areas to dirty areas so that the people and the work performed in the clean areas are not affected by laboratory procedures. The airflow needed to maintain the separation, even during disturbances, is dictated by the types of substances that are being handled and the type and size of the exhaust system. Fume hood placement along with supply diffuser type, throw, and placement are the main causes for disturbances that disrupt the intended airflow direction and maintenance of the desired face velocity of a fume hood.

Purpose for Controlling

The purpose for controlling airflow patterns in a laboratory is to provide a safe working environment for the laboratory personnel and an appropriate environment to ensure the accuracy of the work performed in the laboratory. Specifically, airflow patterns need to be controlled to minimize the spread of contaminants, provide personnel protection, and protect research from contamination.

Control the Spread of Contaminants. Airflow direction is used in laboratories to provide a secondary barrier to control the spread of contaminants to other areas. In laboratories other than cleanrooms, the contaminants contained by the secondary barrier are chemicals that are not removed by direct exhaust equipment, such as fume hoods. By maintaining a negative pressure in a laboratory room, any chemicals that are accidentally released are contained in the room in which they were released and do not spread to surrounding areas. The general exhaust for the room will eventually remove or dilute the accidentally released chemicals. In cleanrooms, the contaminants that are contained by airflow direction are in the less clean, unfiltered air from surrounding areas in the building and outside. By maintaining a positive pressure differential, infiltration from the surrounding areas, which are less clean, can be prevented.

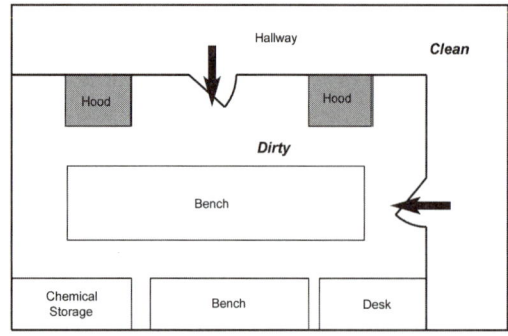 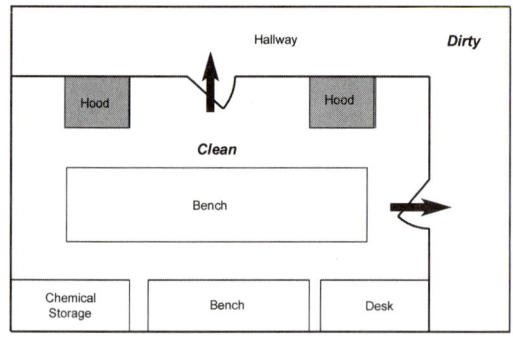

Figure 12-1 *Flow from clean to dirty.*

The integrity of the room envelope is critical in providing the necessary separation needed between contaminants and occupants in a given laboratory space. The envelope is the physical barrier between adjacent spaces and, if it is tight and properly sealed, the correct airflow direction can be effectively maintained with a smaller difference in supply and exhaust air volumes.

Personnel Protection. Airflow patterns need to be maintained to ensure a safe working environment for laboratory personnel. Fume hoods need to maintain a specified face velocity, typically around 100 fpm (0.51 m/s), to keep the materials used in the hood from escaping to the rest of the room. A key factor to the protection of laboratory personnel working near hoods is the location and type of supply air diffuser within the room. Improper placement or selection of a diffuser can result in disturbing the airflow near the hood face, resulting in contaminants from the hood being entrained into the room.

Nonlaboratory personnel are protected from laboratory contaminants through a combination of physical barriers (walls) and airflow direction (nonlaboratory spaces are typically positively pressurized relative to laboratory spaces).

Protection of the Research. In addition to protecting the personnel who use the laboratory, airflow patterns must be maintained to prevent research from being contaminated. This is the case for cleanrooms, where the research inside must be protected from the less clean air surrounding the laboratory. It is also true in animal laboratories, where the animals also may need to be protected from the less clean air surrounding the laboratory or from contaminants generated by the laboratory personnel or other animals. In both cases, less of this protection can ruin the research being conducted. This is especially critical for animal research, where months or years of research are lost when cross-contamination between subjects occur.

Airflow Direction

In order to properly control the airflow direction and protect laboratory personnel and research, the use of recirculated air and the concept of airflow from "cleanest" to "dirtiest" need to be considered.

Use of Recirculated Air. Rooms in a laboratory building can have either once-through airflow or a portion of recirculated air, depending on the uses of the room. Typically, all laboratories, chemical storage rooms, and laboratory support areas use 100% outside air or once-through ventilation. By doing so, contaminants are not spread beyond the initial room in which they are released.

However, most buildings that contain laboratories also contain offices, meeting rooms, and other areas where hazardous materials are not found. These nonlaboratory areas can use recirculated air, although it may be desirable to have higher ventilation rates in these areas than if they were in a typical commercial building. Also, it is advisable to have sufficient barriers between areas where recirculated air is used and areas where hazardous materials are used. For example, offices adjacent to laboratory spaces with hazardous materials would not be a good choice for using recirculated air, as a minor spill or disruption of inward airflow to the nearby laboratory could result in contaminants being transferred to the offices. On the other hand, offices grouped together and separated from laboratory spaces would be an acceptable choice for using recirculated air.

Flow from Cleanest to Dirtiest. Throughout laboratories, pressure differentials are maintained to ensure that airflow is from the cleanest areas to the dirtiest areas. This helps ensure that laboratory personnel and research are protected from the contaminants present in a laboratory building. For cleanroom laboratories, which need to be cleaner that their surroundings, this means maintaining a positive pressure to ensure airflows from the cleanroom to the surrounding areas. For other laboratories, a negative pressure is maintained to ensure that air flows into the laboratory from the cleaner surroundings, thus preventing dangerous chemicals, which were not exhausted by fume hoods, from migrating to the rest of the building. Once air has flowed to the dirtiest area, it is exhausted and new clean air is introduced to the clean areas. Figure 12-1 is a schematic of two types of flow from clean to dirty.

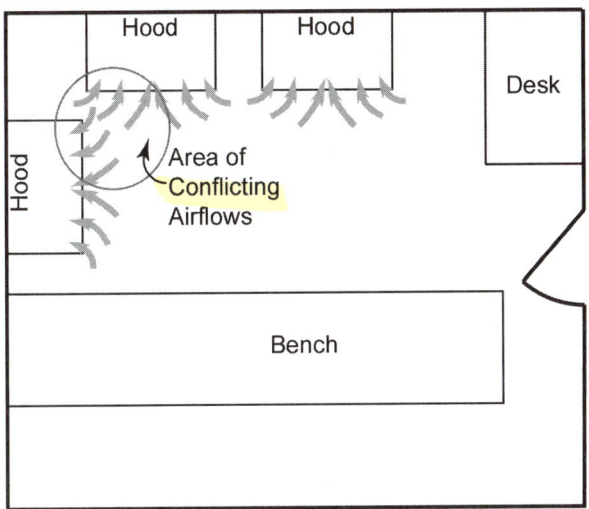

Figure 12-2 *Conflicting airflow to fume hoods.*

Considerations

Numerous variables must be considered when determining the proper airflow pattern for a laboratory building. The general goal of maintaining airflow patterns is to provide safe conditions by controlling the movement of hazardous and unwanted materials. Considerations for selecting airflow patterns include the types of materials handled or created in the laboratory, the type, size, and number of protective exhaust systems, and the acceptable level of air transfer.

Types of Materials Handled or Created. Early in the planning stage, the types of materials handled or created in each space of a laboratory must be determined so that the proper airflow patterns and setpoints for the laboratory can be chosen to prevent the transmission of contaminants. Areas of similar hazard type and level can be grouped together. However, areas containing materials that could adversely interact must be separated (i.e., not connected to the same manifolded exhaust system) to ensure that exhaust airstreams from the rooms will not mix. Also, the face velocity for fume hoods, or biological safety cabinets, may be dictated by the materials that will be used inside of them. Similarly, the materials that are used in the laboratory may determine the ventilation air changes and relative pressure of the laboratory.

Type, Size, and Number of Hoods and Protective Exhaust Systems. Another consideration in determining airflow patterns for a laboratory is the type, size, and number of exhaust equipment. The type of exhaust hood determines many characteristics of the room airflow pattern. First, there is the general choice of constant volume and variable volume systems. Laboratories with variable volume systems will have changing airflow patterns as the variable volume system changes from minimum flow to maximum flow and the proper flow directions must be maintained throughout the operating range, whereas constant volume systems will have a more consistent airflow pattern.

The size of exhaust equipment can also affect airflow patterns. For example, large fume hoods are typically more susceptible to disturbances in face velocity than smaller fume hoods. Therefore, large fume hoods may need to be located farther away from sources of disturbances, such as occupant traffic, doorways, supply diffusers, and other fume hoods.

The number of hoods may present challenges if a large number of hoods are to be placed in one room, as airflow patterns to each hood need to be maintained so that none of them have disturbances in face velocity. Close coordination of the exhaust hood and supply sources during design is needed to ensure that each hood can draw air from the room without drawing in air that is needed to maintain the proper airflow patterns for a nearby exhaust hood. Figure 12-2 is a schematic of conflicting airflow to two different fume hoods.

It is recommended that the fume hoods, in laboratories with multiple fume hoods, be spaced at least 4 feet (1.2 meters) apart when placed on the same wall or on perpendicular walls, and they should be spaced more than 2 feet (0.6 meters) apart when placed on opposite walls. Fume hoods on perpendicular walls, provided adequate spacing is maintained, have been found to perform the best, followed by hoods on opposite walls and hoods on the same wall, respectively (Memarzadeh 1996).

In addition to locating fume hoods an appropriate distance away from each other, fume hoods should also be located away from supply and transfer air diffusers to prevent disturbances to the face velocity of the fume hood. Due to the typical requirement of 100 fpm (0.508 m/s) face velocity for a laboratory hood, it is recommended that the maximum supply air velocity within the occupied space, below approximately 7 ft (2.134 m), be 50 fpm (0.254 m/s) (DiBerardinis et al. 1993). In general, a supply air velocity of 50% of the required hood velocity should be maintained within the occupied space. Supply air velocities at the open face of the hood should be less than 20% of the average face velocity of the hood. To minimize the turbulence within the space, the diffuser supply velocity can be reduced through the addition of multiple diffusers or through the use of a diffuser with multiple orifices. For additional information on diffuser sizing and location, see Chapter 6, "Primary Air Systems."

Acceptability of Air Transfer. A final consideration in determining airflow patterns is the acceptability of air transfer. Air transfer involves allowing air from one area to be transferred to another area in order to maintain pressure relationships. Acceptability of air transfer is determined by the acceptability of the transfer of two types of contaminants: vaporous and particulate. Depending upon the contaminant, if either is acceptable,

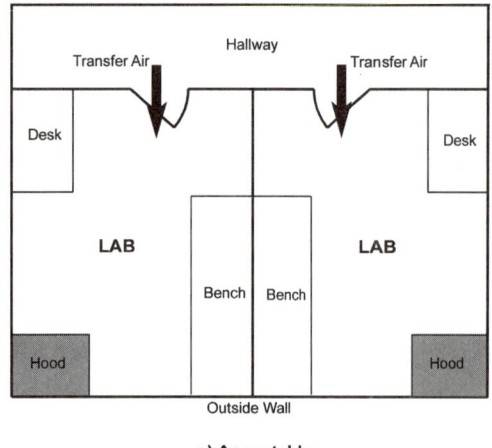
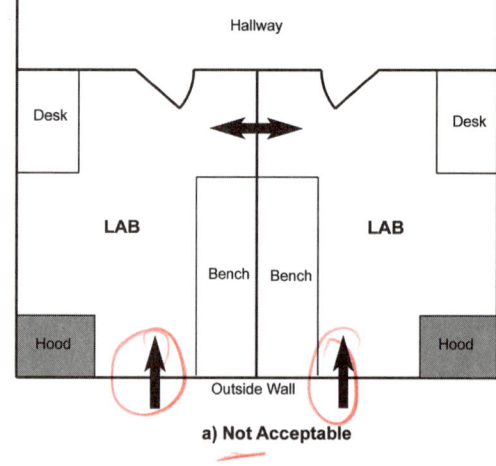

Figure 12-3 *Acceptability of air transfer.*

specific materials and equipment may be required for use in the construction of the laboratory.

In some areas of a laboratory, the use of transfer air is acceptable, for example, between a corridor and a laboratory, between two laboratories with similar hazard levels that use compatible chemicals, and from a laboratory that uses low toxic materials to a laboratory with moderately toxic materials.

In other cases, no air transfer between spaces may be permissible. Examples of cases where air transfer should not be used include: from laboratory to nonlaboratory areas (except for cleanrooms), from areas of high hazard level to areas of low hazard, and between two areas of different types of hazard, such as from a biological laboratory area to a chemistry laboratory area. Figure 12-3 shows acceptable and non-acceptable cases of air transfer.

Air Introduction

The method used to introduce air into a laboratory space is critical to ensure that exhaust equipment is not disturbed. In order to obtain proper air introduction into a laboratory, the concept of challenge velocity must be understood, as well as how type, throw, and placement of diffuser, temperature gradients, turbulence, and cross drafts affect the performance of exhaust equipment. Additional information on supply and exhaust air methods can be found in Chapter 6, "Primary Air Systems."

Concept of "Challenge Velocity." Challenge velocity is the disturbance velocity that interrupts the airflow across the face of a fume hood or biological safety cabinet and causes a disturbance that results in spillage from the hood. Disturbances can include the velocity of air from supply diffusers, occupants walking past the hood, or opening of the door to the laboratory. In order to prevent hood spillage, the allowable velocity of a disturbance at the hood should be at most 20% of the design face velocity of the hood. For example, if the design face velocity of a hood were 110 fpm (0.56 m/s), the maximum velocity of a disturbance would be 22 fpm (0.11 m/s). The throw pattern from the supply diffusers would need to be less than this value where the pattern intersects the hood face.

Types of Diffusers and Diffuser Throw. In laboratories, large quantities of supply air must be introduced to maintain pressure relationships while exhaust equipment is operating. However, the large amount of air must be introduced at a low velocity to prevent creating turbulence in the room, and high-velocity airstreams which intersect fume hood openings, could disturb the airflow into exhaust devices. Therefore, the traditional high-velocity blade diffusers used in most buildings should be avoided in laboratory spaces, as their throw pattern would likely create a disturbance in the exhaust equipment airflow. For this reason, non-aspirating diffusers are recommended for laboratories. Chapter 6, "Primary Air Systems" contains additional information on selecting diffuser types and throws for laboratories.

Diffuser Placement. Diffuser placement in a laboratory differs from that of a typical building. In typical buildings, equal diffuser spacing is used so that the entire room is covered by the throw from the diffusers and the air in the room is thoroughly mixed. This maintains an even temperature distribution and ensures that all areas of the room are ventilated and the occupants are comfortable. Laboratories, on the other hand, need to introduce large quantities of air at a low velocity and maintain safety through proper operation of fume hoods. Diffusers in laboratories should be placed far enough away from exhaust equipment and intakes that they will not cause any interference with the airflow into the exhaust equipment or that their throw pattern does not provide a sufficient challenge velocity to disrupt exhaust airflow patterns under any operating condition. Locations for diffusers should be selected so the velocity profile (both horizontal and vertical) that represents 50% of the face velocity, typically 50 fpm (0.254 m/s), does not intersect the sash area of the fume hood or biological safety cabinet. Chapter 6, "Primary Air Systems," contains additional information on selecting and locating diffusers with respect to exhaust sources.

Temperature Gradients and Equipment Plumes within the Laboratory. Temperature gradients in a laboratory can create unwanted airflow patterns as the warmer air rises and displaces cooler air. In order to minimize possible disturbances to the exhaust equipment airflows created as cool air replaces warm air, temperature gradients should be kept to a minimum. This can be done by providing direct exhaust for high heat producing equipment and proper diffuser selection and location.

Critical Systems

Critical laboratory systems may have special requirements to maintain control of hazardous materials. These requirements include the use of air locks and multiple speed fans.

Application of Air Locks. Air locks are frequently used in laboratories where the disturbances created from opening doors could result in undesirable or unsafe conditions and to contain the spread of contaminants. Examples include a laboratory with hazardous materials that can't be allowed to escape without treatment (e.g., Biosafety Level 3 and 4 laboratories), a laboratory with a high pressure differential to maintain, or the need to prevent contaminants from entering the laboratory (e.g., cleanrooms).

Airlocks consist of a series of two doors separated by a small vestibule. When occupants enter or leave the laboratory, they pass through the first door and wait in the area between the doors until the first door is completely closed. Once the first door has closed, the occupants can open the second door and enter or leave the laboratory. Depending on the nature of the hazards that require the use of airlocks, there can be either a physical mechanism that locks one door while the other is open or the occupants can be responsible for only opening one door at a time, in which case an alarm sounds when both doors are open. Pressure gages are often provided in the airlock to show that the pressure differential is being maintained.

Multiple Speed Fans. Multiple speed fans may be used in critical areas to provide extra airflow in the event of an emergency, such as a large chemical spill. In the event of an emergency, the extra fan capacity can be used to increase the pressure differential of the laboratory and increase the air change rate to remove the spilled chemicals.

TESTING, ADJUSTING, AND BALANCING

The balancing of general HVAC equipment in a laboratory is similar to that for conventional buildings, provided that the piece of equipment is not exposed to hazardous substances from the laboratory and that the laboratory is not in operation. However, if equipment is exposed to hazardous substances, procedures are needed to prevent or control the release of contaminants and ensure the safety of the testing personnel. Also, if the laboratory is in operation, the equipment should only be tested if it will not affect the conditions and safety of the laboratory and if the testing personnel will be safe.

Laboratory health and safety and HVAC equipment have well-defined procedures that should be followed to ensure proper operation. The process for testing, adjusting, and balancing of HVAC systems in a laboratory consists of air and hydronic system balancing, ductwork pressure testing, equipment balancing for both general HVAC equipment and laboratory safety equipment, and the standards that provide various procedures and requirements for testing and balancing.

Air and Hydronic System Balancing

The keys to the successful testing, adjusting, and balancing (TAB) of air and hydronic systems includes the proper design of airflow differentials, adequate access points to obtain accurate measurements, early selection of the TAB contractor, and clearly defined test and balance procedures.

Successful TAB begins during the design stage with the proper selection of airflow differentials (or pressure differentials) needed to contain the spread of contaminants. This includes accounting for variances, accuracy of the control system, and the TAB. As discussed in Chapter 11, "Controls," in a VAV system the accuracy of the central system can easily be ±100 cfm. Since this is for an ideal installation, the variance is likely higher. Therefore, balancing a system to ±10% is very difficult as verification is complicated by the inaccuracy of the control system. Due to the critical nature of laboratories and the inability to maintain differential airflows between spaces, it is recommended that the TAB contractor be required to balance to the following criteria:

- Supply diffusers – The listed airflow is the maximum (minimum for cleanrooms) that can be supplied. Actual acceptable value is the listed value to –10% (+10% for cleanrooms) of the value.

- Exhaust points – The listed airflow is the minimum (maximum for cleanrooms) that can be exhausted. Actual acceptable value is the listed value to +10% (–10% for cleanrooms) of the value.

By using these criteria, the design differential is always maintained. If these are not used and the typical ±10% of design value is used by the TAB contractor, then there will be many instances of laboratories with airflow in the wrong direction, resulting in loss of contaminant control.

Documenting the TAB criteria during design aids in ensuring the building as a whole works properly. Using the previous criteria in a building with 50,000 cfm of exhaust air and 45,000 cfm of supply air to laboratory spaces, the design airflow differential is 5,000 cfm. However, during TAB, the total exhaust airflow could be as high as 55,000 cfm and the supply airflow as low as 40,500 cfm, resulting in am airflow differential of 14,500 cfm. Conversely, using the incorrect TAB criteria could result in an airflow differential of –4,500 cfm (positively pressurized).

The intent of this example is to show why laboratory buildings can become so negatively pressurized. While the system was designed to compensate for 5,000 cfm negative building differential, it was balanced to a negative 14,500 cfm differential. Therefore, the design of the central systems (chillers, boilers, coils, fans, and heat recovery systems) should have been sized for the larger pressure differential and balanced to maintain a positive, or at least a neutral, building pressure.

It is also critical to address the selection of construction materials and methods for the laboratory envelope to ensure they are capable of maintaining the selected airflow and pressure differentials throughout the expected operating conditions of the laboratory.

Another important step in the design of air and hydronic systems, both for laboratories and buildings in general, is to ensure that adequate test access points for making measurements are provided. For air systems, this means providing sufficient lengths of straight ductwork near equipment to make accurate measurements and allowing sufficient space to gain access to the straight length of ductwork. For hydronic systems, this means providing straight lengths of piping with pressure and temperature ports in the necessary locations and ensuring that they are accessible.

Once the design is completed and construction begins, it is important to select a TAB contractor early in the construction phase so that they have an opportunity to review the design documents and HVAC system installation. By including the TAB contractor early in the project, they can provide input on their requirements to be able to successfully and accurately balance laboratory systems. Any modifications needed to be able to balance the systems can be made throughout the construction stage, rather than waiting until construction is complete before seeing if the TAB contractor can actually balance the laboratory systems.

The TAB contractor, once selected, must submit the TAB plan, which should include how it will specifically meet the requirements of the project specifications. This TAB plan must detail the step-by-step procedures to be followed so that the results can be verified and the system can be re-tested and balanced in the future (e.g., yearly) by the same or a separate firm. Review and input from the architect, engineer, laboratory personnel, and the commissioning authority are very valuable in conveying the requirements of the laboratory, which must be verified, and the reasoning behind the requirements to the TAB contractor.

One last consideration in the TAB of exhaust systems is the sealing of test holes downstream of any exhaust fan. At these locations the ductwork is positively pressurized, and inadequate sealing of the test ports may result in contaminating the space. In some instances, such as biological safety cabinets, the test ports should be welded shut to eliminate any possibility of duct leakage. In other systems with low toxicity chemicals, high-quality plugs that won't react with the exhaust airstream can be used.

Ductwork Pressure Testing

Ductwork pressure testing is necessary to ensure high quality, low leakage ductwork was constructed, particularly for exhaust ductwork. Pressure testing helps ensure that the supply and exhaust airflows, needed at the equipment and diffusers, can be provided by the air systems. Generally, the design of ductwork will include an anticipated leakage rate, which must then be compared to the actual leakage rate to determine if the necessary airflows can be provided to the appropriate spaces.

As laboratory exhaust contains hazardous materials and exhaust equipment often needs a relatively high static pressure at the hood or cabinet to operate properly, tightly sealed exhaust ductwork is extremely important to physically and economically provide a high static pressure at the piece of equipment. Also, since large quantities of supply air are used for once-through ventilation, significant energy is needed to condition and move the supply air. Therefore, tightly sealed supply ductwork in a laboratory will help minimize energy losses more dramatically than in a typical building that uses recirculated air.

For both supply and exhaust ductwork testing, care should be taken to ensure that the ductwork isn't damaged in any way during the pressure testing, e.g., develops leaks or overpressurizing reduces its strength. Details on how to pressure test ductwork and the typical acceptable leakage rates can be found in Chapter 6, "Primary Air Systems."

Equipment Balancing

Equipment balancing for laboratories is performed on two categories of equipment: general HVAC equipment (such as chillers, boilers, and air handlers) and laboratory health and safety HVAC equipment (such as fume hoods, biological safety cabinets, etc.). General

HVAC equipment balancing is similar to that done in non-laboratory buildings, with a few additional safety requirements. Laboratory health and safety equipment balancing procedures are often specified in laboratory standards or guidelines and have numerous safety requirements to follow.

General HVAC Equipment. General HVAC equipment in a laboratory can be tested in much the same way as for a typical building, provided a few extra safety steps are followed. For example, testing and balancing should only be performed when the laboratory is unoccupied and all hazardous materials have been safely stored so testing and balancing activities won't disturb them. Also, safety precautions to protect the testing and balancing personnel are required. Testing and balancing personnel should be provided with all necessary safety equipment and informed of safety procedures when entering laboratory areas. Finally, testing and balancing personnel need to be aware of the hazards associated with working with general building equipment that is essential for laboratory safety, such as HEPA filters for air handlers and general building exhaust equipment.

Laboratory Health and Safety HVAC Equipment. It is critical for laboratory safety equipment to be properly tested and balanced to ensure that design values are met and laboratory personnel will be properly protected. Testing and balancing for laboratory equipment has different requirements than for typical HVAC equipment. Therefore, procedures have been incorporated into a variety of standards, such as ASHRAE Standard 110-1995 for fume hood performance and velocity testing and NSF 49-1992 for biological safety cabinet certification for velocity and filter leakage, as well as other tests. In cases where a standard is not available to determine a testing and balancing procedure, there are best practice publications, such as *SEFA 1.2 - Fume Hood Recommended Practices*, which may be used. Also in the case of equipment not covered by a standard, the laboratory designer, safety officer, and occupants should be consulted to determine how the equipment is used, the design values needed for safety, and the procedures to test and balance the equipment to the design values.

Laboratories, as most commercial buildings, rarely operate under full design conditions. However, laboratories have significantly more complex HVAC systems, which must operate in unison, regardless of the current conditions. It is important to consider TAB activities for non-design conditions, especially room pressure differentials, fume hood performance/face velocity testing, and vented biological safety cabinets.

Many actions can cause airflows in a laboratory room to change and create changes in pressure differentials, including leaving doors open, opening and closing fume hood sashes, turning on direct equipment exhaust, and activities in the adjacent laboratory rooms. Therefore, room pressure differentials should be verified under these conditions during testing and balancing to ensure negatively pressurized areas will maintain the appropriate negative pressure, and that positively pressurized areas will maintain the appropriate positive pressure under all operating conditions.

Fume hoods typically operate at either the minimum open position, at a working height, or at the maximum open position, with brief periods to move between these three heights. Each of these three heights will result in different flow characteristics for the fume hood. Therefore, fume hood face velocity testing and fume hood performance testing should be performed at these three heights. Face velocity and performance testing are discussed later in this chapter.

General TAB Standards

There are several standards and documents that provide guidance for general testing and balancing activities and testing of airflow patterns. These include those developed and published by organizations such as the National Environmental Balancing Bureau, Associated Air Balancing Council, and American Society of Heating, Refrigerating and Air-Conditioning Engineers.

National Environmental Balancing Bureau (NEBB). The National Environmental Balancing Bureau is a nonprofit organization that establishes and maintains industry standards for testing, adjusting, and balancing of air and hydronic systems and cleanroom performance testing. They also offer a certification program for testing, adjusting, and balancing firms and have developed several standards that relate to laboratory testing, including:

- *Procedural Standards for TAB Environmental Systems* (NEBB 1998). This standard provides an overview of the NEBB program, contains suggested TAB specifications, and defines instrumentation requirements and TAB reporting forms. It also provides procedures for preliminary TAB work, air system TAB, hydronic system TAB, and TAB of outdoor air ventilation rates.

- *Procedural Standards for Testing Certified Cleanrooms* (NEBB 1996). This publication provides both standards for testing procedures and informational materials for cleanrooms. Some of the testing procedures included in this publication are airflow test procedures, HEPA filter installation leak testing, room cleanliness classification testing, enclosure pressurization and integrity testing, temperature and humidity testing, and lighting level testing. Information on equipment and instrumenta-

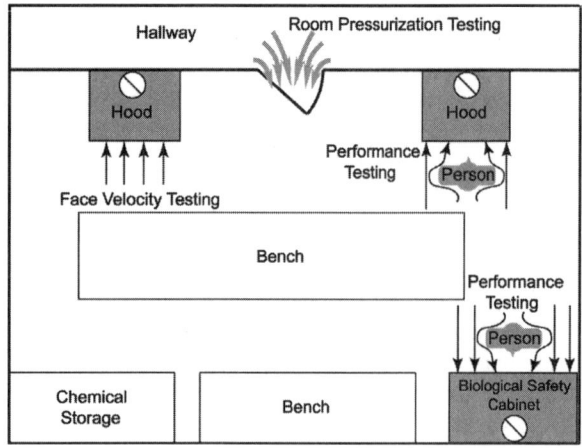

Figure 12-4 *Periodic laboratory equipment and system testing.*

tion, fundamentals of cleanrooms, cleanroom design, and laboratory/health facility cleanrooms is also included.

- *Testing, Adjusting, and Balancing Manual for Technicians* (NEBB 1997). This manual is a field-use manual for TAB contractors. It covers the topics in *Procedural Standards for TAB Environmental Systems* in a how-to format, plus provides information on HVAC fundamentals, HVAC control systems, various HVAC and electrical equipment, TAB mathematics, and system evaluation and troubleshooting.

Associated Air Balancing Council (AABC). The Associated Air Balancing Council has developed requirements for total system balancing and special systems testing. The following describe publications and documents developed by AABC:

- *AABC National Standards* (AABC 1989) define the minimum requirements for total system balancing and contains information on special systems testing such as laboratory fume hoods and cleanrooms. AABC defines total system balance as the process of testing, adjusting, and balancing HVAC systems according to the design intent.
- The *AABC Technician Training Manual* (AABC 1996) provides detailed, in-depth information on total system balancing for training new technicians.
- *AABC Test and Balance Procedures* (AABC 1998) details the minimum procedures required to test and balance air and water systems, HVAC system components, and other specialty systems. It also specifies required instrumentation and test procedures.

American Society of Heating, Refrigerating and Air-Conditioning Engineers (ASHRAE). The American Society of Heating, Refrigerating and Air-Conditioning Engineers has published the following standard that relates to testing, adjusting, and balancing:

- *ANSI/ASHRAE Standard 111-1988, Practices for Measurement, Testing, Adjusting and Balancing of Building Heating, Ventilation, Air-Conditioning and Refrigeration Systems.* The purposes of this standard are
  - to provide systematic and uniform procedures for making measurements in TAB and reporting the performance of building HVAC&R systems in the field,
  - to provide means of evaluating the validity of collected data considering system effects,
  - to establish methods, procedures, and recommendations for providing field-collected data to designers, users, manufacturers, and installers of systems.

LABORATORY TESTING REQUIREMENTS

Fume hoods, biology safety cabinets, and room pressure relationships are the main components of a laboratory that need to be tested on a regular basis, typically annually, in order to ensure that the safety requirements in the laboratory are being met. The varied periodic testing of laboratory equipment and systems is diagrammatically represented in Figure 12-4 and includes fume hood face velocity testing, performance testing of fume hoods and biological safety cabinets, and verification of room pressurization.

Fume Hood Face Velocity Testing

During the design process, a face velocity is chosen for the fume hoods that is believed to prevent the substances in the hoods from entering the laboratory space. Once the building has been constructed, and regularly throughout the life of the laboratory, the face velocity of the fume hood should be tested to ensure that the exhaust system is providing the desired face velocity. Velocity testing directly measures the velocity of the air at the sash opening over a grid pattern to determine if the design face velocity is being met. Adjusting the operation of the fume hood to meet the design face velocity is then accomplished if needed. ASHRAE Standard 110-1995 provides guidance on how to conduct fume hood face velocity testing.

Measurement Grid. An equally divided horizontal by vertical measurement grid should be used to determine the location where the face velocity readings will be taken. Each section of the grid should cover not more than a 1 ft by 1 ft (300 mm by 300 mm) area of the fume hood opening, and velocity readings should be taken at the center of each grid. The grid pattern used for face velocity testing should be sketched for future reference. Figure 12-5 is a schematic of a measurement grid for a fume hood with a 5 ft by 2 ft (0.61 m by 1.5 m) sash opening.

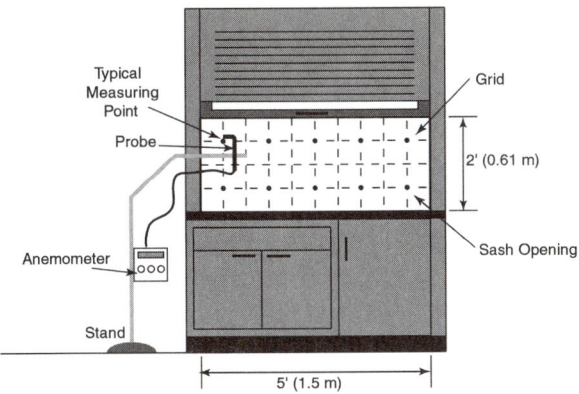

Figure 12-5 *Example measurement grid.*

Avoid Airflow Obstruction by Person Taking Measurements. When measuring fume hood face velocity, it is necessary to prevent disturbances in the airflow pattern that could be caused by the person taking measurements, as a person's profile can create turbulence or block incoming air to the fume hood. To minimize airflow obstructions and to ensure that a steady, consistent reading is attained, a low-profile instrument holder should be used to hold the anemometer, as shown in Figure 12-5.

Average Readings. As the readings of the anemometer will likely vary slightly over time, the airflow readings recorded for each grid section should be an integrated value over a minimum of a five-second period. If the anemometer used can only provide instantaneous readings, at least four readings should be made for each grid section. Once all of the measurements have been made, the average of all the grid measurements should be calculated and the highest and lowest measurements noted.

Maximum Deviation. The maximum allowable deviation between average of all readings and that of a single point is ±20%. Readings that differ by more than this value should be investigated for turbulence caused by disturbances generated in the laboratory or for blockage of airflow by the instrument holder. If the cause of the excessive deviation cannot be found, the outlying readings should not be used in the calculation of the average hood velocity. Repeating the test is recommended.

Sash Position. The measurements to determine average face velocity should be taken with the sash wide open. While the sash may be limited to being less than full open in actual operation, the full open position should be tested to ensure that sufficient airflow will be present in the event that the hood is opened 100%. Also, the fume hood should be tested at the minimum sash position to ensure the proper face velocity is met at both operating extremes. As fume hoods often operate at the minimum sash position, they should be checked to ensure that the necessary face velocity is maintained. Finally, the fume hood should be tested at normal oper-

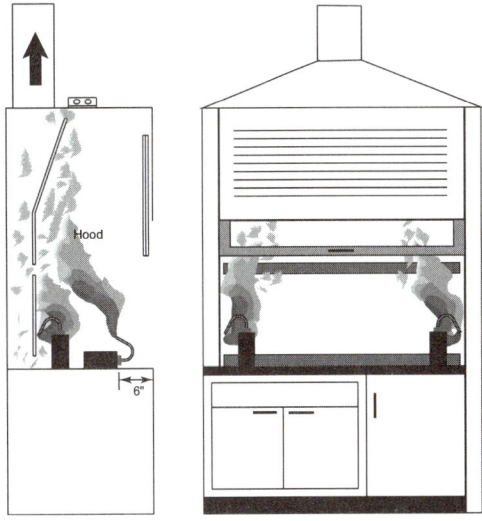

Figure 12-6 *Smoke testing schematic.*

ating height, as this is the height at which the fume hood will operate most often and is the height at which most of the work with hazardous substances will take place.

Smoke Testing. Smoke testing of fume hoods is performed to visually demonstrate if there is any spillage from the hood. Commonly titanium tetrachloride, which generates large quantities of smoke, is released from a smoke bottle along the bottom, side walls, and back of the fume hood and observed for possible spillage and turbulence. In addition to a smoke bottle with titanium tetrachloride, smoke sticks, smoke candles, and other sources that generate a persistent, neutrally buoyant stream of smoke may be used. The air foil on the bottom of hoods should be smoke tested to ensure that when the hood is closed to the minimum position, the air entering the hood is exhausted smoothly to the top of the hood and not trapped by a vortex there. The side walls and bottom of the fume hood are smoke tested 6 in. (15 cm) behind the opening to check for turbulence and spillage from the hood. The back wall of the fume hood is smoke tested to check for reverse airflow, where smoke is carried to the front of the hood, and dead air space, where the smoke lingers and is not removed. If smoke is observed to flow out of the fume hood during any of the smoke tests, the hood fails the test and receives no rating. Figure 12-6 is a schematic showing how smoke testing is performed.

During smoke testing of each surface of the fume hood, the airflow pattern observed should be recorded, along with the time for the hood to clear of smoke. A simple schematic, or a photograph, works well to record the path of the smoke. For each surface of the fume hood that is smoke tested, record the time needed for the smoke to clear and a pass/fail indicating if any smoke was observed to escape the fume hood.

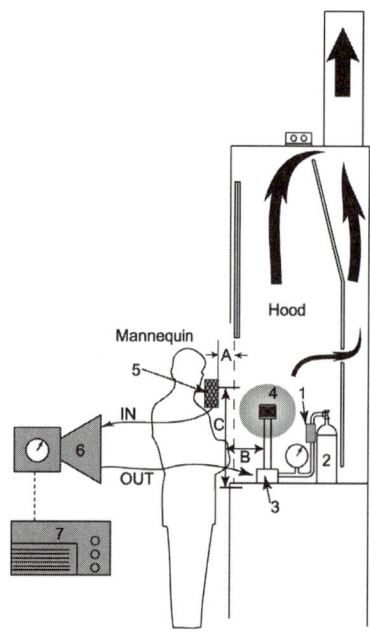

Figure 12-7 *Tracer gas test apparatus.*

Fume Hood Performance Testing

Regular testing is needed to ensure that fume hoods properly protect laboratory workers from the chemicals that they use. The common method of testing fume hoods is specified in *ANSI/ASHRAE Standard 110-1995, Method of Testing Performance of Laboratory Fume Hoods*, in which a tracer gas is used to determine if the fume hood adequately contains contaminants.

Purpose. The purpose of fume hood performance testing is to ensure that the fume hood is operating as designed and is providing protection to the workers. While velocity testing can provide some assurance that the fume hood is operating safely and determine if the operation of the fume hood needs to be adjusted, it only measures a sampling of the face velocity and could miss small vortexes and disturbances in the face velocity. Fume hood performance testing can determine how well the fume hood is containing at a particular location in the lab. The performance of a fume hood cannot be directly linked to worker exposure but is a good indicator that the hood has been installed properly and is operating as a containment device. Requiring performance testing is the best known method for ensuring that the fume hood is providing protection to users. Typically, fume hood performance testing is performed once at the factory, "as manufactured," and once when the hood is installed to verify "as installed" conditions. "As used" testing is dependent on requirements of the particular laboratory in an operating mode with lab equipment inside the fume hood.

Key Points of the Test Procedure. ASHRAE Standard 110 simulates the use of a hazardous material by a laboratory worker through the use of a tracer gas and a mannequin. Similar to velocity testing, a preliminary visual inspection may be performed by using a smoke bottle with titanium tetrachloride, smoke sticks, smoke candles, or other sources that generate a persistent, neutrally buoyant stream of smoke to show the general airflow pattern inside the hood to identify and correct obvious problems.

The mannequin is placed at the opening of a fume hood where the hood user typically would stand, approximately 3 inches (7.5 cm) from the fume hood sash. The mannequin should be 67 inches (170 cm) tall and be dressed in clothing typical for the laboratory workers. A tracer gas, typically sulfur hexafluoride, is released at 1.06 gallons per minute (4.0 liters per minute) on the inside of the fume hood 6 inches (15 cm) behind the sash opening. A detector is placed near the nose/mouth of the mannequin to determine the rate of tracer that escapes from the hood to directly impact the hood's user. The detector for the tracer gas should have a range of at least 0.01 ppm to 100 ppm. Figure 12-7 is a schematic of the tracer gas test apparatus.

The tracer gas ejector system consists of a critical orifice that draws air through the holes in the side of an ejector tube, which is distributed through a wire mesh outlet diffuser that is a 5 inch (12.7 cm) diameter by 3 inch (7.6 cm) high cylinder surrounding the ejector. Standard 110 provides detailed schematics on how to construct the ejector system.

Standard 110 requires that the amount of tracer gas detected should be recorded as AM (as manufactured), AI (as installed), or AU (as used) followed by the detection level in parts per million (ppm). These different labels are required due to the differing conditions present for each test. Generally, the as-manufactured conditions are the most controlled and therefore have the best results (least spillage). The as-installed and as-used test conditions are increasingly less controlled than the as-manufactured conditions and therefore have slightly higher spillage rates.

ASHRAE Standard 110-1995 provides the procedure for fume hood performance testing but does not provide acceptable spillage rates. ANSI/AIHA Standard Z9.5, *Laboratory Ventilation*, provides recommended values of less than 0.05 ppm for "as manufactured" and less than 0.1 ppm for "as used" at the 1.06 gpm (4.0 lpm) diffusion rate.

Biological Safety Cabinet Performance Testing

Biological safety cabinet performance testing is conducted to ensure that cabinets are providing protection from biological aerosols. Testing protocols are specified in *NSF 49-1992 - Class II (Laminar Flow) Biohazard Cabinetry*. Depending on the class of biological safety cabinet, the test parameters may be different because each class of cabinet offers slightly different protective barriers.

Purpose. The purpose of biological safety cabinet performance testing is to ensure that the cabinet meets predetermined criteria and listed requirements established by NSF 49-1992. Assuming that these criteria are met, a particular tested unit is determined to have the ability to provide a defined level of protection when made available for use in the laboratory. Following the acceptance testing, the cabinet unit is then listed and bears an NSF 49 seal.

Key Points of the Certification Procedure. The standard calls for the testing of the following:

- Cabinet structure
- Air velocity and airflow direction
- Pressure decay
- Filter integrity
- Biological aerosol challenge
- Light intensity
- Electrical safety
- Sound and vibration

For biological safety cabinets designed to be vented into the laboratory ventilation system, the strictest velocity testing protocol is required. Safety cabinets must be adjusted to within ±5% of the safety cabinet's design velocity tolerance, which is directly related to the cabinet's ability to contain biological aerosols. Unlike a fume hood, the biological safety cabinet has HEPA filtration, which must be tested for leaks before the unit can be certified for use in a laboratory.

The different classes of biological safety cabinets require different HEPA filtration systems. Class I cabinet filters protect the lab personnel and the environment. Class II and Class III cabinet filters protect personnel, product, and the environment. NSF 49-1992 discusses the different tests required for the different cabinet classes available and maintains a listing of authorized safety cabinets and field certifiers by geographical location.

For open face safety cabinets connected to the building exhaust system, the TAB contractor should consult with the cabinet manufacturer regarding the acceptable airflow range that should be used to verify the exhaust rate using a flow hood or the procedure required using a discrete velocity meter. Often a generic airflow is specified on a mechanical plan, which ultimately could require precise tuning to safety cabinet tolerances to be within acceptance limits for the field certification test.

Verification of Room Pressurization

Room pressurization is used to either prevent substances not captured by a fume hood from migrating to the rest of the building or to maintain a cleaner environment in the laboratory than the surrounding areas. Supply and exhaust airflows may change over time due to normal wear of the HVAC system. The building envelope and pressure sensor calibration may also change over time and affect room pressurization. Therefore, room pressurization should be tested and verified that it meets the necessary requirements. Both instrument testing and (visual) flow direction testing are used to verify room pressurization.

Instrument Testing. Laboratories need to maintain a pressure differential to maintain airflow direction from cleanest to dirtiest. For most laboratories, a minimum of 0.01 in. w.g. (2.5 Pa) is needed to provide this airflow. Laboratories that use highly hazardous materials will likely require greater pressure differentials, up to 0.05 in. w.g. (12.5 Pa) or higher. A pressure differential of 0.01 in. w.g. (2.5 Pa) is intended to provide a 100 fpm (0.51 m/s) velocity through a 0.125 in. (3.2 mm) crack (e.g., under a door), while the actual velocity will vary with the integrity of the envelope construction.

Instrument testing of room pressurization uses a manometer to verify the pressurization of the laboratory. The manometer should be accurate to ±1%, and the precision should be ±0.005 in. w.g. (1.25 Pa). The room pressurization should be tested under the maximum open, minimum open, and typical working positions of the fume hood sashes (ASHRAE Standard 41.2-1987 (RA 92)).

Flow Direction. In addition to instrument testing, flow direction in laboratories can be visually verified. For example, smoke testing can be used to visually demonstrate that the pressure differential is being maintained for the room and that there is no air leaving a laboratory space and migrating to another part of the building. Flow direction testing can be accomplished by using smoke candles around openings to the laboratory to verify flow direction. The typical openings to be verified under the various operating conditions listed previously are:

- Door closed
- Door open
- Electrical outlets
- Window
- Other penetrations (piping, wiring, duct)

REFERENCES

American National Standards Institute (ANSI). 1992. *Standard Z9.5, Laboratory Ventilation, American National Standard Practices for Respiratory Protection.* New York: ANSI.

American Society of Heating Refrigerating and Air conditioning Engineers (ASHRAE). 1992. ASHRAE Standard 41.1-1987 (RA 92): Standard Methods for Laboratory Airflow Measurement. Atlanta, GA.

American Society of Heating Refrigerating and Air conditioning Engineers (ASHRAE). 1995. ASHRAE Standard 110-1995: Method for Testing Performance of Laboratory Fume Hoods. Atlanta, GA.

American Society of Heating Refrigerating and Air conditioning Engineers (ASHRAE). 1988. ASHRAE Standard 111-1988: Practices for Measurement, Testing, Adjusting, and Balancing of Building Heating, Ventilation, Air-Conditioning, and Refrigeration Systems. Chapter 10. Atlanta, GA.

Associated Air Balance Council (AABC), *Test and Balance Procedures*. 1998. Washington DC.

Associated Air Balance Council (AABC), *Technician Training Manual*. 1996. Washington DC.

Associated Air Balance Council (AABC), National Standards. 1989. Washington DC, 1989.

DiBerardinis, L.J., J.S. Baum, M.W. First, G.T. Gatwood, E. Groden, and A.K. Seth. 1993. *Guidelines for Laboratory Design: Health and Safety Considerations*, 2nd ed. New York: John Wiley & Sons, Inc.

Memarzadeh, F. 1996. *Methodology for Optimization of Laboratory Hood Containment, Volume I*, Chapter 4, pages 41-73. National Institutes of Health.

National Environment Balancing Bureau (NEBB). 1998. *Procedural Standards for TAB of Environmental Systems,* 6th ed. Gaithersburg, Md.

National Environment Balancing Bureau (NEBB). 1997. *Testing, Adjusting, and Balancing Manual for Technicians,* 2nd ed. Gaithersburg, Md.

National Environment Balancing Bureau (NEBB). 1996. *Procedural Standards for Certified Cleanrooms,* 2nd ed., Chapter 8. Gaithersburg, Md.: NEBIB.

National Sanitation Foundation International (NSF). 1992. *Class II (Laminar Flow) Biohazard Cabinetry.* Ann Arbor, MI: NSF Publications.

Scientific Equipment and Furniture Association (SEFA). 1996. *SEFA 1.2 – Laboratory Fume Hoods, Recommended Practices.* Maclean, Va.: SEFA.

Chapter 13
Operation and Maintenance

OVERVIEW

Proper operation and maintenance of laboratories ensures the safety of occupants, the ability of the HVAC system to operate in the harsh environment to which it is exposed, and the efficient operation of the various laboratory processes and equipment. Rather than waiting for major problems to occur, maintenance measures such as monitoring, testing, calibrating, cleaning, replacements, and repairs need to be performed on a regular basis (weekly, monthly, and annually). This ensures that the laboratory will maintain precise conditions for an extended period of time.

DECONTAMINATION OF EXISTING LABORATORIES

During its lifetime, a laboratory will most likely need to be decontaminated. Common reasons for decontamination include:

- Changes in the type of work performed
- Renovation or remodeling of the laboratory workspace
- Maintenance of laboratory equipment

Decontamination procedures should be established and documented to assist maintenance staff in providing proper maintenance to the laboratory equipment. Coordination of decontamination is needed to minimally impact the work being performed in the laboratory.

MAINTENANCE OF EQUIPMENT AND SYSTEMS

Routine maintenance measures are needed for various pieces of laboratory equipment and systems. Some of these include:

- Fume hoods
- Biological safety cabinets
- Ventilation systems
- Exhaust systems

Established standards, codes, and guidelines provide stringent requirements and recommendations that should be adhered to for ensuring the safe operation of the equipment and systems. The following sections present justification for routine maintenance. Table 13-1 through Table 13-5 summarize requirements stipulated by several applicable codes and standards.

Fume Hoods

To continually protect the occupants of laboratories, fume hoods need to be inspected, tested, and calibrated on a regular basis. Over time, regular use of fume hoods can cause the face velocity to deviate from the design value due to inaccurate sensor readings, buildup of dirt and debris inside the fume hood and ductwork, and wear and stretching of belts on exhaust fans. This is especially critical with variable volume fume hoods, as they rely on sensor readings to determine the proper airflow needed to maintain a safe face velocity. If the sensor becomes damaged or goes out of calibration, a variable volume fume hood can expose the laboratory occupants to dangerous levels of chemicals. Also, if the fume hood exhausts at a higher rate than is needed, then energy will be wasted on providing excess supply air for the laboratory. Table 13-1 presents requirements for operation and maintenance of fume hoods.

Table 13-1: O&M Requirements for Fume Hoods

O&M Requirement	Frequency	Reference
Each hood should have a continuous monitoring device to allow convenient confirmation of adequate hood performance.	Continuous	US OSHA, 29 CFR Part 1910, p. 3332, 4(b)
A flow monitor shall be installed on each new laboratory hood.	Continuous	National Fire Protection Association, Standard NFPA 45, 6-8.7.1
A flow monitor shall be installed on existing laboratory hoods whenever any modifications or changes are made that can affect laboratory unit ventilation or the airflow through existing laboratories.	Continuous	National Fire Protection Association, Standard NFPA 45, 6-8.7.2
When furnished, a monitoring device shall, by visible and audible signal, or both, give warning when the airflow through the hood falls below a predetermined level of adequate hood performance.	Continuous	Scientific Equipment & Furniture Association – Laboratory Fume Hoods Recommended Practices, SEFA 1/1992, 4.1.10
Quality and quantity of ventilation should be evaluated. These should be regularly monitored and re-evaluated whenever a change in ventilation devices is made.	At installation and at least quarterly thereafter.	US OSHA, 29 CFR Part 1910, p. 3332, 4(h)
All fume hoods should be tested and their performance certified.	Annually	1999 ASHRAE Handbook—HVAC Applications, p. 13.5, Fume Hood Performance Criteria
A routine performance test shall be conducted on every fume hood.	At least annually or whenever a significant change has been made to the operational characteristics of the system.	American National Standard for Laboratory Ventilation (ANSI/AIHA Z9.5) p. 13, 5.6.2
The following performance tests must be performed: Test for face velocity Test for exhaust airflow rate Test for exhaust fan performance Test for controls Test for entrainment of auxiliary air supply (auxiliary air supply fume hoods only)	At least once per year	Canada Public Works, Standards and Guidelines – MD 151218 Laboratory Fume Hoods, A.2 Preventative Maintenance Programs, A.2.1.4

Table 13-2: O&M Requirements for Biological Safety Cabinets

O&M Requirement	Frequency	Reference
To assure that all cabinet operating criteria continues to be met, each cabinet should be field tested. The following physical tests should be performed: 1. Downflow velocity profile test 2. Inflow velocity test 3. Airflow smoke patterns 4. HEPA filter leak test 5. Cabinet leak test 6. Electrical leakage and ground circuit resistance and polarity tests 7. Lighting intensity test 8. Vibration test 9. Noise level test 10. Decontamination (if necessary)	At installation and at least annually thereafter.	National Sanitary Foundation, NSF 49-1992, Class II (Laminar Flow) Biohazard Cabinetry, Annex F
Recertification of cabinets should be performed.	Whenever HEPA filters are changed, maintenance repairs are made on internal parts, or cabinets are relocated. More frequently for more hazardous or critical applications or workloads	National Sanitary Foundation, NSF 49-1992, Class II (Laminar Flow) Biohazard Cabinetry, Annex F

Biological Safety Cabinets

Biological safety cabinets, like fume hoods, need to be inspected, tested, and recalibrated periodically in order to be certified and provided the seal of approval from the National Sanitation Foundation (NSF). These cabinets protect laboratory users from biological aerosols that range from known substances of the Biosafety Level I category to the more exotic unknown substances of Biosafety Level IV category. A strict maintenance program is essential to the proper containment of infectious agents that can be malicious to other projects, personnel, and the environment. Table 13-2 presents requirements for operation and maintenance of biological safety cabinets.

Ventilation Systems

The laboratory ventilation system ensures that occupied spaces get the correct amount of outside air while maintaining the appropriate temperature and humidity levels for acceptable thermal comfort and experimental quality. Also, the precise control of air change rates will preserve the correct pressure relationships between adjacent spaces throughout the laboratory. However, without periodic maintenance of the key components of the ventilation system, the requirements for safe and reliable operation will not be met. Table 13-3 presents requirements for operation and maintenance of ventilation systems.

Exhaust Systems

Fume hoods and biological safety cabinets are an intimate part of the exhaust system of any laboratory. Whereas this exhaust equipment protects personnel or product by containing fumes, aerosols, and particles, the exhaust system, which includes components such as fans, stacks, ductwork, control equipment, dampers, and monitoring/alarm devices, continues the protection process by expelling the initially contained substances to the outside of the building. The exhaust system also works in tandem with the ventilation system to provide the correct pressure relationships and total air change rates for laboratory spaces. Therefore, for the exhaust system to operate in a safe and dependable manner, continuous maintenance is a high priority. Table 13-4 presents requirements for operation and maintenance of exhaust systems.

Table 13-3: O&M Requirements for Ventilation Systems

O&M Requirement	Frequency	Reference
Quality and quantity of ventilation should be regularly monitored.	At least quarterly	US OSHA, 29 CFR Part 1910, p. 3332, 4 (h)
Centralized monitoring of laboratory variables (e.g., pressure differentials and face velocity of fume hoods, supply flows, and exhaust flows) are useful in predictive maintenance of equipment and in assuring safe conditions.	Continuous	ASHRAE, HVAC Applications Handbook, 1999, p. 13.16, Operation and Maintenance
Quality and quantity of ventilation should be evaluated and regularly monitored.	At installation and at least quarterly and whenever a change in local ventilation devices is made.	US OSHA, 29 CFR Part 1910, p. 3332, 4 (h)
Fans, blowers, and drive mechanisms shall be visually inspected. Observe for abnormal noise or vibration, bearing noise, excessive temperature of motors, lubricant leaks, etc. Records should be maintained for all inspections and maintenance.	Weekly	American National Standard for Laboratory Ventilation (ANSI/AIHA Z9.5), p. 11– 4.14.7.1 and 4.13.4
V-belt drives shall be stopped and inspected for belt tension and signs of belt wear of checking.	Monthly	American National Standard for Laboratory Ventilation (ANSI/AIHA Z9.5), p. 11 – 4.14.7.1
Installed air system flow detectors shall be inspected and tested.	Annually (increase frequency appropriately where potentially corrosive or obstructive conditions exist).	National Fire Protection Association, Standard NFPA 45, Par: 6-13
Air supply and exhaust fans, motors, and components shall be inspected.	Quarterly	National Fire Protection Association, Standard NFPA 45, Par: 6-13

Table 13-4: O&M Requirements for Exhaust Systems

O&M Requirement	Frequency	Reference
Provision should be made for maintenance of adequate suction in the (exhaust) manifold. This requirement would be satisfied by providing an installed spare manifold exhaust fan that can be put into service rapidly by energizing its motor and switching a damper.	Continuous	American National Standard for Laboratory Ventilation (ANSI/AIHA Z9.5), p. 21
Laboratory exhaust systems should be designed for high reliability and for ease of maintenance. One way to achieve this design is to provide redundant exhaust fans and a means of sectionalizing equipment so that an individual exhaust fan can be maintained while the system is operating.	N/A	1999 ASHRAE Handbook—HVAC Applications, p. 13.9, Exhaust Systems
Controls and dampers where required for balancing or control of the exhaust system shall be of a type that in the event of failure will fail open to assure continuous draft.	N/A	National Fire Protection Association, Standard NFPA 45, Par. 6.5.7

Others

Other equipment and components that require routine maintenance for efficient laboratory operations include:

- Filters
- Dampers (fire, smoke, and ceiling)
- Ductwork
- Plenums
- Fans and fan motors
- Controls

Appendix B of NFPA (1996) provides guidelines for periodic maintenance of the above equipment. Table 13-5 summarizes these guidelines and frequency. The frequency of maintenance may vary widely depending on the system operation period, fresh air conditions, return air dust quantity, and other factors. Intervals should be shortened if system conditions warrant.

COST INFORMATION

In addition to the initial cost of designing and constructing a laboratory, there are several costs associated with using the laboratory after completion, including the cost of maintenance, energy, and operation. Maintenance costs include the cost of repair/replacement parts and materials for the HVAC system. The energy cost of a laboratory is the cost needed to purchase electricity, natural gas, and other fuels to operate the laboratory. The cost of operation includes the salary of the maintenance staff (or the cost to subcontract maintenance assignments) and other expenses for running the laboratory, such as the cost for non-energy utilities and general building supplies.

Since laboratory HVAC systems are generally more complex and energy intensive than those of typical buildings and their continued operation is essential for safety, it is important to budget appropriate funds for the operation and maintenance of a laboratory, which includes the provision for sufficient qualified personnel. Coordination between laboratory designers, personnel, and maintenance staff can reduce costs by making the laboratory easier to operate and to maintain.

Table 13-5: O&M Guidelines for HVAC Equipment and Components

Equipment	O&M Guideline	Frequency
Filters	All air filters should be kept free of dust and combustible material. Unit filters should be renewed or cleaned. Filters designed to be thrown away after use should not be reused.	When the resistance to airflow has increased to two times the original resistance, or when the resistance has reached a value of recommended replacement by the manufacturer.
Dampers	Each damper should be examined to ensure that it is not rusted or blocked, giving attention to hinges and other moving parts. It is recommended to operate dampers with normal system airflow to ensure that they close and are not held open by the airstream. Care should be exercised to ensure that such tests are performed safely and do not cause system damage.	Biennially
Ductwork	Inspections to determine the amount of dust and waste material in the ducts (both supply and return) should be made.	Quarterly (if, after several inspections, such frequency is determined to be unnecessary, the interval between inspections can be adjusted to suit the conditions).
Plenums	Apparatus casing and air-handling unit plenums should be inspected for dirt buildup conditions.	Monthly (if, after several inspections, such frequency is determined to be unnecessary, the interval between inspections can be adjusted to suit the conditions).
Fans and fan motors	Fans and fan motors should be inspected and checked for proper alignment and, when necessary, cleaned and lubricated. Care should be exercised in lubricating fans to avoid allowing lubricant to run onto the fan blades.	Quarterly
Controls	Controls should be examined and activated to ensure that they are in operable condition.	At least annually

Operations Cost

Operation costs include the costs associated with the operation of the equipment and systems of a laboratory. Energy costs and maintenance costs are major categories of the cost of operation. These costs include wages and non-energy utility costs.

Energy Cost. Due to the large amounts of exhaust air and the typical use of 100% outside air for laboratory spaces, energy costs are higher than those of typical buildings. The operations and maintenance staff should closely monitor the monthly energy costs, as large increases or decreases may identify possible HVAC system problems. There are many steps that maintenance personnel can take to reduce energy costs, including closing fume hood sashes during unoccupied periods, regularly changing filters, and keeping air passages clean and open. A DDC control system can greatly simplify the optimization of laboratory operation and energy minimization.

Maintenance Cost. As laboratories have high concentrations of HVAC equipment, the time and cost to provide maintenance can be significant. Therefore, a carefully planned and executed preventive maintenance program should be used, as it will reduce time and costs for maintenance and ensure a safe working environment in the laboratory. Documentation during design and construction, in the form of the basis of design, as-built drawings, and O&M manuals, is essential to ensuring that maintenance personnel have the necessary resources to maintain the laboratory. Additional costs invested in maintenance can result in significant savings in energy costs.

TRAINING

Training is required throughout the life of any building to keep the maintenance staff and occupants of the building informed of the procedures for safe, energy-efficient, and cost-effective use of the building's HVAC system. As laboratories use many hazardous substances, use more energy than a typical commercial building, and have HVAC systems that can be dangerous or costly to fix if used improperly, training is especially important for the maintenance staff and occupants. Training is also important for laboratories with a frequent change in occupants, such as a teaching or research laboratory.

Maintenance Staff

In order to safely and cost-effectively operate a laboratory, the maintenance staff needs to be properly trained in the operation of building equipment and systems. Maintenance staff should be aware of the hazardous materials used in the laboratory and how that affects maintaining the HVAC system. Training should be tailored to meet the specific needs of the maintenance staff, allow for trainee interaction with equipment, provide documentation and demonstrate its use, and provide guidance for preventive maintenance.

Training should be customized to provide the level of detail that the maintenance staff needs, as they may or may not already be familiar with operating the equipment installed in the laboratory. If they have already used the same type of equipment, general background information on the equipment is not needed.

Trainees should also be allowed to work with the equipment during training so that they can actually see what the trainer is talking about and have their questions answered. It is unlikely that the maintenance staff will be able to remember or comprehend all of the information that is presented to them during training. Therefore, they should be made aware of the reference materials available to them, such as O&M manuals, and have a chance to use them.

Finally, maintenance staff should be informed of the preventive maintenance that is required for the laboratory equipment. Maintenance procedures, including work to perform, notification of laboratory users, and documentation of work and testing procedures, should be discussed.

Laboratory Users

Laboratory users, both new and experienced, need to know how to safely operate a laboratory, know emergency procedures, realize the limitations of the HVAC system, and understand the impact their work has on energy usage.

To operate a laboratory safely, laboratory users must know the proper procedures for operating laboratory equipment to minimize the risk of accidents. Items such as fume hood operation, keeping doors closed to maintain pressure differentials, and limits on where various materials can be used must be discussed. Laboratory users should also be aware of the procedures for reporting a potentially dangerous condition so that maintenance personnel can take corrective action.

Laboratory users should know the proper emergency procedures to minimize the impacts of an accident. Regular reviews of emergency procedures should be performed to sharpen emergency actions, to determine possible improvements, and to develop new procedures when the type of work performed in the laboratory changes. This includes use of emergency announcement systems and evacuation procedures.

Also, laboratory users need to be aware of limitations in the HVAC system, such as the use of diversity factors and modification limits. For example, if a diversity factor of 75% was used for fume hoods, laboratory users need to know that it is not physically possible to conduct large experiments simultaneously that would require 85% of the fume hoods. They should be aware of the reasons for the limits, which in the case of the above example would be that the supply air system could not provide the needed air volume and there might not be sufficient air exhausted through the fume hood to keep them safe.

Finally, laboratory users need to be aware of the impact their actions have on energy usage. For example, they should be aware that unnecessarily leaving fume hood sashes open will use a significant amount of energy to condition and to move air.

REFERENCES

American Society of Heating, Refrigerating and Air-Conditioning Engineers (ASHRAE). 1999. *1999 ASHRAE Handbook—HVAC Applications*, Chapter 13, Laboratories. Atlanta: ASHRAE.

American National Standards Institute (ANSI). 1992. *Standard Z9.5, Laboratory Ventilation*. New York: American National Standard Practices for Respiratory Protection.

National Fire Protection Association (NFPA). 2000. *NFPA 45—Fire Protection for Laboratories Using Chemicals*. Quincy, Mass: NFPA Publications.

National Sanitation Foundation International (NSF). 1992. *Class II (Laminar Flow) Biohazard Cabinetry*. Ann Arbor, MI: NSF Publications.

Occupational Safety and Health Administration (OSHA). 1990b. *OSHA 29 CFR Part 1910.1450– Occupational Exposure to Hazardous Chemicals in Laboratories*. Washington, D.C.: U.S. Government Printing Office.

Occupational Safety and Health Administration (OSHA). 1990c. *OSHA 29 CFR Part 1910.1330– Occupational Exposure to Bloodborne Pathogens*. Washington, D.C.: U.S. Government Printing Office.

Occupational Safety and Health Administration (OSHA). 1991. *OSHA 29 CFR Part 1910.35– Health and Safety Standards*, Chapter 12, Subpart E. Washington, D.C.: U.S. Government Printing Office.

Office of Biosafety, Laboratory Centre for Disease Control. 1990. Canada Public Works, *Standards and Guidelines – MD 151218 Laboratory Fume Hoods*. A.2 Preventative Maintenance Programs, A.2.1.4 Performance Tests. Ottawa, Ontario, Canada.

Scientific Equipment and Furniture Association (SEFA). 1996. *SEFA 1.2—Laboratory Fume Hoods, Recommended Practices*. Maclean, VA.

Chapter 14
Laboratory Commissioning Process

OVERVIEW

Laboratories are considerably complicated in that they must include ventilation for comfort and for safety and have other special requirements such as cleanrooms and strict temperature and humidity control. These requirements necessarily increase the amount of components as well as control strategies. The corrosive and sometimes explosive properties of gases or particles in the exhaust system, health hazards, high efficiency filtration, and energy efficiency must also be considered.

To ensure that all of the diverse requirements for laboratories are met, a quality control process must be instituted for the planning, design, construction, and operation of laboratories. The process recommended to consistently meet the requirements is called *commissioning*. Commissioning, in general terms, is the process adopted owners for the planning, design, construction, and operation of their facilities to ensure they get at the end of a project what they wanted at the beginning. As commissioning is "just the way things are done," the key requirements for proper commissioning were integrated into the previous chapters of this guide. Therefore, this chapter is intended to provide a general overview of the commissioning process and to provide some detailed guidance for laboratory systems. The first section provides an overview of the commissioning process, followed by sections that detail the commissioning tasks for each phase of a typical laboratory project. The final section details how the commissioning process can be applied to existing laboratory facilities.

THE COMMISSIONING PROCESS

Commissioning is defined by ASHRAE as:

> The process of ensuring that systems are designed, installed, functionally tested, and capable of being operated and maintained to conformity with the design intent.

While commissioning can start in any phase of a building project, the greatest benefits are achieved by starting in the planning phase of a project. In so doing, maximum benefits are achieved throughout the lifetime of the building. The general phases of the commissioning process are:

- Planning
- Design
- Construction
- Acceptance
- Operation

ASHRAE Guideline 1-1996 was published to provide guidance on implementing the commissioning process. It is important to understand that these are just guidelines and not "hard and fast" rules. As with any quality control process, commissioning is continuously changing and improving. The practitioners of commissioning are responsible for maintaining the principles of this process.

PLANNING PHASE

The goals and expectations of the owner are documented during the planning phase in a significant document known as the *design intent*. The *design intent* provides a measure for the success of the laboratory building process. It is a living document and must be updated as the project progresses. Also completed during this phase of the project is the drafting of the commissioning plan, which is discussed in a later section.

Establishing Project Goals and Expectations

The building project goals must be stated in clear and concise terms as early in the project as possible to ensure that all the work performed meets the goals of the owner. While the project goals change during the project, as the owner becomes aware of the implications of the initial project goals, documenting changes and maintaining a current design intent enables everyone involved in the project to work toward the same goals. There are a number of tools available to document the design intent, including workshops, surveys, and interviews.

Items of particular interest when developing the design intent for a laboratory building include:

- Risk assessment
- Safety features
- Flexibility and reliability
- Economics
- Functional performance

Risk Assessment. The planned usage of the laboratory must be detailed as thoroughly as possible to ensure that the designed and constructed laboratory system can meet the requirements. These details can, for example, be the expected use of chemicals, the mixture of chemicals, the biological hazard, the level of cleanliness required, types of animals, etc. The more details provided at this stage will help ensure that the risks are determined accurately and that the design of the system can handle these risks. Chapter 3, "Laboratory Planning" discusses risk assessment in more detail.

Safety Features. Based on the risk assessment, the safety features to handle emergencies such as spills, fires, explosions, malfunctioning of fans or motors, and room pressure problems can be identified. These features help ensure that the laboratory system will be able to handle situations that are abnormal without excessive risk to the users. The object of the safety features is to ensure that the users are protected from release of contaminants even in extreme situations.

Flexibility and Reliability. The flexibility of the design is important to be able to meet changing criteria. The flexibility of the system may require additional cost, but good planning can considerably ease the process of adding to the system or making a change to the system.

Reliability is an important issue for laboratory systems, as failure of the system can have dramatic consequences. In critical systems that deal with highly toxic materials or where a shutdown could be extremely expensive, redundant system or component configurations are often necessary. This redundancy will ensure that maintenance can be done without shutting down the system and that the system will function even in a case of component failure. The level of reliability must be determined since systems can only approach 100% reliability.

Economics. As in any project, the economics is a constraint that determines what can actually be accomplished. The owner always has a limited amount of funds available, and the project must meet this criterion as well as the other criteria. Often, the cost of the requirements will exceed the funds available. While compromises are then made, the safety of the users must be maintained. It is important to update the design intent every time something is compromised and to explain the reason for the compromise so that all project participants understand what is changed and why it was changed to avoid future misunderstandings or discussion (lawsuits) about changes.

Functional Performance. The functional performance to be met by the system must be documented to ensure that all the participants clearly know what the design and installation has to achieve. Functional performance includes the flexibility, reliability, and safety functions with a main focus on the operational requirements. The functional requirements should be as detailed as possible to ensure that the design and construction team can meet the requirements throughout the project.

Commissioning Plan

The commissioning plan is a roadmap of how to meet the owner's requirements using quality techniques. It is intended to provide background information for the team members not familiar with the commissioning process.

DESIGN PHASE

During the design phase, the commissioning process ensures that the construction documents meet the design intent. This includes the development of the basis of design and the review of all construction documents as well as the integration of the commissioning requirements into them.

Basis of Design

The basis of design states what codes and standards the design must meet and guidelines that are used to design the system. In addition, the basis of design includes the reasoning for the chosen system configuration and calculation that documents and explains how the components and systems meet the requirements set by the codes, standards, and guidelines.

Construction Documents

Drawings and specifications are the main construction documents developed during the design phase. These documents are used for bidding and the construction of the project and are based on the basis of design. The commissioning requirements for the laboratory building project must be integrated into these documents so that all parties involved will be aware of the quality process used to achieve the goals and expectations of the owner.

CONSTRUCTION PHASE

Commissioning during the construction phase focuses on the quality of the installations to be in accordance with the design intent. While similar to the design phase, the procedures are different. The emphasis during the construction is on accomplishing the installation correctly the first time. Guidance for the installers of what is important for the installation to meet the design intent is clearly provided. Commissioning tools and techniques used during construction include mock-ups, construction checklists, and system verification.

Mock-Ups

The use of mock-ups is one way to ensure quality during construction. The contractor is required to construct a small section of the installation that has to be approved before the remaining construction sections are allowed. This procedure catches installation faults that otherwise would not have been detected until a significant portion of the installation was complete. Mock-ups increase the quality and reduce the cost of the installation for the contractor, as it ensures the individual workers know what is required to meet the owner's design intent.

Construction Checklists

The installation of the individual components of the various systems must be verified to meet the design intent. Individual workers accomplish this through continuous spot-checking of work and through the use of construction checklists that are developed by the commissioning authority.

System Verification

Once all systems are installed, a verification of the control system sequences is accomplished. This entails the review of control panel wiring and labeling to ensure that they match the as-built drawings. The individual control loops are also checked. After the control acceptance, all air and water systems can then be tested, adjusted, and balanced. The system TAB is then verified to ensure that the laboratory has the proper flows under all conditions to maintain the health and safety of the occupants and to meet the owner's design intent.

ACCEPTANCE PHASE

The acceptance phase commissioning focuses on the functional performance of the equipment. Ideally, the commissioning process has eliminated the majority of problems before the functional performance of the system is tested. Minor failures during the test are corrected. Major failures require a retest. This acceptance phase section focuses on functional performance tests of HVAC equipment that are critical for laboratories, such as pressurization controls and exhaust hoods and systems.

Functional Testing Overview

Functional testing of the equipment ensures that the installed equipment and systems perform as intended. All the normal operating conditions are tested. If an operating condition cannot be tested (e.g., space cooling during the winter season) and there is no way to simulate the necessary conditions, the functional performance test should be scheduled for a later time when it is expected that the necessary conditions are present. In addition to the normal operating conditions, emergency, safety, start-up, shutdown, and all other foreseeable abnormal operating conditions are also tested.

Controls

The control system is the "brain" of the system and determines the sequence of operation and how the individual pieces of equipment interact. The control system receives information from sensors located at strategic locations throughout the system. A control system can only function correctly if it receives correct information from all the sensors, the control sequences are correct, and the equipment performs the tasks the way the control system instructed it.

Exhaust Hoods and Systems

Exhaust hoods and systems often have requirements that require complex controls as well as stringent equipment performance requirements. These requirements depend on the type of use of the laboratory and the specific use of the exhaust hood. Face velocity and proper hood containment can be tested by smoke test under all normal operating conditions. The acceptance criteria for these tests should be determined before projects are bid and included in the documents, with the actual test procedures determined when the contractor's submittals have been accepted and the manufacturer determined. The extent of functional performance tests for the laboratory systems include:

- Part-load and full-load performance
- Fume hoods
- Biological safety cabinets
- Miscellaneous exhausts

OPERATION PHASE

The last phase of the commissioning process focuses on actions needed to follow up on warranties and making sure the expected performance is maintained for the life of the building. The requirements of training and documentation are addressed in this section, as well as the need for audits of laboratory systems.

Documentation

Documentation should include all the necessary information to safely operate and maintain the laboratory systems. These include operating and maintenance (O&M) manuals, record drawings, material safety data sheets (MSDS), and any key references for specific laboratory procedures. The documentation must be well organized and only include information on the installed components. No information on components that were not actually installed should be allowed. The information should be short and concise and meet the design intent. This allows the O&M staff to have the correct information to perform their required tasks. The documentation should be compiled early during the construction process with details such as the record drawings and the commissioning report added.

Training

The training of the O&M staff is a key factor that determines if the laboratory systems will be operated as intended when the building is occupied. While the system may be capable of meeting the owner's and occupants' requirements, the systems will fail if the O&M staff do not understand the system. If the maintenance of systems is insufficient, dangerous conditions as well as a shorter lifetime of individual components are inevitable. It is an advantage if the training focuses on teaching the O&M staff to use the O&M manual so they are confident about being able to work in a situation that resembles the way they will have to work during day-to-day operation.

The user of laboratories must also be trained to be able to take advantage of the system's capabilities, know the system's limitations, and understand safety procedures.

Ongoing Laboratory Audits

The purpose of laboratory audits is to systematically review the existing state of the laboratory facility and determine any code compliance issues that are related to the safety of personnel and product, energy usage, and the storage, use, and handling of chemicals and biohazardous materials. Laboratory audits should be performed periodically even if all the steps of a quality control process such as commissioning were followed in the procurement of a laboratory. A report of the findings and recommendations of the audit is an invaluable source for focused improvements. Laboratory personnel should meet, discuss, and rank the recommendations in order of priority. A HAZOP style format can be used for these meetings. See Chapter 3, "Laboratory Planning" for a discussion on risk assessment and various other hazard analysis methods.

COMMISSIONING OF EXISTING BUILDINGS

Commissioning of systems in existing buildings is divided into two separate sections:

- *Retrofitting* - Commissioning to achieve the expected performance from an existing laboratory system
- *Remodeling* - Commissioning of a project to remodel an existing building that did not have a particular laboratory system(s) before remodeling

Retrofitting

Retrofitting concentrates on upgrading the system to perform as designed or to accommodate slight changes to the existing system. Training of O&M staff, improved documentation, replacement of failing equipment, recalibration of sensors, and testing and balancing of air and water flows are usually the focus of these projects. Laboratory systems can cause significant losses if the laboratory system does not perform as designed. Therefore, many of these tasks might be considered maintenance tasks, but a more thorough retrofitting will in many cases be required at yearly intervals.

Remodeling

Remodeling of existing buildings will usually follow the same process as a new building process. However, if certain parts of systems are reused, considerations such as cleanliness, integration with new components, etc., must be considered. Remodeling also requires that contaminated equipment be handled according to safety and health standards and codes during demolition.

REFERENCES

American Society of Heating, Refrigerating and Air-Conditioning Engineers (ASHRAE). 1996. *Guideline 1-1996, The HVAC Commissioning Process*. Atlanta: ASHRAE.

Chapter 15
HVAC System Economics

OVERVIEW

Cost analysis for buildings can be divided into two categories: initial and life-cycle cost. Generally, the initial price is emphasized due to budget constraints on a project. Because of the high density of HVAC equipment in laboratories, the initial price to purchase equipment can be a considerable percentage of the total building price (between 30% and 50%). Therefore, accurate initial price estimates are necessary to complete the laboratory within the budget constraints.

First cost analysis, while important in maintaining the budget, ignores the investment required to operate the building, such as utility costs and maintenance cost, over the mechanical system's life span of 15 to 30 or more years. Due to the high amounts of HVAC equipment (and, thus, high energy consumption) in laboratories that must be rigorously maintained to ensure the occupants' safety, life-cycle cost is highly important in designing a laboratory.

INITIAL COST OF SYSTEM

The HVAC system choice and layout, and thus the initial cost of the laboratory, is affected by many factors, such as the usage patterns of the laboratory, the substances handled in the laboratory, the possibility for future expansion, and the potential for changes in layout.

The usage patterns of a laboratory may dictate which system options are more attractive. For example, in a laboratory where most fume hoods are continuously in use, they may not benefit from the added initial cost of using variable volume fume hoods, as the need for exhaust would be relatively constant.

The substances used in the laboratory may also dictate several design requirements, such as the material to use for exhaust ductwork. Using similar chemicals, which don't interact, throughout the laboratory could allow manifolded exhaust to be used, whereas chemicals that interact would require separate exhaust stacks.

The possibility for future expansion and changes in layout may affect sizing factors such as the diversity factor. A laboratory that plans to expand or reconfigure laboratory space in the future may opt to use a higher, or no, diversity factor to allow for additional capacity for future work. In some cases, the diversity factor may equal one, in that the central equipment has the same capacity as the summed distributed systems.

Each of the factors mentioned above (usage patterns, substances handled in the laboratory, the possibility for future expansion, and the potential for changes in layout) influence the initial price of the central air-handling equipment and exhaust equipment systems.

Central Air-Handling Equipment

Central air-handling equipment for laboratories must be able to respond accurately and quickly to changes in load. In the case of a variable volume air handler, this means making rapid changes in supply air volume when a fume hood sash is opened to maintain room pressure differentials. For constant volume air handlers, this may mean rapidly adjusting heating or cooling valves to prevent overheating or overcooling the laboratory space. The equipment selected needs to be able to perform these rapid changes and not be selected only on lowest cost or familiarity alone. Last, central air-handling equipment may have special filtering and treatment requirements that require additional equipment or add-on options for the equipment.

Air Distribution System. The air distribution system for laboratories often has special design requirements. For example, diffuser type and location are important in preventing disturbances, which affect fume hood operation. Also, temperature variations and air currents should be minimized. When large quantities of air need to be supplied to laboratories to make up for the exhausted air, special methods are needed to prevent temperature variations and air currents that may disturb exhaust equipment.

Constant Volume Reheat. Constant volume air-handling equipment with reheat may require less initial cost to purchase the equipment, but it can use significantly more energy in laboratories with fluctuating equipment usage or wide changes in outdoor air temperature throughout the year.

Variable Air Volume. While variable air volume equipment requires additional initial equipment, such as variable frequency drives and controls, and thus costs more when compared to constant volume equipment, this additional initial cost can be offset by the energy cost savings over the life of the building.

Exhaust Systems

To determine the initial cost of exhaust systems, the cost of individual versus manifolded exhaust, ductwork material options, and fume hood control options should be considered.

Individual Exhausts. The use of individual exhausts can be required for certain materials or if multiple materials that interact are to be used in the laboratory. It is also possible that building layout makes individual exhaust attractive, as in the case of a single-story building, or impractical, such as a tall multi-story laboratory, where individual exhaust stacks could occupy a significant portion of the floor area on upper floors. The cost-effectiveness of individual exhausts is determined by the number of sources to be exhausted (each will need a fan and a stack), the length of ductwork to the exhaust stack, and space available in the building for individual duct runs and exhaust stacks.

Manifolded Exhausts. The use of manifolded exhausts can be prohibited for certain materials or if the materials used in different areas of the laboratory will interact. The cost-effectiveness of manifolded exhaust is determined by the grouping of exhaust sources, the building layout, and possibilities for future changes in materials used in the laboratory (may need to convert to individual exhaust if different materials that interact are later used).

Material Options. As with the choice of individual or manifolded exhaust, some chemicals used in a laboratory require special ductwork materials and connection methods. When a choice is available for ductwork material and connection methods, the material costs and installation time and cost for various options should be considered. Also, the potential for changing the chemicals that are exhausted should be considered, as some ductwork materials limit the types of chemicals that can be handled by the exhaust system.

Fume Hood Controls. Fume hood controls are either constant volume or variable volume. While constant volume controls are simpler and less expensive for initial cost, constant volume control may not be economical over the life of the building. Variable volume controls may be more expensive for initial cost but provide additional features, such as monitoring the face velocity of the fume hood. For additional information on fume hood controls, see the "Equipment Control" section of Chapter 11, "Controls."

LIFE-CYCLE COST ANALYSIS

To determine the life-cycle cost, the costs, both capital and operating, associated with a building need to be converted to a common cost per time period using an appropriate interest rate. Typically, the time period is for a 30-year life expectancy. Numerous cost factors must be estimated or calculated during life-cycle cost analysis. Often, these cost factors are very project specific; changes in usage patterns of the laboratory or the building location can dramatically affect which cost factor is dominant in determining the most cost-effective system option.

Method of Analysis

In order to perform a life-cycle cost analysis, consultations between the design team and owner are needed to determine the appropriate cost factors and the common time increment to use for the life-cycle cost. Once the cost factors have been determined, they are used along with the interest rate to calculate the life-cycle cost.

Cost Factors

There are many cost factors that are needed to calculate the cost of a building over its lifetime. These factors can be divided into three groups: design factors, economic factors, and performance factors.

Design Factors. Design factors include items selected or determined during the design stage of a project that can greatly affect the materials and equipment used in the laboratory and thus the initial cost.

The heat gain from laboratory equipment may be the dominant factor in sizing equipment for some laboratories; for other laboratories, it is the rate of exhaust needed to provide separation between hazardous materials and laboratory occupants.

The climate where the laboratory is located can significantly affect the sizing of the heating and cooling systems for a building. Due to the need to use 100% outside air, wide temperature variations throughout the year have a significant impact on laboratory equipment sizing and heating and cooling costs. In order to compensate for wider variations in temperature, additional design features may need to be considered, such as using variable volume fume hoods, improving the building envelope, and using energy recovery.

The design face velocity for fume hoods (and other exhaust equipment) may be dictated by the chemicals used in the hood. In this case, a designer may not have a choice in selecting a face velocity. In other cases, a designer may have a range of values to choose from. While safety is the primary design objective, the cost to operate a laboratory must also be considered. Therefore, design face velocities should be chosen to be the minimum to ensure the safety of the occupants, as face velocity is directly proportional to the rate of airflow exhausted from the fume hood for a given sash opening, which is a cubic relationship to energy use.

Economic Factors. Several economic factors are used to calculate the life-cycle cost of a laboratory, including interest rate, initial system cost, cost of maintenance, and operating/energy costs.

The interest rate is used to determine the time value of money, such as determining the current value of electric utility costs over the life of the building. It should be chosen after consultation with the building owner, as organizations often have fixed requirements on interest rates to use for cost analysis.

The initial system cost includes the purchase cost for equipment and materials, the cost of installation, and overhead costs for the installing contractors. The whole building does not necessarily need to be included in the cost analysis. For example, if the building envelope is the same for all options under consideration, its initial cost may be excluded.

The cost of maintenance includes the costs associated with keeping the laboratory equipment functioning as originally constructed. This includes replacement parts for items that normally wear out, such as fan belt and filters, and for items that may wear out prematurely, such as a fan or pump motor. Laboratories have higher maintenance costs than typical buildings as there are many critical systems that require regular, preventive maintenance to ensure that they do not fail and create unsafe working conditions. Typical maintenance costs for commercial buildings should not be substituted for careful consultation between the building owner, O&M staff, and designers to develop a preventive maintenance program and identify the costs of the program.

Operating and energy costs are the costs for items such as electricity, natural gas, and water and sewer. The cost for these items will likely vary in the future, so estimation of an inflation rate may be needed. Reviewing the utility costs for a similar laboratory, if possible, may be useful in estimating energy use. Some building analysis programs used for determining heating and cooling loads can also be used to determine energy usage.

Performance Factors. Performance factors include diversity factor, service life, and average heat gains. In most laboratories, not all hoods will be used at the same time. If this is the case, a diversity factor, which represents the percentage of equipment that will be used at any one time, may be used to downsize central equipment, such as air handlers, boilers, and chillers. There are some exceptions, though, when diversity factors should not be used. This may include teaching laboratories, which may have students using all of the fume hoods at once during a class. Also, a diversity factor may not be practical for laboratories, which are often reconfigured when expansion is planned when a diversity factor is not used during the initial design, the additional equipment installed when the laboratory is reconfigured or expanded accounts for the diversity. The owner and users of the laboratory should be consulted when determining if using a diversity factor is appropriate.

The service life for the equipment must be considered when performing life-cycle cost analysis. Life-cycle cost analysis requires that a time period be chosen for the analysis. If the expected life of equipment being analyzed is less than the time period for the analysis, then the cost of replacing the equipment for as many times as needed to equal the time period of analysis should be included in life-cycle cost analysis.

Average heat gain is the average thermal load supplied to the space, not the connected electric load, as it is rare for all the equipment to be used at once. An average heat gain, however, may not accurately represent a large heat-producing piece of equipment in a laboratory without much other heat-producing equipment. A more accurate method of determining the heat gain in a laboratory is to determine the heat gain from each piece of equipment and use the expected usage, which can be obtained from the building owner, laboratory users, or equipment manufacturers. However, this may also present challenges, as heat gain values from equipment may not be available or the number of pieces of equipment may change.

As life-cycle cost considers the life span of the building, the amount of adaptability for the laboratory should be considered, as it is likely the type of work performed in the laboratory will change at some point during the life of the building. The building owner and occupants should be consulted to determine the level of adaptability needed to allow for future changes in laboratory uses.

Example. The following example is intended to provide guidance on the procedure to use in determining the life-cycle cost of a laboratory.

In the planning stage, the design team suggested two different approaches for the design of the mechanical system for the laboratory. One system emphasized an energy conservation design and the other was of a more traditional design. The owner wanted to know which system design option had the lower life-cycle cost. The lifetime for both mechanical systems was the same and inflation was not considered in this example.

- Given economic factors:
  - Interest rate = 8%
  - Service life = 20 years
- System 1
  - Mechanical system initial cost = $1,250,000
  - Energy cost = $65,000/year
  - Maintenance cost = $150,000/year
- System 2
  - Mechanical system initial cost = $1,100,000
  - Energy cost = $95,000/year
  - Maintenance cost = $145,000/year

The economic formula that can be used to calculate the life-cycle cost for this scenario is shown in Equation 15-1:

$$LC = IC + AC \cdot \frac{(1+i)^n - 1}{i \cdot (1+i)^n} \quad (15\text{-}1)$$

where

- LC = life-cycle cost
- IC = initial cost
- AC = annual cost (energy cost + maintenance cost)
- i = interest rate in absolute value (e.g., 8% = 0.08)
- n = number of years

System 1 life-cycle cost:

$$LC = 1,250,000 + (65,000 + 150,000) \cdot \frac{(1 + 0.08)^{20} - 1}{0.08 \cdot (1 + 0.08)^{20}}$$

$$= 3,411,000$$

System 2 life-cycle cost:

$$LC = 1,100,000 + (95,000 + 145,000) \cdot \frac{(1 + 0.08)^{20} - 1}{0.08 \cdot (1 + 0.08)^{20}}$$

$$= 3,456,000$$

The energy-efficient System 1 had the lower life-cycle cost with these parameters.

What would the life-cycle costs of the systems be if the interest rate was 12%?

System 1 life-cycle cost:

$$LC = 1,300,000 + (65,000 + 150,000) \cdot \frac{(1 + 0.12)^{20} - 1}{0.12 \cdot (1 + 0.12)^{20}}$$

$$= 2,905,000$$

System 2 life-cycle cost:

$$LC = 1,100,000 + (95,000 + 145,000) \cdot \frac{(1 + 0.12)^{20} - 1}{0.12 \cdot (1 + 0.12)^{20}}$$

$$= 2,892,000$$

The traditional system 2 has the lower life-cycle cost with the higher interest rate.

Chapter 16
Microbiological and Biomedical Laboratories

OVERVIEW

This chapter contains information on the classifications of biological contaminants and discusses laboratory animals and their requirements to provide a basis for the design of these types of laboratories. As it is very important to contain the biological material within the biological containment laboratory (BCL) and laboratory animal areas, issues such as space pressurization, system reliability, proper controls, and redundancy, discussed and illustrated in earlier chapters, apply here. Specific details about biological safety cabinets used for biological containment are covered in Chapter 5, "Exhaust Hoods."

BIOLOGICAL CONTAINMENT

In this section, the classification of biosafety levels along with key guidelines for biological containment in microbiological and biomedical laboratories are discussed. Adhering to these guidelines will reduce the health and safety risks posed by these kinds of laboratory facilities.

Guidelines

Prudent guidelines for managing microorganisms are provided by the Centers for Disease Control (CDC) and National Institutes of Health (NIH) in the publication, *Biosafety in Microbiological and Biomedical Laboratories*. Its objective is to offer information that will reduce the spread of diseases from laboratories handling microorganisms by recommending practices for the design and use of these laboratories. The guideline describes four different biosafety levels summarized in the following section, which range from microorganisms that are not known to cause diseases in humans to microorganisms that can cause life-threatening diseases. The four biosafety levels are parallel to NIH *Guidelines for Research Involving Recombinant DNA* and are consistent with the general criteria originally used in assigning agents to Classes I to IV in *Classification of Etiologic Agents on the Basis of Hazards*.

Classifications

The following are classifications of the four biosafety levels as defined by CDC/NIH (1999):

Biosafety Level 1 (BL-1). BL-1 practices, safety equipment, and facilities are appropriate when working with microorganisms that are not known to cause disease in healthy humans. Laboratories for undergraduate, secondary educational training, and teaching are facilities where BL-1 can be a sufficient measure of caution. BL-1 does not require any primary or secondary barrier except a sink for hand washing. Very few biomedical research laboratories operate at the BL-1 level.

Biosafety Level 2 (BL-2). BL-2 practices, safety equipment, and facilities are appropriate when working with indigenous moderate risk agents that are known to cause human diseases with varying severity. Good laboratory practices allow the use of these agents on an open bench provided the potential for splashing and aerosols are minimal. Laboratories for clinical, diagnostic, and teaching are facilities where BL-2 can be a sufficient measure of caution. The primary hazard is accidents where the agent gets in contact with the percutaneous (skin) or mucous membranes, or is ingested. BL-2 requires extreme precaution when working with contaminated needles or sharp objects. Procedures that can cause splashing or aerosols must be conducted in primary containment enclosures such as certified biological safety cabinets or safety centrifuge cups. Gloves, face masks, splash shields, and gowns should be used when appropriate. A sink for hand washing and facilities for decontamination must be available.

Biosafety Level 3 (BL-3). BL-3 practices, safety equipment, and facilities are appropriate when working with indigenous or exotic agents with a potential for respiratory transmission, which may cause serious and potentially lethal infection. Laboratories for clinical, diagnostic, research, production, and teaching are facilities where BL-3 can be a sufficient measure of caution. The primary hazards relate to autoinoculation, ingestion, and exposure to infectious aerosols. BL-3 places emphasis on primary enclosures, and all laboratory manipulation should be done in certified biological safety cabinets or similar enclosures. The ventilation system should be made to minimize the release of aerosols and the laboratories should have limited access.

Strict attention should be given to the secondary barrier construction. These labs operate under a defined negative pressure, which must be monitored and controlled precisely. Their electrical and plumbing penetrations must be sealed and they should have anteroom entrance and egress. Some facilities will require shower out and there must be areas reserved for personal protective clothing. Sinks, lights, and phones should be automatic or handless and floors seamless to minimize contamination. All windows, if any, should be permanently sealed and doors should be installed with automatic closure and magnetic interlocks. Windows should be installed in all doors to allow entering personnel to view the lab for normal and emergency conditions. Utilities should be easily accessible from outside the containment rooms. These labs generally have dedicated exhaust and supply systems. If building supply air is used, a controlled shutoff supply damper should be installed.

Biosafety Level 4 (BL-4). BL-4 practices, safety equipment, and facilities are appropriate when working with very dangerous and exotic agents that pose a high individual risk of life-threatening disease, may be transmitted via the aerosol route, or for which there are no available vaccines or treatment. Agents with a close or similar antigenic relationship to BL-4 agents should be handled at this level. The primary hazards relate to autoinoculation, mucous membrane exposure to infectious droplets, and exposure to infectious aerosols. All manipulation of BL-4 agents poses a danger to the personnel, the community, and the environment. All work with BL-4 agents must be in certified Class III biological safety cabinets, Class II, Type B2 biological safety cabinets, or with full body positive pressurized suits. BL-4 laboratories should, in general, be in separate buildings or completely isolated zones with precisely operated ventilation and waste management systems. The laboratory director is responsible for the safe operation of the laboratory. All BL-3 requirements apply to BL-4.

ANIMAL OVERVIEW

There are many types of animals commonly used in laboratories for experiments. Each species has individual requirements detailed in guidelines, standards, and codes. These requirements, including temperature, humidity, sound, and light, are required for healthy and humane housing of animals. This section provides an overview of animal requirements that are the basis for the design of such areas.

Standard Types of Animals

Standard types of laboratory animals are:

- Mouse
- Rat
- Rabbit
- Dog
- Cat
- Small farm animal
- Large farm animal

The use of these species in microbiological and biomedical laboratories and special considerations for each of these categories of species will be discussed.

Environmental Requirements

Each species of animal has a unique range of preferred environmental conditions that must be present to survive and reproduce. Even with a specific species, there is variability among subclasses. For example, a hairless mouse requires a warmer temperature than one with hair for optimal conditions. Some animals, such as squirrels and cottontail rabbits, can survive in captivity but will not reproduce. In general, an animal does better when the environmental conditions are close to its natural habitat.

In addition to the environmental conditions that are controlled by HVAC and electrical systems (temperature, humidity, acoustics, and light) there are other conditions such as space requirements, diets, and social and sexual relations that can be related to the environmental conditions and must be accounted for. It is apparent that maintaining the proper conditions is expensive and sometimes impractical resulting in less than ideal conditions.

The HVAC system is responsible for maintaining the temperature and humidity in the animal cage. The specific temperature and humidity to maintain is dependent upon the animal's metabolism and its insulation level. Changes in temperature and humidity affect the amount of food consumed and the weight of the animal.

Table 16-1 (NRC 1996) lists appropriate environmental conditions for typical laboratory animals.

Noise and acoustics can have a direct impact on the animal's response to an experiment. This is due to the fact that some animals rely on sound more than others for courtship, mating behavior, and predator detection while others are more visually dependent. Feeding and cleaning operations, heating and ventilation equipment, and the animals themselves are sources of noise in an animal laboratory. Typical noise levels range from 40 to 100 dBA, with levels below 44 dBA considered acceptable for normal conditions and 55 to 68 dBA during feeding. Extremely high noise levels can cause temporary and permanent hearing loss and stress reactions for both the animals and their handlers.

Lighting levels and cycling preferred by an animal depend on their visual capability and circadian rhythm (circadian describes a periodicity that is close to one day). The level of lighting should be kept similar to the animal's natural environment. Inappropriate light levels can result in damage to their vision and cause undo stress on the animals. Cycling the lighting on and off influences when the animal is active. Many animals are active and reproduce at night. In some instances, lights are never used to either match natural conditions or to keep animals active longer.

Minimum space requirements are generally dependent on the size of the animal. Table 16-2 and Table 16-3 list appropriate space requirements for typical laboratory animals.

Table 16-1:
Recommended Temperature, Relative Humidity, and Ventilation for Common Laboratory Animals

Species	Temperature °F (°C)		Relative Humidity (%)		Minimum Room Air Exchange Rate (ACH)	
	ILAR*	CCAC**	ILAR	CCAC	ILAR	CCAC
Mouse	64.4 – 78.8 (18 – 26)	71.6 – 77.0 (22 – 25)	40 – 70	50 – 70	15	8 – 12
Rat	64.4 – 78.8 (18 – 26)	68.0 – 77.0 (20 – 25)	40 – 70	50 – 55	15	10 – 20
Hamster	64.4 – 78.8 (18 – 26)	69.8 – 75.2 (21 – 24)	40 – 70	45 – 65	15	6 – 10
Guinea pig	64.4 – 78.8 (18 – 26)	64.4 – 71.6 (18 – 22)	40 – 70	50 – 60	15	4 – 8
Rabbit	60.8 – 69.8 (16 – 21)	60.8 – 71.6 (16 – 22)	40 – 60	40 – 50	10	10 – 20
Cat	64.4 – 84.2 (18 – 29)	68.0 – 71.6 (20 – 22)	30 – 70	45 – 60	10	10 – 18
Dog	64.4 – 84.2 (18 – 29)	64.4 – 69.8 (18 – 21)	30 – 70	45 – 55	10	8 – 12
Nonhuman primate	64.4 – 84.2 (18 – 29)	69.8 – 78.8 (21 – 26)	30 – 70	45 – 60	10 – 15	12 – 16
Chicken	60.8 – 80.6 (16 – 27)	64.4 – 71.6 (18 – 22)	45 – 70	45 – 70	10 – 15	5 – 15

* Institute of Laboratory Animal Resources
** Canadian Council on Animal Care

Table 16-2:
Recommended Space for Commonly Used Group-Housed Laboratory Rodents

Animals	Weight, g (lb)	Floor Area/Animal, in.^{2} (cm^2)	Height,* in. (cm)
Mice	Less than 10 (0.022)	6 (39)	5 (13)
	Up to 15 (0.033)	8 (52)	5 (13)
	Up to 25 (0.055)	12 (77)	5 (13)
	Greater than 25** (0.055)	Greater than 15 (97)	5 (13)
Rats	Less than 100 (0.22)	17 (110)	7 (18)
	Up to 200 (0.44)	23 (148)	7 (18)
	Up to 300 (0.66)	29 (187)	7 (18)
	Up to 400 (0.88)	40 (258)	7 (18)
	Up to 500 (1.1)	60 (387)	7 (18)
	Greater than 500** (1.1)	Greater than 70 (452)	7 (18)
Hamsters	Less than 60 (0.13)	10 (65)	6 (15.24)
	Up to 80 (0.18)	13 (84)	6 (15.24)
	Up to 100 (0.22)	16 (103)	6 (15.24)
	Greater than 100** (0.22)	Greater than 19 (123)	6 (15.24)
Guinea pigs	Less than 350 (0.77)	60 (387)	7 (18)
	Greater than 350** (0.77)	Greater than 101 (651)	7 (18)

* From cage floor to cage top.
** Larger animals might require more space to meet the performance standards.

Table 16-3:
Recommended Space for Rabbits, Cats, Dogs, Nonhuman Primates, and Birds

Animals	Weight, kg (lb)	Floor Area / Animal ft^2 (m^2)	Height,* in. (cm)
Rabbits	Less than 2 (4.4)	1.5 (0.14)	14 (36)
	Up to 4 (8.8)	3.0 (0.27)	14 (36)
	Up to 5.4 (11.9)	4.0 (0.36)	14 (36)
	Greater than 5.4** (11.9)	Greater than 5.0 (0.45)	14 (36)
Cats	Less than 4 (8.8)	3.0 (0.27)	24 (61)
	Greater than 4** (8.8)	Greater than 4.0 (0.36)	24 (61)
Dogs***	Less than 15 (33)	8.0 (0.72)	—
	Up to 30 (66)	12.0 (1.08)	—
	Greater than 30** (66)	Greater than 24 (2.16)	—
Monkeys[1,2] (baboons included)			
Group 1	Up to 1 (2.2)	1.6 (0.14)	20 (51)
Group 2	Up to 3 (6.6)	3.0 (0.27)	30 (76)
Group 3	Up to 10 (22)	4.3 (0.39)	30 (76)
Group 4	Up to 15 (33)	6.0 (0.54)	32 (81)
Group 5	Up to 25 (55)	8.0 (0.72)	36 (91)
Group 6	Up to 30 (66)	10.0 (0.90)	46 (117)
Group 7	Greater than 30** (66)	15.0 (1.35)	46 (117)
Apes (Pongidae)[2]			
Group 1	Up to 20 (44)	10.0 (0.90)	55 (140)
Group 2	Up to 35 (77)	15.0 (1.35)	60 (152)
Group 3	Greater than 35[3] (77)	25.0 (2.25)	84 (213)
Pigeons[4]	—	0.8 (0.072)	—
Quail[4]	—	0.25 (0.023)	—
Chickens[4]	Less than 0.25 (0.55)	0.25 (0.023)	—
	Up to 0.5 (1.1)	0.50 (0.045)	—
	Up to 1.5 (3.3)	1.00 (0.09)	—
	Up to 3.0 (6.6)	2.00 (0.18)	—
	Greater than 3.0** (6.6)	Greater than 3.00 (0.27)	—

* From cage floor to cage top.
** Larger animals might require more space to meet the performance standards.
*** These recommendations might require modification according to body conformation of individual animals and breeds. Some dogs, especially those toward the upper limit of each weight range, might require additional space to ensure compliance with the regulations of the Animal Welfare Act. These regulations (CFR 1985) mandate that the height of each cage be sufficient to allow occupant to stand in "comfortable position" and that the minimal square feet of floor space be equal to "mathematical square of the sum of the length of the dog in inches (measured from the tip of its nose to the base of its tail) plus 6 inches; then divide the product by 144."

[1] Callitrichidae, Cebidae, Cercopithecidae, and Papio. Baboons might require more height than other monkeys.

[2] For some species (e.g., Brachyteles, Hylobates, Symphalangus, Pongo, and Pan), cage height should be such that an animal can, when fully extended, swing from the cage ceiling without having its feet touch the floor. Cage-ceiling design should enhance brachiating movement.

[3] Apes weighing over 50 kg are more effectively housed in permanent housing of masonry, concrete, and wire-panel structure than in conventional caging.

[4] Cage height should be sufficient for the animals to stand erect with their feet on the floor.

Caging Systems

Typical cage and housing systems define the boundaries of an animal's immediate environment. These include:

- Cage racks
- Pens
- Indoor and outdoor runs
- Built-in units
- Ventilated racks
- Cubicles

The cage material has a significant influence on the animal's heat loss (thereby the temperature requirements of the animal), the noise reduction, and the light transmission. Open cages are more susceptible to cross-contamination through the transfer of particles from one cage to another cage. In general, it is acceptable practice, according to NRC (1996), for caging systems to allow for the following:

- Normal physiologic and behavioral needs of the animals, including urination and defecation, maintenance of body temperature, normal movement and postural adjustments, and reproduction, where indicated.
- Animals to remain clean and dry (as consistent with the requirements of the species).
- Adequate ventilation.
- Access to food and water with easy filling, refilling, changing, servicing, and cleaning of food and water utensils.
- A secure environment that does not allow escape of or accidental entrapment of animals or their appendages between opposing surfaces or by structural openings.
- Absence of sharp edges or projections that could cause injury to the animals.
- Observation of the animals with minimal disturbance to them.

Guidelines

Prudent guidelines for the care and use of laboratory animals, which will prove invaluable to designing laboratory HVAC systems, are provided by the Institute of Laboratory Animal Resources, National Research Council. The guidelines are made available in the publication *Guide for the Care and Use of Laboratory Animals*, which has the distinct purpose to "assist institutions in caring for and using animals in ways judged to be scientifically, technically, and humanely appropriate" (NRC 1996). The guide is divided into four chapters that cover institutional policies and responsibilities; animal environment, housing, and management; veterinary medical care; and physical plant. It also includes additional information in its appendices on organizations, federal laws, and public health service policies relevant to the care and use of animals.

DESIGN OF LABORATORY ANIMAL AREAS

Laboratory animal areas within facilities have special requirements to maintain the comfort and health of the laboratory personnel and animals and to avoid contamination between and within individual rooms. The intent of this section is to provide an overview of the conditions and requirements that must be considered when designing laboratory animal areas.

Ventilation Rates

The primary function of the ventilation system is to provide clean air, remove thermal loads, adjust the humidity, and remove odors, smoke, and particles. The ventilation system is also used to control cross-contamination by pressurizing all cages.

While considerably less is known about the ventilation rate requirement for animals compared to humans, the generally accepted ventilation rate is between 10 and 15 supply air changes per hour for passive rooms with open cages (NRC 1996). In general, the ventilation rate will depend on population densities and refuse removal schedules. Typically 100% outdoor air is used for animal laboratories. However, in some instances recirculated air can be used, as long as the following criteria are followed:

- Minimum 50% outdoor air
- Sufficient cleaning practices to keep the toxic and odor level down
- The return air must come from the same room
- The thermal and humidity requirements must be met
- HEPA filtration (filters must be certified annually for airflow and leak test)

Filtered cage systems, to ensure adequate ventilation, must use a forced ventilation system to provide sufficient air to keep the contamination and temperature below the required levels. The diffusers and exhaust openings used with these systems should be located and constructed so that they prevent creating drafts but provide for a uniform low-velocity ventilation throughout the occupied spaces. In addition they should be easy to clean. Care should be taken to adequately screen the room and outside openings of the ventilation system to prevent the escape of animals as well as the entry of vermin. Screens should be easily removed and cleaned periodically to avoid blocking the airflow.

It is important to understand that increasing the air circulation will not guarantee that the animal cages receive more air. However, animal cages that have only the top open, will usually have an increased amount of air circulation with increased amount of air supplied to the room.

Air Distribution

The primary function of the air distribution device is to distribute the supply air to all areas of the room without causing draft or temperature discomfort. For rooms with open cages, these devices will not distribute the air equally to the cages due to the arrangement of the cages relative to each other. The intent of the air distribution device with the help of the exhaust is to move as much air as possible from the cleaner areas to the more contaminated areas within a room. In rooms with dedicated supply and exhaust to the cage racks, the intent of the air distribution is to maintain the comfort of the laboratory personnel.

Types of diffusers available include linear diffusers, ceiling (round/rectangular) diffusers, sidewall grilles, floor diffusers, and perforated ceilings. Linear and ceiling diffusers are mounted in the ceiling and perform similarly with the exception of a longer throw for the slot diffuser. The sidewall grilles are usually located under a window or close to the ceiling. The object of these three diffuser types is to mix the incoming air with room air to maintain a comfortable temperature and air speed within the occupied space.

Perforated ceilings do not try to mix the air, but rather they try to "push" the contaminated air out of the room with supply air. This usually requires a larger volume of low-velocity air to ensure that the supply air does not mix with the contaminated air. Objects, animals, and humans cause turbulence that mixes the contaminated air with the clean air or creates stagnant areas.

Unlike perforated ceilings, floor diffusers supply a plug of airflow from the bottom to the top and utilize the stratification principle. Stratification occurs when cool supply air is supplied at the floor and rises as it warms. For this to work, it is assumed that the contaminated air is also heated and rises to the ceiling, where it is exhausted. Turbulence from animals and occupants can cause problems for this distribution strategy. Chapter 6, "Primary Air Systems" and Chapter 12, "Airflow Patterns and Air Balance" provide more details on air distribution.

Separation of Areas

Separation of human areas and animal areas is required to ensure human comfort and health protection. Physical separation in addition to separation of ventilation and sanitation supply is required to avoid contamination of the human areas. In addition, animals themselves might have to be separated from other animals.

Physical. Physical separation can be achieved by housing the animal in a separate building, wing, floor, or room. Good design should ensure that animal housing areas are close to the laboratories but separated by a barrier.

Space Pressurization and Airflow. Space pressurization is a method to reduce infiltration and airflow from dirty and contaminated areas to cleaner areas. This is usually accomplished by exhausting a different volume of air than is supplied. If more air is exhausted than supplied, the space is negatively pressurized and if less air is exhausted than supplied, the space is positively pressurized. Another control method is the use of a variable air volume system with a pressure sensor to maintain a constant pressure differential in the room/area relative to surroundings.

A similar approach is usually applied within a room where the design of the air distribution system is made to have the air flowing from cleaner areas to dirtier/contaminated areas or from sensitive animals to less sensitive, or infected animals. However, movement of personnel and opening of doors disturbs the flow pattern and this strategy should not be used where it is critical that airborne contaminants are not spread to sensitive areas within the room.

Special Animal Conditions. The spread of pathogens must be controlled to avoid infecting disease-free animals and humans (if the disease can spread to humans). Pathogens can spread by different vectors, such as airborne, direct contact, indirect contact, through intermediate hosts, placental transmission, or a combination of these methods. The sensitivity of the pathogen to temperature, pressure, pH, and chemicals must be determined to be able to inactivate the pathogen in rooms and on equipment that have been contaminated. The pathogen is less sensitive to inactivating methods as its associated risk of spread increases. This can result in an increased severity of the disease and thereby the stringency of precautions. Specially stringent precautions must be taken if the pathogen can cause health risks for humans.

Some of the precautions to be taken include:

- Separation of infected animals
- Cremation of dead animals
- Delivery of food and water with reduced/eliminated contact with the infected animals
- Disinfection of water and urine
- Burning of feces
- Negatively pressurized contaminated areas
- HEPA filtration of air leaving the contaminated areas
- Showering and disinfecting the personnel that have been in contact with the contaminated animals
- Separate clothing for working inside and outside of the contaminated areas
- Metal bins to carry infected equipment to and from the contaminated areas

- Ensuring that the intermediate host (flies, mice, birds) can not enter or leave the contaminated areas
- Disinfection of the areas and equipment (including filters) contaminated by the animals after the animals have been removed from the containment areas

The actions taken depends on risk factors and considerations, which include:

- Disease infectious to humans
- Disease infectious to other animals
- Specific pathogen free (SPF)
- Immuno-compromised animals

Sanitation

Sanitation and the cleanability of the animal areas are the most significant factors that determine the local environmental conditions for the animals. Poor sanitary conditions can cause toxic buildup and unacceptable odor levels that result in both animal and human health problems. The ventilation system cannot compensate for poor sanitation procedures beyond its design.

Spaces. Design issues that are important for the cleanability of the laboratory spaces include:

- Moisture resistant, nonabsorbent, impact resistant, chemically resistant, and relatively smooth (unless a textured surface is required) floors.
- Adequate drainage pipes (if required).
- Washable paint on ceilings and walls.
- Walls that can be cleaned with chemicals and water under high pressure.
- Walls, ceiling, floor, and junctions between these; doors and windows should be free of cracks and crevices.

HVAC Components. Design issues that are important for the cleanability of the laboratory HVAC components include:

- Piping, wires, and ductwork should not be exposed unless easily cleanable.
- The exhaust duct of the HVAC system is likely to get excessively dirty or get contaminated by pathogens. Therefore, the exhaust and supply duct (if air is reused) will require periodic internal cleaning. The design of the duct must be made to accommodate cleaning by installing access panels to allow easy access for reaching every part of the internal ductwork. The surface of the ductwork that is to be cleaned must be cleanable.

Equipment and Materials. Sanitation procedures for materials exiting the containment are essential to safe operation of laboratories. Each biosafety level has its own requirements for sanitation of materials exiting and used in the laboratory or containment. Biosafety level 1 (BSL-1) requires decontamination of cultures, stocks, and other regulated wastes before disposal and that the work surface of the laboratory be decontaminated at least once a day and after each spill. BSL-2 requires in addition that all equipment and surfaces be decontaminated with an approved disinfectant on a routine basis. BSL-3 requires in addition to BSL-2 that all potentially contaminated waste material, such as gloves, lab coats, and spills, be decontaminated before disposal or reuse. BSL-4 requires that all material that leaves the laboratory be decontaminated and that equipment be routinely decontaminated as well as decontaminated after contact with infectious material.

These actions are usually required at some point when operating an animal laboratory area:

- *Sterilization* – Autoclaving (procedure where highly pressurized steam-heated vessel is used) is the procedure of choice for sterilization (WHO 1983). Alternative methods are boiling for 30 minutes in water containing sodium bicarbonate or the use of a pressure cooker at the highest possible pressure. Materials that cannot be autoclaved can be decontaminated within an airlock by gaseous or vapor methods.
- *Incineration* – Waste can be burned instead of being sterilized. However, the waste must be contained in a leakproof container if incineration takes place outside the containment zone.
- *Waste Handling* – Waste should be separated into containers for each category of waste. Examples of these categories are noncontaminated waste, sharps, contaminated material for autoclaving and reuse, and contaminated material for disposal.

System Redundancy

System redundancy measures usually depend on how critical the continuous operation of the HVAC system is for the animal laboratory to function safely. They also depend on how extensive the economic losses will be if the HVAC system or one of its components fails or is shut down due to maintenance. Although the implementation of these measures, which includes backup components and emergency power, will decrease the chance of system failure, the chance of system failure cannot be completely eliminated. The actual cost of installing redundant systems is also an important factor and must be considered.

Exhaust fan failure will cause the most problems since the designed pressurization of a given space will be jeopardized. This will result in contaminants being able to flow to cleaner adjacent areas, as well as allow contamination within a space to build up beyond safe and permissible thresholds. Failure of cooling equipment and systems can be critical if it is for an extended period of time. This can result in unacceptably high temperatures and humidity.

In addition to installing equipment that can handle abnormal operating conditions, control systems must be able to handle all conditions that are expected to occur. The system must be tested in all modes of operation to ensure that the desired capabilities function properly. See Chapter 11, "Controls" for details on laboratory control systems.

Emergency Power. Emergency power equipment will prevent the laboratory system from shutting down when the electricity supply fails. The emergency power system can be designed to supply enough energy to keep all systems running or to only supply enough power to keep the vital systems and components running. The highest priority should be given to the continued operation of the ventilation system.

Multiple or Redundant Fan Systems. Failure or shutdown of the exhaust fan in dirty and contaminated areas will usually cause the most problems or health hazards. Without the exhaust fan, the negative pressure in the dirty and contaminated areas will disappear. This can cause odors and pathogens to infiltrate the cleaner areas and thereby create unpleasant or unhealthy environments. Infiltration from the contaminated areas will begin immediately after the exhaust system fails or is shut down. Installation of backup fans and motors in addition to an emergency power generator will ensure operation of the exhaust fan in most circumstances.

Operational Issues

Proper maintenance and operation of the laboratory systems will ensure a long lifetime, reduce or eliminate the number of emergency situations, and ensure that the systems work according to the design intent. Topics of special importance to systems used in animal laboratory areas include:

- Fan system failure
- Component services
- Caulking and sealing

Fan System Failure. A fan system failure can cause a health hazard for the occupants. Therefore, there is a need for monitoring the critical fans and there should be a plan for how to handle the situation when the emergency occurs.

Component Service. Filters that likely will get contaminated should be installed in such a way that the filters can be removed without exposure to laboratory or maintenance personnel. All other components that are likely to be contaminated should be installed for easy access and safe handling.

Caulking and Sealing. Ductwork and HVAC components that will be contaminated should be properly sealed to avoid leakage and contamination of areas through which the ductwork passes or contamination of the supply air. The contaminated parts of the HVAC system should be kept under negative pressure while the clean parts should be kept under positive pressure. After cleaning and maintenance, it is important to make sure the access panels and other openings or connections are properly sealed. All electrical conduits and plumbing penetrations should be sealed to prevent room-to-room cross-over contamination.

Support Areas

Some of the rooms required for the care and housing of animals have unique demands. These include:

- Cage wash areas (hoods, exhaust, and drains)
- Food and bedding storage
- Pathology-necropsy areas (general and table exhaust)

Cage Wash Areas. Cage wash areas must be designed for easy cleaning and sanitation. The air supply and exhaust must be located to prevent water intrusion, which can cause growth of microorganisms, and be possible to clean. The floor must be tilted toward a drain and have no local low points that cause water accumulation.

Food and Bedding Storage. Food must be stored in such a way that it will last until it is used. A dry location will be sufficient for many types of foods (e.g., hay), while subfreezing temperatures are required for lengthy storage of other types of food (e.g., meat). Bedding requires a dry area with no access to animals and should be stored off the floor on carts, racks, or pallets so that contamination is minimized and quality can be maintained. Both food and bedding storage areas must have a sufficient capacity for convenient management of the supply.

Pathology-Necropsy Areas. Pathology-necropsy areas are used to perform gross postmortem evaluation and testing on animal carcasses to examine fresh and fixed tissues. These areas must be designed for easy cleaning and sanitation. Washing facilities should provide proper drainage and there should be adequate storage space for disinfectant, cleaning supplies, and waste disposal equipment.

Routine postmortem examinations should be performed on down-draft tables, but necropsies on animals that hosts agents classified as biological safety level II or higher, which contain certain radionuclides, should be performed in an appropriate biological safety cabinet or chemical fume hood. Automated tissue processor units that use toxic or volatile chemicals should be provided localized exhaust to expel the fumes or be enclosed in ventilated cabinets.

REFERENCES

Centers for Disease Control and Prevention (CDC) and National Institutes of Health (NIH). 1999. *Biosafety in Microbiological and Biomedical Laboratories*, 4th ed. U.S. Bethesda, Md: Department of Health and Human Services.

Centers for Disease Control and Prevention (CDC), Office of Biosafety. 1974. *Classification of Etiologic Agents on the Basis of Hazard*, 4th ed. Bethesda, Md: U.S. Department of Health and Human Services.

National Research Council (NRC). 1996. *Guide for the Care and Use of Laboratory Animals*. Washington, D.C.: Institute of Laboratory Animal Resources, Commission on Life Science.

World Health Organization. *Laboratory Biosafety Manual*. 1983. Washington, D.C.: WHO Publications Center.

BIBLIOGRAPHY

Federal Register. 1976. *Recombinant DNA Research Guidelines*. 41:27902-27943.

Federal Register. 1986. *Guidelines for Research Involving Recombinant DNA molecules*. 51:16958-16968.

Hare, Ronald, and P.N. O'Donoghue. 1968. Laboratory animal symposia. 1. *The design and function of laboratory animal houses*. Geerings of Ashford, Ltd. Ashford, England.

Maghirang, R.G., G.L. Riskowski, and L.L. Christianson. 1996. Ventilation and Environmental Quality in Laboratory Animal Facilities. *ASHRAE Transactions* 102(2): 186-194.

Office of Research Safety, National Cancer Institute, and the Special Committee of Safety and Health experts. 1978. *Laboratory Safety Monograph*. A supplement to the NIH Guidelines for Recombinant DNA Research. Bethesda, Md.: National Institutes of Health.

Appendix

**Table A-1:
Laboratory Equipment**

Equipment	Description	Nameplate (W)***	Operational Mode-Power Consumption (W)				Fan	Usage
			Peak	Average	Radiant	Convective		
Analytical Balance	Mettle Toledo, PR 8002	7	7	7			N	Intermittent
Centrifuge	IEC HN-SII Centrifuge	288	136	132 41*	14* (34%)	27* (66%)	N	Intermittent
Centrifuge	International Clinical Centrifuge	138	89	87			N	Intermittent
Centrifuge	IEC B-20A Centrifuge	5500	1176	730			N	Intermittent
Electrochemical Analyzer	Voltammetric Analyzer CV-50W	50	45	44			N	Intermittent
Electrochemical Analyzer	Voltammetric Analyzer BAS 100B	100	85	84			N	Intermittent
Fluorescent Microscope	Leica MZ 12	150	144	143			N	Intermittent
Fluorescent Microscope	Axioplan2	200	205	178			N	Intermittent
Function Generator	Interstate High Voltage Function Generator-F41	58	29	29			N	Intermittent
Incubator**				83	47 (57%)	36 (43%)	N	Intermittent
Incubator	Lab-line Orbit Environ-Shaker No.3527	600	479	264 220*	68* (31%)	152* (69%)	N	Intermittent
Incubator	Controlled Environment Incubator Shaker G-28	3125	1335	1222			N	Intermittent
Incubator	National Incubator 3321	515	461	451			N	Intermittent
Orbital Shaker	VWR Scientific Orbital Shaker	100	16	16			N	Intermittent

Table A-1: Laboratory Equipment (Continued)

Equipment	Description	Nameplate (W)***	Operational Mode-Power Consumption (W)				Fan	Usage
			Peak	Average	Radiant	Convective		
Oscilloscope	BK Precision 20MHZ2120	72	38	38			N	Intermittent
Oscilloscope	Nicolet Instrument Corp 201	345	99	97	10 (10%)	87 (90%)	Y	Intermittent
Rotary Evaporator	BUCHI Rotary	75	74	73			N	Intermittent
Rotary Evaporator	BUCHI-RE121 Rotavapor Spectronics	94	29	28			N	Intermittent
Spectronics	Spectronics20	36	31	31	15 (49%)	16 (51%)	N	Intermittent
Spectrophotometer	Hitachi U-2000	575	106	104			N	Intermittent
Spectrophotometer	Double Beam Spectrophoto-meter Hitachi Perkin-Elmer	200	122	121			N	Intermittent
Spectrophotometer	Infrared Perkin-Elmer 1310	N/A	127	125			N	Intermittent
Flame Photometer	Atomic Absorption Spectometer 3110	180	107	105			N	Intermittent
Spectro Fluorometer	Spectro Fluorometer Model 430	340	405	395	64 (16%)	331 (84%)	N	Intermittent
Tissue Culture	Fisher Scientific CO_2 Incubator	475	132	46			N	Intermittent
Tissue Culture	SteriGRAD HOOD VBM-600	2346 Peak & Ave. in Idel mode at 53 W	1178	1146			Y	Intermittent
Thermocycler	Neslab RTE-221	1840	965	641[a] 479[b]	24 (5%)	455 (95%)	Y	Intermittent
Thermocycler	DNA Thermo-cycler 480	N/A	233	198			Y	Intermittent

Notes:
* 400 rpm, i.e., one-fourth of the maximum rotation speed for centrifuge (1600 rpm), 200 rpm for shaker (i.e., one-fifth of maximum speed), and one-half maximum temperature.
** Equipment tested in phase 1 of ASHRAE Research Project 1055-RP.
*** If the nameplate rating is given in terms of voltage and amperage, the wattage is listed as the product, i.e., power factor is approximately unity.
[a] Measured at 40°C
[b] Measured at 30°C

REFERENCES

Hosni, M.H., B.W. Jones, and H. Xu. 1999. Measurement of Heat Gain and Radiant/Convective Split from Equipment in Buildings, Final Report—ASHRAE 1055-RP. Atlanta: American Society of Heating, Refrigerating and Air-Conditioning Engineers, Inc.

References

American Conference of Governmental Industrial Hygienist (ACGIH). 1999. *Documentation of the Threshold Limit Values and Biological Exposure Indices*, 6th ed. Cincinnati, Ohio: ACGIH.

American Conference of Governmental Industrial Hygienist (ACGIH). 1998. *Threshold Limit Values for Chemical Substances and Physical Agents and Biological Exposure Indices*. Cincinnati, Ohio: ACGIH.

American Conference of Governmental Industrial Hygienists (ACGIH). 1998. *Industrial Ventilation—A Manual for Recommended Practices*, 23rd ed., Cincinnati, Ohio: ACGIH.

Alereza, T., and J. Breen. 1984. Estimates of Recommended Heat Gains Due to Commercial Appliances and Equipment. *ASHRAE Transactions* 90(2A): 25-58.

American Chemical Society (ACS). 1991. *Design of Safe Chemical Laboratories: Suggested References*, 2nd ed. Committee on Chemical Safety. Washington, D.C.

American Institute of Chemical Engineers (AIChE). 1985. *Guidelines for Hazard Evaluation Procedures*. New York.

American Industrial Hygiene Association (AIHA). 1993. *Standard Z9.5-93, Laboratory Ventilation*. Fairfax, Va.

American National Standards Institute (ANSI). 1982. *Standard Z88.2-1980*. New York: American National Standard Practices for Respiratory Protection.

American National Standards Institute (ANSI). 1992. *Standard Z9.5, Laboratory Ventilation*. New York: American National Standard Practices for Respiratory Protection.

American Society of Heating, Refrigerating and Air-Conditioning Engineers (ASHRAE). 1999. *1999 ASHRAE Handbook—HVAC Applications*. Atlanta: ASHRAE.

American Society of Heating, Refrigerating and Air-Conditioning Engineers (ASHRAE). 1999. *1999 ASHRAE Handbook—HVAC Applications*, Chapter 13, Laboratories. Atlanta: ASHRAE.

American Society of Heating, Refrigerating and Air-Conditioning Engineers (ASHRAE). 1999. *1999 ASHRAE Handbook—HVAC Applications*, Chapter 47, Water Treatment. Atlanta: ASHRAE.

American Society of Heating, Refrigerating and Air-Conditioning Engineers (ASHRAE). 1999. *1999 ASHRAE Handbook—HVAC Applications*, Chapter 50, Evaporative Cooling Applications. Atlanta: ASHRAE.

American Society of Heating, Refrigerating and Air-Conditioning Engineers (ASHRAE). 1997. *1997 ASHRAE Handbook—Fundamentals*. Atlanta: ASHRAE.

American Society of Heating, Refrigerating and Air-Conditioning Engineers (ASHRAE). 1997a. *1997 ASHRAE Handbook—Fundamentals*, Chapter 13, Odors. Atlanta: ASHRAE.

American Society of Heating, Refrigerating and Air-Conditioning Engineers (ASHRAE). 1997b. *1997 ASHRAE Handbook—Fundamentals*, Chapter 12, Air contaminants. Atlanta: ASHRAE.

American Society of Heating, Refrigerating and Air-Conditioning Engineers (ASHRAE). 1997. *1997 ASHRAE Handbook—Fundamentals*, Chapter 15, Airflow around Buildings. Atlanta: ASHRAE.

American Society of Heating, Refrigerating and Air-conditioning Engineers (ASHRAE). 1997. *1997 ASHRAE Handbooks—Fundamentals*, Chapter 33. Atlanta: ASHRAE.

American Society of Heating, Refrigerating and Air-Conditioning Engineers (ASHRAE). 1996. 1996

ASHRAE Handbook—HVAC Systems and Equipment. Atlanta: ASHRAE.

American Society of Heating, Refrigerating and Air-conditioning Engineers (ASHRAE). 1996. *1996 ASHRAE Handbook—HVAC Systems and Equipment*, Chapter 12, Hydronic Heating and Cooling System Design. Atlanta: ASHRAE.

American Society of Heating, Refrigerating and Air-conditioning Engineers (ASHRAE). 1996. *1996 ASHRAE Handbooks—HVAC Systems and Equipment*, Chapter 19, Evaporative Air Cooling. Atlanta: ASHRAE.

American Society of Heating, Refrigerating and Air-Conditioning Engineers (ASHRAE). 1996. *1996 ASHRAE Handbook—HVAC Systems and Equipment*, Chapter 24, Air Cleaners for Particulate Contaminants. Atlanta: ASHRAE.

American Society of Heating, Refrigerating and Air-Conditioning Engineers (ASHRAE). 1996. *1996 ASHRAE Handbook—HVAC Systems and Equipment*, Chapter 25, Industrial Gas Cleaning and Air Pollution Control. Atlanta: ASHRAE.

American Society of Heating, Refrigerating and Air-Conditioning Engineers (ASHRAE). 1996. *1996 ASHRAE Handbook—HVAC Systems and Equipment*, Chapter 42, Air-to-Air Energy Recovery. Atlanta, GA.

American Society of Heating, Refrigerating and Air-Conditioning Engineers (ASHRAE). 1995. *ANSI/ASHRAE Standard 110-1995, Method for Testing Performance of Laboratory Fume Hoods*. Atlanta: ASHRAE.

American Society of Heating, Refrigerating and Air-Conditioning Engineers (ASHRAE). 1992. *ANSI/ASHRAE Standard 41.1-1987 (RA 92), Standard Methods for Laboratory Airflow Measurement*. Atlanta: ASHRAE.

American Society of Heating, Refrigerating and Air-Conditioning Engineers (ASHRAE). 1992. *ANSI/ASHRAE Standard 52.1-1992, Gravimetric and Dust Spot Procedures for Testing Air Cleaning Devices Used in General Ventilation for Removing Particulate Matter*. Atlanta: ASHRAE.

American Society of Heating, Refrigerating and Air-Conditioning Engineers (ASHRAE). 1992. *ANSI/ASHRAE Standard 55-1992, Thermal Environmental Conditions for Human Occupancy*. Atlanta: ASHRAE.

American Society of Heating, Refrigerating and Air-Conditioning Engineers (ASHRAE). 1989. *ANSI/ASHRAE Standard 62-1989, Ventilation for Acceptable Indoor Air Quality*. Atlanta: ASHRAE.

American Society of Heating, Refrigerating and Air-Conditioning Engineers (ASHRAE). 1988. *ANSI/ASHRAE Standard 111-1988, Practices for Measurement, Testing, Adjusting, and Balancing of Building Heating, Ventilation, Air-Conditioning, and Refrigeration Systems*, Chapter 10. Atlanta: ASHRAE.

American Society of Heating, Refrigerating and Air-Conditioning Engineers (ASHRAE). 1985. *ANSI/ASHRAE Standard 51-1985, Laboratory Methods of Testing Fans for Rating*. Atlanta: ASHRAE.

American Society of Heating, Refrigerating and Air-Conditioning Engineers (ASHRAE). 1980. *ANSI/ASHRAE Standard 90A-1980, Energy Conservation in New Building Design*. Atlanta: ASHRAE.

Associated Air Balance Council (AABC). 1998. *Test and Balance Procedures*. Washington, D.C.

Associated Air Balance Council (AABC). 1996. *Technician Training Manual*. Washington, D.C.

Associated Air Balance Council (AABC). 1989. National Standards. Washington, D.C.

Berglund, B., and T. Lindvall. 1986. Sensory Reactions to Sick Building. *Environment International*, Vol. 12, pp. 147-159.

Bretherick, L. 1981. Hazards in the Chemical Laboratory, 3rd ed. London: Royal Society of Chemistry, Burlington House.

Building Officials and Code Administrators International, Inc. (BOCA). 1987b. *The BOCA National Mechanical Code, 6th ed*. Country Club Hills, Ill.: BOCA Publications.

Building Officials and Code Administrators International, Inc. (BOCA) 1987a. *The BOCA National Building Code 10th Edition—Model building regulations for the protection of public health, safety and welfare*. Country Club Hills, Ill: BOCA Publications.

Centers for Disease Control and Prevention (CDC) and National Institute of Health (NIH). 1999. *Biosafety in Microbiological and Biomedical Laboratories*, 4th Edition. Bethesda, Md: U.S. Department of Health and Human Services.

Centers for Disease Control and Prevention (CDC), Office of Biosafety. 1974. *Classification of Etiologic Agents on the Basis of Hazard*, 4th edition. Bethesda, Md: U.S. Department of Health and Human Services.

Department of Justice. 1994. 28 CFR Part 36. ADA Standards for Accessible Design, July 1 Revision.

DiBerardinis, L.J., J.S. Baum, M.W. First, G.T. Gatwood, E. Groden, and A.K. Seth. 1993. *Guidelines for Laboratory Design: Health and Safety Considerations, 2nd ed*. New York.: John Wiley & Sons, Inc.

Dorgan, C.B., and C.E. Dorgan. 1996. *Ventilation Best Practices Guide*. Palo Alto: Electric Power Research Institute (EPRI).

Dorgan, C.B., S.P. Leight, and C.E. Dorgan. 1995. *Application Guide for Absorption Cooling/Refrigeration Using Recovered Heat.* Atlanta: ASHRAE.

Federal Register. 1976. *Recombinant DNA Research Guidelines.* 41:27902-27943.

Federal Register. 1986. *Guidelines for Research Involving Recombinant DNA molecules.* 51:16958-16968.

Hare, Ronald, and P.N. O'Donoghue. 1968. Laboratory animal symposia. 1. *The design and function of laboratory animal houses.* Ashford, England: Geerings of Ashford, Ltd.

Huchingson, R.D. 1981. *New Horizons for Human Factors in Design.* New York: McGraw-Hill.

Hosni, M.H., B.W. Jones, and H. Xu. 1999. *Measurement of Heat Gain and Radiant/Convective Split from Equipment in Buildings. Final Report—ASHRAE 1055-RP.* Atlanta: ASHRAE.

Illumination Engineering Society of North America (IESNA). 1993. *IESNA Lighting Handbook*, 9th ed. New York: IESNA.

Illumination Engineering Society of North America (IESNA). 1989. *IES Lighting Ready Reference.* New York: IESNA.

Institute of Laboratory Animal Resources. 1985. *Guide for the Care and Use of Laboratory Animals.* Washington, D.C.: National Academy of Sciences.

Kawamura, P.I., and D. MacKay. 1987. The evaporation of volatile liquids. *Journal of Hazardous Materials*, Vol. 15, pp. 343-364.

Lawley, H.G. 1974. Operability Studies and Hazard Analysis. *Loss Prevention*, Vol. 8, pp. 105-116.

Maghirang, R.G., G.L. Riskowski, and L.L. Christianson. 1996. Ventilation and Environmental Quality in Laboratory Animal Facilities. *ASHRAE Transactions* 102(2): 186-194.

Mayer, Leonard. 1995. Design and Planning of Research and Clinical Laboratory Facilities. New York: John Wiley and Sons, Inc.

McQuaid, J., and B. Roebuck. 1984. *Large Scale Field Trials on Dense Vapor Dispersion.* Final report to sponsors on the heavy gas dispersion trials at Thorney island, 1982-84. Sheffield, UK: Health and Safety Executive Safety Engineering Laboratory.

Memarzadeh, Farhad. 1996. *Methodology for Optimization of Laboratory Hood Containment, Volume I.* National Institutes of Health, Chapter 4, pp 41-73.

Monger, Samuel C. 1994. Designing or Renovating Fume Hood Laboratories. *TAB Journal*, pages 13-19. Associated Air Balance Council.

National Environment Balancing Bureau (NEBB). 1998. *Procedural Standards for TAB of Environmental Systems,* 6th Edition. Gaithersburg, Md: NEBB.

National Environment Balancing Bureau (NEBB). 1997. *Testing, Adjusting, and Balancing Manual for Technicians,* 2nd Ed. Gaithersburg, Md: NEBB.

National Environment Balancing Bureau (NEBB). 1996. *Procedural Standards for Certified Clean Rooms,* 2nd Ed., Chapter 8. Gaithersburg, Md: NEBB.

National Fire Protection Association (NFPA). 2000. *NFPA 12—Carbon Dioxide Extinguishing Systems.* Quincy, Mass.: NFPA Publications.

National Fire Protection Association (NFPA). 2000. *NFPA 30—Flammable and Combustible Liquids Code.* Quincy, Mass.: NFPA Publications.

National Fire Protection Association (NFPA). 2000. *NFPA 45—Fire Protection for Laboratories Using Chemicals.* Quincy, Mass.: NFPA Publications.

National Fire Protection Association (NFPA). 2000. *NFPA 101—Life Safety Code,* Chapter 5, Section 13-2, Section 14-2. Quincy, Mass.: NFPA Publications.

National Fire Protection Association (NFPA). 1999. *NFPA 70—National Electrical Code. NFPA Publications.* Quincy, Mass.: NFPA Publications.

National Fire Protection Association (NFPA). 1998. *NFPA 801—Facilities handling radioactive materials.* Quincy, Mass.: NFPA Publications.

National Fire Protection Association (NFPA). 1996. NFPA 30—Flammable and Combustible Liquids Code. Quincy, Mass.: NFPA Publications.

National Institute for Occupational Safety and Health (NIOSH). 1994. *Pocket Guide to Chemical Hazards Department of Health and Human Services.* Cincinnati, Ohio.

National Research Council (NRC). 1996. *Guide for the Care and Use of Laboratory Animals.* Washington, D.C.: Institute of Laboratory Animal Resources, Commission on Life Science.

National Research Council (NRC). 1995. *Prudent Practices in the Laboratory, Handling and Disposal of Chemicals.* Washington, D.C.: National Academy Press.

National Research Council (NRC). 1991. *Standards for Protection against Radiation.* U.S. Nuclear Regulatory Commission, 10 CFR 20.

National Sanitation Foundation International (NSF). 1992. *Class II (Laminar Flow) Biohazard Cabinetry.* Ann Arbor, MI: NSF Publications.

Neuman, V.A. 1989. Disadvantages of Auxiliary Air Fume Hoods. *ASHRAE Transactions* Part 1, p. 73.

Nielson, D.S., O. Platz, and B. Runge. 1975. A Cause-Consequence Chart of a Redundant Protection System. *IEEE Transactions on Reliability*, Vol. R-24, pp. 8-13.

Office of Biosafety, Laboratory Centre for Disease Control. 1990. Canada Public Works, *Standards and Guidelines – MD 151218 Laboratory Fume Hoods.* A.2 Preventative Maintenance Programs, A.2.1.4 Performance Tests. Ottawa, Ontario, Canada.

Office of Research Safety, National Cancer Institute, and the Special Committee of Safety and Health experts. 1978. *Laboratory Safety Monograph.* A supplement to the NIH Guidelines for Recombinant

DNA Research. Bethesda, Md: National Institute of Health.

Occupational Safety and Health Administration (OSHA). 1991. *OSHA 29 CFR Part 1910.35—Health and Safety Standards*, Chapter 12, Subpart E. Washington, D.C.: U.S. Government Printing Office.

Occupational Safety and Health Administration (OSHA). 1991. *OSHA 29 CFR Part 1990—Identification, Classification, and Regulation of Potential Occupational Carcinogens*. Washington, D.C.: U.S. Government Printing Office.

Occupational Safety and Health Administration (OSHA). 1990a. *OSHA Regulated Hazardous Substances—Health, Toxicity, Economic and Technological Data, Vol. 1 & 2*. New Jersey: Noyes Data Corporation.

Occupational Safety and Health Administration (OSHA). 1990b. *OSHA 29 CFR Part 1910.1450—Occupational Exposure to Hazardous Chemicals in Laboratories*. Washington, D.C.: U.S. Government Printing Office.

Occupational Safety and Health Administration (OSHA). 1990c. *OSHA 29 CFR Part 1910.1330—Occupational Exposure to Bloodborne Pathogens*. Washington, D.C.: U.S. Government Printing Office.

Petersen, R.L., and M.A. Ratcliff. 1991. An Objective Approach to Laboratory Stack Design. *ASHRAE Transactions* 97(2): 553-562.

Ruth, J.H. Odor Thresholds and Irritation Levels of Several Chemical Substances: A Review. *Journal of American Industrial Hygienists Association*, Vol. 47, p. A-142.

Ruys, Theodorus. 1990. *Handbook of Facilities Planning, Volume 1, Laboratory Facilities*. New York: Van Nostrand Reinhold.

Scientific Equipment and Furniture Association (SEFA). 1999. *SEFA 8—Laboratory Furniture, Recommended Practices*. Maclean, Va.

Scientific Equipment and Furniture Association (SEFA). 1997. *SEFA 2.3—Installation of Scientific Furniture and Equipment, Recommended Practices*. Maclean, Va.

Scientific Equipment and Furniture Association (SEFA). 1996. *SEFA 1.2—Laboratory Fume Hoods, Recommended Practices*. Maclean, Va.

Sheet Metal and Air Conditioning Contractors National Association (SMACNA). 1987. *HVAC Systems Applications*. Vienna, Va.: SMACNA, Inc.

Sheet Metal and Air Conditioning Contractors National Association (SMACNA). 1985. *HVAC Duct Construction Standards-Metal and Flexible*. Vienna, Va.: SMACNA, Inc.

Simmons C.G., and R. Dvoodpour. 1994. Design Considerations for Laboratory Facilities Using Molecular Biology Techniques. *ASHRAE Transactions* 100(1): 1266-1274.

Streets and Setty. 1983. Energy Conservation in Institutional Laboratory and Fume Hood Systems. *ASHRAE Transactions* 89(2B): 542-551.

U.S. Department of Health and Human Services (DHHS), Public Health Service, Centers for Disease Control and Prevention (CDC), National Institutes of Health (NIOSH). 1999. *Biosafety in Microbiological and Biomedical Laboratories*, 4th ed. J.Y. Richmond and R.W. Mckinney, Eds. Washington D.C.: U.S. Government Printing Office.

U.S. Department of Justice (DOJ). 1994. *American with Disabilities Act (ADA) Regulation for Title III—Non-discrimination on the basis of disability by public accommodations and in commercial facilities*. Washington, D.C.: DOJ.

U.S. Environmental Protection Agency (EPA). 1995. *Testing of Meteorological and Dispersion Models for Use in Regional Air Quality Modeling*. Washington, D.C.: EPA.

U.S. Environmental Protection Agency (EPA). 1992a. *Standards for the Tracking and Management of Medical Waste.*, 40 CFR 259.

U.S. Environmental Protection Agency (EPA). 1992b. *General Regulations for Hazardous Waste Management.*, 40 CFR 260.

U.S. Environmental Protection Agency (EPA). 1992. *User's Guide for the Industrial Source Complex (ISC2) Dispersion Models. Volume II – Description of Model Algorithms*. Washington, D.C.: EPA.

Varley, J.O. 1998. Applying Process Hazard Analysis to Laboratory HVAC Design. *ASHRAE Journal*, Vol. 40, No. 2, pp. 54-57.

Vesley, W.E., et. al. 1981. *Fault Tree Handbook*. Report (NUREG-0492). Springfield, Va.: National Technical Information Service.

Walters, Douglas, B.Ed. 1980. *Safe Handling of Chemical Carcinogens, Mutagens, Teratogens, and Highly Toxic Substances, Vol. 1*. Ann Arbor, MI: Ann Arbor Science.

Wilson, D.J., I. Fabris, and M.Y. Ackerman. 1998. Measuring Adjacent Building Effects on Laboratory Exhaust Stack Design. *ASHRAE Transactions* 104(2): 1012-1028.

Wilson, D. J., and G. Winkel. 1982. The Effect of Varying Exhaust Stack Height on Contaminant Concentration at Roof Level. *ASHRAE Transactions* 88(1): 515-533.

Wisconsin Department of Administration (WI-DOA), Division of Facilities Development. 2000. *Fume Hood Performance Test and Life Cycle Cost Analysis*, pp. 7and 19. Milwaukee, WI.

Wisconsin Department of Commerce (WI-DOC). 1998. *Commercial Building Energy Conservation/HVAC Code Handbook*. Madison, WI: WI-DOC.

World Health Organization. 1983. *Laboratory Biosafety Manual*. Washington, D.C.: WHO Publications Centre.

Wunder, J.S. 2000. Personal Communication from Operating Experiences with Laboratory Equipment. University of Wisconsin–Madison.

Annotated Bibliography

American Conference of Governmental Industrial Hygienist (ACGIH). 1999. *Documentation of the Threshold Limit Values and Biological Exposure Indices*, 6th ed. Cincinnati, Ohio: ACGIH.

This publication provides the basic rationale for the development of TLVs for chemical substances and physical agents and of BEIs for selected chemicals. In the chemical substances TLV section, "timelines" present a historical overview of the TLV adoption process. Information is also provided on the OSHA PELs, NIOSH RELs, and NTP studies; carcinogen designations from various sources; and values from selected other countries.

American Conference of Governmental Industrial Hygienists (ACGIH). 1998. *Industrial Ventilation—A Manual for Recommended Practices*, 23rd ed. Cincinnati, Ohio: ACGIH.

This manual of recommended practices is the outgrowth of years of experience by ACGIH committee members and a compilation of research data and information on design, maintenance, and evaluation of industrial exhaust ventilation systems. The manual attempts to present a logical method of designing and testing these systems. It has found wide acceptance as a guide for official agencies, as a standard for industrial ventilation designers, and as a textbook for industrial hygiene courses.

American Society of Heating, Refrigerating and Air-Conditioning Engineers (ASHRAE). 1999. *1999 ASHRAE Handbook—HVAC Applications*, Chapter 13, Laboratories. Atlanta: ASHRAE.

This chapter of the Handbook was prepared by ASHRAE Technical Committee TC 9.10 and addresses biological, chemical, animal, and physical laboratories. Within these generic categories, some laboratories have unique requirements. This chapter provides an overview of the heating, ventilating, and air-conditioning (HVAC) characteristics and design criteria for laboratories, including a brief overview of architectural and utility concerns. It does not cover pilot plants, which are essentially small manufacturing units.

American Society of Heating, Refrigerating and Air-Conditioning Engineers (ASHRAE). 1999. *1999 ASHRAE Handbook—HVAC Applications*, Chapter 47, Water Treatment. Atlanta: ASHRAE.

This chapter of the Handbook was prepared by ASHRAE Technical Committee TC 3.6 and covers the fundamentals of water treatment and some of the common problems associated with water in heating and air-conditioning equipment. Specific topics include water characteristics, corrosion control, scale control, biological growth control, suspended solids and depositation control, startup and shutdown of cooling towers, and selection of water treatment.

American Society of Heating, Refrigerating and Air-Conditioning Engineers (ASHRAE). 1999. *1999 ASHRAE Handbook—HVAC Applications*, Chapter 50, Evaporative Cooling Applications. Atlanta: ASHRAE.

This chapter of the Handbook was prepared by ASHRAE Technical Committee TC 5.7 and covers topics such as indirect evaporative cooling, booster refrigeration, residential or commercial cooling, exhaust requirements, two-stage cooling, industrial and other applications, economic factors, psychrometrics, and entering air concerns.

American Society of Heating, Refrigerating and Air-Conditioning Engineers (ASHRAE). 1997. *1997 ASHRAE Handbook—Fundamentals*, Chapter 15, Airflow around Buildings. Atlanta: ASHRAE.

This chapter of the Handbook was prepared by the ASHRAE Technical Committee TC 2.5 and contains information for evaluating flow patterns, estimating wind pressures and air intake contamination, and solving problems caused by the effects of wind on intakes, exhausts, and equipment.

American Society of Heating, Refrigerating and Air-Conditioning Engineers (ASHRAE). 1997b. *1997 ASHRAE Handbook—Fundamentals*, Chapter 12, Air Contaminants. Atlanta: ASHRAE.

This chapter of the Handbook was prepared by ASHRAE Technical Committees TC 2.3 and TC 2.4 and includes topics such as classification of air contaminants, nature of airborne contaminants, suspended particulates, industrial air contaminants, flammable gases and vapors, combustible dusts, air pollution, radioactive contaminants, atmospheric pollen, bioaerosols, and indoor air quality.

American Society of Heating, Refrigerating and Air-Conditioning Engineers (ASHRAE). 1996. *1996 ASHRAE Handbook—HVAC Systems and Equipment*, Chapter 12, Hydronic Heating and Cooling System Design. Atlanta: ASHRAE.

This chapter of the Handbook was prepared by ASHRAE Technical Committee TC 6.1 and discusses only closed water systems and their design considerations. Specific topics covered are piping circuits, capacity control of load system, sizing control valves, low-temperature heating systems, chilled water systems, dual-temperature systems, design procedures, and antifreeze solutions.

American Society of Heating, Refrigerating and Air-Conditioning Engineers (ASHRAE). 1996. *1996 ASHRAE Handbook—HVAC Systems and Equipment*, Chapter 24, Air Cleaners for Particulate Contaminants. Atlanta: ASHRAE.

This chapter of the Handbook was prepared by ASHRAE Technical Committee TC 2.4 and discusses the cleaning of both ventilation air and recirculated air for the conditioning of building interiors. Complete air cleaning may require the removal of airborne particles, microorganisms, and gaseous pollutants. This chapter addresses only the removal of airborne particles.

American Society of Heating, Refrigerating and Air-Conditioning Engineers (ASHRAE). 1996. *1996 ASHRAE Handbook—HVAC Systems and Equipment*, Chapter 25, Industrial Gas Cleaning and Air Pollution Control. Atlanta: ASHRAE.

This chapter of the Handbook was prepared by ASHRAE Technical Committee TC 5.4 and covers each generic type of industrial gas-cleaning equipment is discussed on the basis of its primary method for gas or particulate abatement. Specific topics discussed are regulations and monitoring, particulate contaminant control, gaseous contaminant control, incineration of gases and vapors, auxiliary equipment, and operations and maintenance.

American Society of Heating, Refrigerating and Air-Conditioning Engineers (ASHRAE). 1996. *1996 ASHRAE Handbook—HVAC Systems and Equipment*, Chapter 42, Air-to-Air Energy Recovery. Atlanta: ASHRAE.

This chapter of the Handbook was prepared by ASHRAE Technical Committee TC 5.5 and covers both air-to-air energy recovery applications and equipment. The topics covered in the applications section are economic considerations, technical considerations, and air-to-air energy recovery examples. The topics covered in the equipment section are: fixed plate exchangers, rotary air-to-air energy exchangers, coil energy recovery (runaround) loops, heat pipe heat exchangers, twin tower enthalpy recovery loops, and thermosiphon heat exchangers.

Bretherick, L. 1981. *Hazards in the Chemical Laboratory*, 3rd ed. London: Royal Society of Chemistry, Burlington House.

This handbook is intimately concerned with the prevention of injury from fire, explosion or from exposure to hazardous substances. There are nine chapters that include an introduction, health and safety at work act of 1974, safety planning and management, fire protection, reactive chemical hazards, chemical hazards and toxicology, health care and first aid, hazardous chemicals, and precautions against radiations.

Building Officials and Code Administrators International, Inc. (BOCA). 1987b. The BOCA National Mechanical Code, 6th ed. Country Club Hills, Ill: BOCA Publications.

This code book was prepared for the safe installation and maintenance of all mechanical equipment in order to protect the public safety, health and welfare. It sets forth regulations for the safe installation and maintenance of mechanical facilities where great reliance was previously placed on accepted practice and engineering standards.

Building Officials and Code Administrators International, Inc. (BOCA) 1987a. *The BOCA National Building Code 10th Edition—Model building regulations for the protection of public health, safety and welfare*. Country Club Hills, Ill: BOCA Publications.

This code book states regulations in terms of measured performance rather than in rigid specification of materials and, in this way, makes possible the acceptance of new materials and methods of construction which can be evaluated by accepted standards, without the necessity of adopting cumbersome amendments for each variable condition.

Centers for Disease Control and Prevention (CDC) and National Institute of Health (NIH). 1999. *Biosafety in Microbiological and Biomedical Laboratories*. 4th Edition. Bethesda, Md: U.S. Department of Health and Human Services.

This publication describes combinations of standards and special microbiological practices, safety equipment and facilities that constitute biosafety levels 1-4, which are recommended for working with a variety of infectious agents in various laboratory settings. These recommendations are advisory and are intended to provide a voluntary guide and code of practice as well as a goal for upgrading operations. Furthermore, the recommendations are offered as a guide and reference in the construction of new laboratory facilities and in the renovation of existing facilities.

DeLuga, Greg. 1994. Key Issues to Consider in Laboratory Control Systems. *TAB Journal,* pages 7-11. Associated Air Balance Council.

This article discusses the following five key issues that must be constantly identified and addressed through the life of a laboratory facility: Ensuring operational safety for the operators and other occupants; Maintaining compliance with applicable regulations; Satisfying occupant comfort and process constraints; Minimizing operational costs, and Providing the flexibility needed to accommodate continuous change

DiBerardinis, L.J., J.S. Baum, M.W. First, G.T. Gatwood, E. Groden, A.K. Seth. 1993. *Guidelines for Laboratory Design: Health and Safety Considerations*, 2nd ed. New York: John Wiley & Sons, Inc.

Chapter 1 of this book has a good overview of the common elements of laboratory design. The purpose of this book is to provide reliable design information related to specific health and safety issues that should be considered when planning new and renovated laboratories. The objective is approached within the framework of other important facts such as efficiency, economy, energy conservation, and design flexibility.

Dorgan, C.B., and C.E. Dorgan. 1996. *Ventilation Best Practices Guide.* Palo Alto: Electric Power Research Institute (EPRI).

Productivity losses and health costs resulting from indoor air quality problems in U.S. commercial buildings amount to $63 billion per year. Indoor air quality problems, however, can be avoided when the interaction between a heating, ventilating, and air conditioning (HVAC) system and the building occupants are understood. This guide employs six case studies to educate the reader on indoor air quality terminology, sources of problems, and mitigation strategies.

Hare, R., and P.N. O'Donoghue. 1968. Laboratory animal symposia. 1. *The design and function of laboratory animal houses*. Ashford, England: Geerings of Ashford, Ltd.

Compilation of papers presented at a meeting organized jointly by The Laboratory Animals Center and The Laboratory Animal Science Association, held at the Zoological Society of London, 1967.

Haugen, R.K., and R. Poblete. 2000. *A Containment Study of Two Dynamic Barrier Low Constant Volume Fume Hoods*. Statesville, NC: Kewaunee Scientific Corporation.

This study covers research into a new type of low constant volume fume hood—the dynamic barrier fume hood, which uses a low constant volume approach coupled with air input vectors designed to maximize containment by creation of a low vapor "buffer zone" between the hood interior and the sash plane. Energy conservation and fume containment are the two factors used to evaluate this low volume technology. ASHRAE 110 is the test methodology used to evaluate this technology.

Hosni, M.H., B.W. Jones, and H. Xu. 1999. Measurement of Heat Gain and Radiant/Convective Split from Equipment in Buildings. Final Report—ASHRAE 1055-RP. Atlanta: ASHRAE.

This final report for the ASHRAE Research Project 1055-RP sponsored by Technical Committee TC 4.1 (Load Calculation Data and Procedures) discusses the experimental test facility, measurement equipment, and test procedures used to obtain heat gain data for various office, laboratory, and hospital equipment. General guidelines are given for the use of the experimental results. The report also includes pictures of various laboratory and hospital equipment.

Kawamura, P.I., and D. MacKay. 1987. The evaporation of volatile liquids. *Journal of Hazardous Materials*, Vol. 15, pp. 343-364.

This study develops two models to estimate the evaporation rate of volatile and non-volatile liquids resulting from ground spills. Steady-state heat balances form the bases of both models and the effects of solar insolation, evaporative cooling, and heat transfer from the ground is also included. Whereas, one of the models estimates the evaporation rate directly from chemical-physical data, the other determines the surface temperature of the evaporating pool by an iterative method and then uses this temperature to estimate the evaporation rate.

Lawley, H.G. 1974. Operability Studies and Hazard Analysis. *Loss Prevention*, Vol. 8, pp. 105-116.

This paper presents examples of two loss prevention methods. One method called the "operability study" was developed by the Petrochemicals Division of Imperial Chemical Industries. It is based on the claim that most problems are missed due to a system being complex instead of the design team's lack of knowledge. This method can be used to examine preliminary process design flow sheets at the beginning of a project or intricate instrument and piping diagrams at end of the design phase.

The other method called "hazard analysis" provides a full quantitative examination after the identification of a severe hazard. A fault tree forms the basis of this method, where the events or coincidence events leading up to the specific hazard are related and then quantified.

Bell, G., and E. Mills. 1996. *A Design Guide for Energy Efficient Laboratories*. Berkeley, Calif.: Lawrence Berkeley National Laboratory (LBNL).

This guide, available online at http://ateam.lbl.gov/design-guide/, is intended to assist facility owners, architects, engineers, designers, facility managers, and utility demand-side management specialists in identifying and applying advanced energy-efficiency features in laboratory-type environments. It focuses comprehensively on laboratory energy design issues with a "systems" design approach. Although a laboratory-type facility includes many sub-system designs, e.g., the heating system, the authors believe that a comprehensive design approach should view the entire building as the essential "system." This means the larger, macro energy-efficiency considerations during architectural programming come before the smaller, micro component selection such as an energy-efficient fan.

Maghirang, R.G., G.L. Riskowski, and L.L. Christianson. 1996. Ventilation and Environmental Quality in Laboratory Animal Facilities. *ASHRAE Transactions* 102(2): 186-194.

This paper describes the environmental conditions which laboratory animal ventilation systems should be designed to provide. It presents a study based on a detailed literature review conducted to evaluate ventilation strategies for improving thermal and air quality conditions in laboratory animal facilities.

Mayer, L. 1995. *Design and Planning of Research and Clinical Laboratory Facilities*. New York: John Wiley and Sons, Inc.

This book is intended to be a primer for administrators, researchers, facility managers, architects, and engineers. It presents biomedical, environmental, physical, and basic sciences research, laboratory programming, planning, and design criteria based on the experience of the author. The book focuses on what may be considered as wet laboratory environments, such as chemistry and biology including related specialties. However, the basic programming and planning principles indicated are equally applicable to all research, clinical, teaching, and production-type laboratory facilities.

Memarzadeh, F. 1996. *Methodology for Optimization of Laboratory Hood Containment, Volume I*, Chapter 4, pp. 41-73. National Institutes of Health.

This research investigates the effect of room airflow on hood containment. It demonstrates that the performance of the hood is strongly dependent on the flow in the laboratory. It is the combination of the parameters that define the resulting flows and which in turn can result in good or bad containment. This document details the work undertaken and its findings, identifying design methodologies that enable the designer to evaluate the likely hood containment performance of proposed designs as well as verifying the performance of current installations either by comparison with the data provided in this document, experimentally, or by simulation for the proposed design. Most importantly, this document provides specific recommendations that can be expected to improve the design in terms of hood containment.

Monger, Samuel C. 1994. Designing or Renovating Fume Hood Laboratories. *TAB Journal*, pp. 13-19. Associated Air Balance Council.

The author mentions the lag time that exists between publications, guidelines, and standards for laboratories and the lack of specificity in the standards that has created some confusion and discomfort for safety and industrial hygienists, operations and maintenance staff, and laboratory workers. In an attempt to clear up some of this confusion, this article poses and answers several questions related to the designing and renovating of laboratories that use fume hoods. Some of the many questions posed include (1) Do you need a laboratory management program? (2) What to ask about the laboratory ventilation system? (3) Should you have an open or closed loop system? (4) What is a laboratory fume hood? (5) What face velocities are required? and (5) What other questions must be answered before choosing or installing a system?

National Environment Balancing Bureau (NEBB). 1996. *Procedural Standards for Certified Testing of*

Clean Rooms, Chapter 8, 2nd Edition. Gaithersburg, Md: NEBB.

This two-part publication contains the latest data and information on cleanroom testing methods and procedures and sample test report forms. This text covers cleanroom design, cleanliness classes based on Federal Standard 209E, construction, airflow system design, filtration, airflow patterns, testing instruments, and equipment. This manual also contains a comprehensive glossary, engineering data and metric conversion and provides the basic and fundamental knowledge needed for certified testing of cleanrooms.

National Institute for Occupational Safety and Health (NIOSH). 1994. *Pocket Guide to Chemical Hazards Department of Health and Human Services.* Cincinnati, Ohio: NIOSH.

This pocket guide is a source of general industrial hygiene information on over 600 chemicals in the workplace (including substances from Tables Z-1-A and Z-2 of the OSHA General Industry Air Contaminants Standard). The guide helps users recognize and control occupational chemical hazards with information including NIOSH RELs, OSHA PELs, IDLH values, and odor thresholds; protective equipment recommendations; incompatibilities and reactivity information; measurement methods; chemical and physical properties; as well as health hazards, target organs, symptoms, and routes of exposure.

National Research Council (NRC). 1996. *Guide for the Care and Use of Laboratory Animals.* Washington, D.C.: Institute of Laboratory Animal Resources, Commission on Life Science.

The purpose of the guide is to assist institutions in caring for and using animals in ways judged as to be scientifically, technically and humanely appropriate. The guide is also intended to assist investigators in fulfilling their obligation to plan and conduct animal experiments in accord with the highest scientific, humane and ethical principles. The guide is organized into four chapters on the major components of an animal care and use program: institutional policies and responsibilities; animal environment, housing, and management; veterinary medical care; and physical plant.

National Research Council (NRC). 1995. Prudent Practices in the Laboratory, Handling and Disposal of Chemicals. Washington, D.C.: National Academy Press.

This books constitutes a strong recommendation to workers in laboratories to exercise prudence in designing and carrying out their studies so as to maintain a safe workplace and safe operational procedures. Mentions Sara Title III (chapter 9), discusses hazards and safety practices, and laboratory ventilation (chapter 8).

Neuman, V.A. 1989. Disadvantages of Auxiliary Air Fume Hoods. *ASHRAE Transactions*, Part 1, p. 73.

This paper presents a review of recent research as well as original analysis to show the disadvantages of auxiliary air chemical fume hoods and several design alternatives to increase the safety and cost effectiveness of fume hoods.

Occupational Safety and Health Administration (OSHA). 1990a. *OSHA Regulated Hazardous Substances – Health, Toxicity, Economic and Technological Data, Vol. 1.* New Jersey: Noyes Data Corporation.

This book is the first of a two-volume series that provides industrial exposure data and control technologies for more than 650 substances currently regulated, or candidates for regulation, by Occupational Safety and Health Administration (OSHA). The health, toxicity, economic and technological data provided are intended to serve as a reference for those who are potentially exposed to one or more of these substances in their workplace, or for those who have supervisory or management responsibility for workers potentially exposed. OSHA "permissible exposure" limits" (PEL) for the 650 substances reflect all updates and changes as presented in the January 19, 1989 Federal Register. This volume contains substance starting with A-I.

Occupational Safety and Health Administration (OSHA). 1990a. *OSHA Regulated Hazardous Substances—Health, Toxicity, Economic and Technological Data, Vol. 2.* New Jersey: Noyes Data Corporation.

This book is the second of a two-volume series that provides industrial exposure data and control technologies for more than 650 substances currently regulated, or candidates for regulation, by Occupational Safety and Health Administration (OSHA). The health, toxicity, economic and technological data provided are intended to serve as a reference for those who are potentially exposed to one or more of these substances in their workplace, or for those who have supervisory or management responsibility for workers potentially exposed. OSHA "permissible exposure" limits" (PEL) for the 650 substances reflect all updates and changes as presented in the January 19, 1989 Federal Register. This volume contains substance starting with K-Z.

Petersen, R.L., and M.A. Ratcliff. 1991. An Objective Approach to Laboratory Stack Design. *ASHRAE Transactions* 97(2): 553-562.

This paper states that the quality of air that enters a laboratory depends a great deal on the height of the stacks on top of the building and the location of the fresh air intakes relative to the stack location and the prevailing winds. Also, if the exhaust and ventilation systems are poorly designed, routine or accidental emissions from the stacks can be re-circulated into the building. The paper presents and objective approach used to evaluate the effect of stack design on concentration levels at fresh-air intakes and pedestrian areas. The method examines local meteorology, expected chemical emissions, and their odor and health thresholds. Screening equations for estimating maximum concentrations are then employed. If thresholds are exceeded, wind tunnel measurements are then employed to refine concentration estimates and to improve the stack design. Examples of the methods are also presented.

Rankin, J.E., and G.O. Tolley. 1978. *Safety Manual No. 8—Fault Tree Analysis*. U.S. Washington, D.C.: Government Printing Office.

This pamphlet provides easy to read information on fault tree analysis, a technique that can be effectively used in the prevention and investigation of accidents and in the recognition, evaluation and control of hazards. A list of references is included for those interested in additional information on the topics discussed in this pamphlet.

Ruys, Theodorus. 1990. *Handbook of Facilities Planning, Volume 1, Laboratory Facilities*. New York: Van Nostrand Reinhold.

This handbook is a hands-on guide to many different topics including: similarities and differences between physical science and biological science laboratories; laboratory planning issues (programming, documentation, space guidelines, etc.); and a coverage of HVAC electrical, structural, and waste handling and disposal systems. This presentation also covers energy conservation cost issues and important codes and regulations.

Sheet Metal and Air Conditioning Contractors National Association (SMACNA). 1987. *HVAC Systems Applications*. Vienna, Va.: SMACNA, Inc.

This manual was developed to present information and data on the application of both hydronic and air systems for contractors, system designers, and others involved in industrial and commercial heating, ventilating, and air conditioning (HVAC) systems.

Chapters I and II contain a general discussion of basic central HVAC system and equipment installations. Electric and pneumatic control systems are covered in depth in Chapter III. The characteristics of individual types of air and water systems are discussed in Chapters IV through VIII. Special types of air systems (lab exhaust and smoke control, dehumidification, and cleanroom) can be found in Chapter IX. Chapters X through XIII specific and general information on refrigeration and hydronic systems. Finally, the Chapter XIV provides engineering tables and charts and a comprehensive glossary follows.

Sheet Metal and Air Conditioning Contractors National Association (SMACNA). 1985. *HVAC Duct Construction Standards-Metal and Flexible*. Vienna, Va.: SMACNA, Inc.

This publication was produced in the hope that duct system designers and contractors will apply their separate skills more cost effectively and responsibly. There are seven sections which cover the following topics (1) basic duct construction, (2) fittings and other construction, (3) round, oval and flexible duct, (4) hangers and supports, (5) exterior components, (6) casings, and (7) functional criteria for demonstrating equivalency. There are 33 Appendices included.

Siemens Building Technologies, Inc. 1999. *Laboratory Control and Safety Solutions Application Guide*. Buffalo Grove, Ill.

This 80-page paperback guide provides information to assist in properly configuring laboratory ventilation and their associated control systems. It covers laboratory ventilation requirements beginning with the quantity of ventilation air needed, desirable room airflow patterns to maximize effectiveness, preventing cross contamination by room pressurization, and the unique ventilation needs of the different types of laboratory rooms. It also addresses how to minimize the potentially high energy costs associated with these systems and how to ensure continued proper system functionality.

Siemens Building Technologies. 1999. *Laboratory Ventilation Standards and Codes*. Buffalo Grove, Ill.

This 24-page guide covers the requirements of ventilation systems that serve laboratory facilities. Each individual aspect that a laboratory ventilation system must commonly address is listed along with the applicable national standards and regulatory requirements. Informative comments are also included to provide further explanation where applicable and to provide the preferred means of satisfying the requirements.

Simmons C.G., and R. Dvoodpour. 1994. Design Considerations for Laboratory Facilities Using Molecular Biology Techniques. *ASHRAE Transactions* 100(1): 1266-1274.

This paper reviews the architectural and HVAC requirements of the different biosafety level (BL-1 to BL-4) laboratories. It briefly explains the typical procedures used in these laboratories so that engineers may better understand what is involved.

The discussion also covers the necessity of directional airflow, the requirement for a "once-through" air system, the use of anterooms, the use of biological safety cabinets, exhaust requirements for autoclaves, the heat loads for typical scientific equipment, the practicality of local supplemental cooling, and the need for redundant exhaust and alarm systems.

Streets and Setty. 1983. Energy Conservation in Institutional Laboratory and Fume Hood Systems. *ASHRAE Transactions* 89(2B): 542-551.

This paper states that in the design of fume hoods and their associated mechanical systems, the most pivotal concern is safety. Opportunities for cost reduction rarely exist if the safety of working personnel is threatened by a proposed design trade-off. Through successful integration of energy saving air systems, however, operating costs can be reduced without sacrificing the performance or reliability of the fume hood, thereby ensuring the continued safety of personnel. This paper discusses energy conservation features of laboratory fume hood systems and includes and overview of the primary energy conservation elements of a laboratory/classroom complex under design in central Oklahoma.

Varley, J.O. 1998. Applying Process Hazard Analysis to Laboratory HVAC Design. *ASHRAE Journal*, Vol. 40, No. 2, pp. 54-57.

This paper states that regardless of the nature of the hazards (chemical or biological) contained in the laboratory, the engineers and operations personnel responsible for laboratory design, construction and maintenance can benefit from using process hazard analysis (PHA) techniques, as authored by the Occupational Safety and Health Administration (OSHA). The paper also points out that although not intended to be used as a design tool, PHA offers designers a systematic approach to engineering safety and proceeds to review the PHA requirements and offer ideas for applying them to HVAC system design.

Wilson, D.J., I. Fabris, and M.Y. Ackerman. 1998. Measuring Adjacent Building Effects on Laboratory Exhaust Stack Design. *ASHRAE Transactions* 104(2): 1012-1028.

Unlike current methods for designing exhaust stacks, which are based on avoiding contamination of the roof, walls, and nearby ground surface of the building on which the stack is located, this paper takes into account the effect of adjacent buildings that add turbulence and increase dispersion if they are located upwind and may be contaminated themselves if they are located downwind of the emitting building. To account for such adjacent effects, water channel simulation of the atmosphere is used to evaluate rooftop contamination in more than 1,700 different configurations of adjacent building height, width, spacing, stack location, stack diameter, height and exit velocity. The implications of the study for developing practical stack design are discussed.

Wilson, D. J., and G. Winkel. 1982. The Effect of Varying Exhaust Stack Height on Contaminant Concentration at Roof Level. *ASHRAE Transactions* 88(1): 515-533.

This paper presents the development of a simple theory, which uses the results of a systematic wind tunnel study of a large number of building, stack and air intake configurations to predict the effect of exhaust stack height modifications on air intake contamination by exhaust gases. Only flat roofed buildings, typical of industrial situations were used in this study. Tests show that although there are many cases where complex roof flow patterns prevent reliable predictions, the effect of stack height on roof level intake contamination can be predicted in many situations.

Wisconsin Department of Administration (WI-DOA), Division of Facilities Development. 2000. *Fume Hood Performance Test and Life Cycle Cost Analysis*, pp. 7 and 19. Milwaukee, WI.

The purpose of this report is to present and economic and engineering assessment between chemical fume hoods from two manufacturers who submitted bids for the University of Wisconsin - Milwaukee Chemistry Fume Hood Replacement Project (DFD #98303). The project scope called for installing a total of 94 new fume hoods. The report shows that the low-flow fume hood technology of one of the manufacturers was economically viable and gave a greater degree of safety and containment than today's conventional fume hoods.

Index

A

acceptability of air transfer 159, 160
acceptance phase 179
access and egress 40
ACGIH 5, 27, 36, 43, 55, 65, 79, 81, 117–119, 123
acquisition 13, 33
activity level 29, 30, 64, 155
ADA 4, 35, 44, 45
adjacent building effects 129, 135
administrative spaces 109
aesthetics 37, 42, 52, 125, 130
AIHA 5, 27, 32, 33, 35, 43, 80, 167, 170, 172
air and hydronic system balancing 161
air change requirements 32, 33
air distribution system 26, 49, 85, 86, 88, 103, 182, 191
air intake location 37
air introduction 160
air locks 31, 61, 161
air quality 1, 28, 29, 32–35, 44, 47, 51, 95, 102, 123, 135
air recirculation 32, 75, 109
air treatment 2, 32, 34, 47, 51, 99, 101, 117, 122, 133
airflow direction 42, 52, 96, 154, 157, 158, 167
airflow obstruction 165
airflow patterns 2, 60, 128, 157–161, 163
air-handling unit 97, 98, 101, 102, 104, 132, 173

air-to-air heat recovery 141
allergens 26, 28
allowable concentrations 117
ambient lighting 29
animal laboratory 7, 9, 25, 53, 58, 59, 143, 187, 192, 193
ANSI 5, 33–35, 69, 80, 81, 94, 104, 109, 123, 146, 156, 164, 166–168, 170, 172, 175
anteroom 31, 32, 42, 48, 61, 97, 186
appliance loads 29, 47
application disadvantages 72
ASHRAE 1, 2, 26, 28, 50, 72, 92, 94, 113, 118, 126, 137, 149, 163, 170, 177, 178, 180, 194
ASHRAE 110-1995 testing 72
ASHRAE dilution equations 134
Associated Air Balancing Council 163, 164
auxiliary air 51, 72, 154, 155, 170
auxiliary air fume hood 72, 73, 81, 154, 155
average readings 165

B

baffle fume hood 70
baseboard 87, 88
basis of design 48, 49, 53, 174, 179
biological containment 2, 9, 53, 58, 185
biological containment laboratory 7, 9, 53, 58, 185
biological safety cabinet 2, 4, 9, 10, 12, 29, 40, 42, 52,

56, 58, 60, 67, 74–77, 80, 84, 89, 145, 149, 150, 159, 162, 163, 164, 167, 169, 171, 180, 185, 186

biological safety cabinet performance testing 167

Biosafety Level 1 35, 90, 185, 192

Biosafety Level 2 75, 91, 186

Biosafety Level 3 76, 103, 161, 186

Biosafety Level 4 186

budget 42, 43, 173, 181

budgeting 37, 42

building 52, 145

building concept 37, 40

Building Officials and Code Administrators International, Inc. 3, 34, 40, 44

building pressurization 2, 155

bypass fume hood 69, 70

C

cage wash area 193

caging systems 190

canopy hood 77, 79, 151

caulking and sealing 193

cause-consequence analysis 23

Centers for Disease Control and Prevention 3, 17, 44, 194

central air system 51

central air-handling equipment 181, 182

central system 30, 37, 48, 50, 51, 73, 96, 97, 98, 99, 100, 109, 145, 155, 161, 162

centrifuges 29, 111, 112, 151

certification requirements 47, 52

CETA 5

challenge velocity 160

chemical and fume exhaust 109

chemical fume hoods 8, 9, 56, 67, 77, 79

climate 72–74, 122, 127, 129, 131, 137, 140, 142, 143, 183

commissioning integration 47, 52

commissioning plan 53, 178

commissioning process 2, 43, 52, 53, 177, 178, 179, 180

component service 193

compound pumping 114, 115

computational fluid dynamics 133, 134

computer simulation 130, 133, 134

condenser water heat recovery 141

condensing 34, 117, 123

constant volume reheat 182

construction checklists 179

construction documents 52, 178, 179

construction phase 162, 179

contaminants 29, 35, 42, 60, 96, 102, 103, 108, 117, 119, 121, 123, 127, 131, 132, 134, 149, 150, 152, 157–159, 161, 166, 178, 185, 191, 192

contaminated airstreams 131

continuous monitoring 54, 80, 81, 170

control strategies 33, 47, 52, 71, 102, 145, 153, 154, 177

control strategy 96, 102, 151, 154

controls 2, 8, 35, 37, 50, 52, 61, 71, 73, 81, 92, 95, 96, 98, 102, 104, 115, 146, 148–150, 152, 153–155, 161, 170, 172, 173, 179, 182, 185, 193

conventional fume hood 67, 68, 69, 70, 71

cooling load 29, 33, 32, 49, 50, 103, 115, 141, 151, 183

corridor ceiling distribution 38

cost factors 182, 183

critical spaces 2, 145, 154, 155

critical systems 38, 43, 161, 178, 183

D

decontamination 9, 54, 58, 59, 62, 64, 65, 76, 87, 96, 103, 169, 171, 186, 192

dedicated equipment exhaust 77, 79

design factors 183

design of laboratory animal areas 190

design phase 52, 53, 63, 104, 133, 178, 179

device types 84

diffuser 29, 40, 42, 51, 53, 56, 59, 80, 83–86, 157–159, 160–162, 166, 190, 191

direct equipment exhaust 8, 145, 149, 151, 163

direct pressure control 148, 152, 153

directional airflow 10, 40, 42, 50, 61, 80, 153

dispersion modeling 133

diversities 29, 30, 48

document assumptions 48

documentation 13, 25, 48, 52, 64, 123, 147, 174, 180

drainage 95, 125, 131, 192, 193

duct construction 2, 74, 83, 102, 110

duct materials 74, 102, 105, 107, 108

duct system components 102, 103

ductwork pressure testing 161, 162

E

economic evaluation 47, 53
economic factors 183, 184
economics 2, 36, 43, 53, 116, 142, 143, 178
electronics/instrumentation laboratory 7, 10, 53, 61, 62
elevated receptors 132, 133
emergency power 43, 59, 192, 193
emergency situations 35, 52, 55, 133, 145, 155, 193
emissions characterization 133
energy cost 137, 173, 174, 182–184
energy efficiencies 2, 83, 109
energy recovery options 47, 52
energy use 102, 105, 107, 108, 125, 130, 146, 183
EPA 5, 8, 133, 134, 135
equipment balancing 161–163
equipment control 145, 182
equipment plumes 161
evaporative cooling 137, 138, 140, 141, 143
event tree analysis 18, 23
exhaust air system 37, 51, 81, 83, 88, 96, 103, 108
exhaust ducts 76, 96, 108, 139
exhaust stack design 2, 37, 52, 89–91, 131, 134, 135
exhaust stack velocity 95
exhaust systems 2, 25, 27, 29, 37, 51, 74, 76, 81, 84, 88–91, 95, 98, 103–105, 107, 108, 131, 159, 162, 169, 171, 172, 182
exhaust/supply requirements 47, 48
experimental quality 25, 28, 29, 171

F

failure mode, effects, and criticality analysis 19
fan system failure 193
fan types 94, 102
fan-powered dilution 117, 119
fault tree analysis 18, 20, 45
filter retaining system 122
filtration 3, 34, 54, 59, 61, 74–78, 90, 101, 109, 117, 120, 121, 123, 138, 149, 167, 177, 190
fire separations 40
fixed 137
fixed offset 152, 154
fixed plate 139, 142, 143

flammable and solvent storage cabinets 12, 149–151
flexibility and reliability 178
flow direction 154, 167, 168
flow tracking control 148, 153, 154
food and bedding storage 193
fume hood 2, 10, 27–30, 32, 35, 37, 38, 40, 42, 43, 45, 52, 54, 55, 61, 64, 67–74, 78–81, 89, 93, 94, 103–105, 129–132, 145–150, 153–160, 162–166, 168–172, 174, 175, 180, 181, 183
fume hood controls 182
fume hood face velocity testing 163, 164
functional performance 53, 178, 179
functional testing overview 179

G

gas phase filters 120, 121
general chemistry laboratory 7, 8, 10, 53, 55
general HVAC equipment 161–163
general laboratory exhaust 149–151
general TAB standards 163
glove box 54, 77, 78, 145, 149

H

hazard analysis methods 13, 18, 180
hazard and operability studies 18
hazards 8, 13–15, 17, 18, 20, 21, 44, 54, 62, 73, 78, 80, 81, 111, 123, 131, 133, 138, 161, 163, 177, 185, 193
heat 137
heat pipe 138, 139, 140, 142
heat recovery 2, 90, 101, 109, 139–143, 162
heat wheel 139, 143
heating and cooling coils 101
hood construction requirements 73
hood safety certification 80
hospital and clinical laboratory 7, 9, 53, 56
hot water waste heat recovery 141, 142
human error analysis 18, 23, 24
humidification 29, 87, 88, 101, 102
humidity 2, 8–10, 25, 26, 28, 35, 49, 56, 58–61, 63, 72, 73, 79, 87, 98, 100, 101, 122, 137, 139, 145, 151, 154, 155, 163, 171, 177, 186, 187, 190, 192
hydronic system balancing 161

I

identification and understanding of hazards 14

individual exhausts 182

indoor air quality 1, 13, 25, 26, 29, 32—35, 44, 47, 51, 95, 102, 123

Institute of Environmental Sciences and Technology 3

Institute of Laboratory Animal Resources 3, 44, 187, 190, 194

instrument testing 167

insufficient plume rise 132, 133

interstitial space 39

inventory 25, 33, 34, 64, 133

isolation/cleanrooms 7, 9, 53, 60

L

laboratory codes 13, 34

laboratory health and safety HVAC equipment 162, 163

laboratory layout approaches 37, 39

laboratory volumes 40

laminar flow clean air stations 77, 79

lasers 111, 113

layout of ducts and rooms 47, 51

life-cycle cost analysis 53, 182, 183

lighting levels and quality 27

liquid 137

liquid desiccant 139, 140

load calculation 29, 47, 48, 50

loads for comfort 48

local air system 51

local exhaust scoop 63, 77, 79

M

maintenance 1, 2, 10, 19, 23, 35–37, 51, 52, 72, 73, 76, 87, 94, 96, 99, 101, 102, 122, 131, 138, 148, 153, 155, 157, 169, 170–175, 178, 180, 181, 183, 184, 190, 192, 193

maintenance cost 173, 174, 181, 183, 184

maintenance staff 54, 169, 173, 174

manifolded exhausts 81, 182

material options 182

materials testing laboratory 7, 10, 53, 61

maximum deviation 165

measurement grid 164, 165

minimum exhaust air changes 32

minimum outdoor air changes 32

minimum supply air changes 32

minimum total air changes 32

mock-ups 179

modular design 39

multiple exterior shafts 38

multiple interior shafts 38, 39

multiple or redundant fan systems 193

multiple speed fans 161

N

National Environmental Balancing Bureau 163

National Fire Protection Association 3, 12, 14, 15, 16, 34, 44, 45, 65, 110, 143, 170, 172, 175

National Institutes of Health 3, 9, 17, 34, 44, 168, 185, 194

National Research Council 3, 4, 44, 65, 81, 190, 194

National Sanitation Foundation International 4, 44, 81, 168, 175

negative pressure room 31

neutral pressure room 31, 32

noise 18, 26, 28, 49, 51, 59, 92, 104, 125, 130–132, 171, 172, 187, 190

O

O&M requirements 47, 52, 170, 171, 172

occupant densities 29

occupants 1, 2, 17, 20, 25, 26, 28–30, 33, 38, 48–51, 53, 67–69, 71, 84, 88, 96, 107, 117, 119, 143, 145, 151, 153, 154, 157, 160, 161, 163, 169, 170, 174, 179, 180, 181, 183, 184, 191, 193

occupational health limit 117, 118

Occupational Safety and Health Administration 4, 34, 40, 44, 117, 175

odors 26–28, 31, 59, 63, 79, 107, 117–119, 123, 150, 190, 193

off-peak loads 84, 86

ongoing laboratory audits 180

operation and maintenance 1, 2, 19, 37, 52, 169, 170–173

operation phase 2, 180

operations cost 174

oxidization 123

P

parallel pumping 114
particulate filters 120, 121
pathology-necropsy areas 193
perchloric acid hood 67, 73, 74, 91
performance factors 183
performance testing 102, 104, 154, 163, 164, 166, 167
personnel comfort 25, 26
personnel protection 157, 158
photographic darkroom 62
physics laboratory 53–55
planning documents 13, 43
planning phase 1, 2, 13, 24, 25, 30, 33, 38, 42, 47, 48, 50, 177, 178
positive pressure room 32
potential pollutants 32, 33
pressure mapping 47, 50
pressure relationship 13, 27, 29, 30, 47, 50, 58, 71, 91, 96
primary air systems 2, 38, 47, 51, 81, 83, 111, 120, 149, 159, 160, 191
primary and secondary barriers 40, 42
project goals and expectations 178
protection of the research 158
protective exhaust systems 159
pumping subsystems 113, 114

R

radiant panel 87, 88
radiation laboratory 53
radio-chemistry laboratory 7, 8
radioisotope fume hood 77, 80, 89
recirculated air 26, 32, 48, 76, 148, 149, 157, 158, 162, 190
recirculation considerations 33
recycling 33, 34, 64
refrigeration machine heat recovery 141
relative device location 84
relative static pressures around buildings 128
remodeling 169, 180
research laboratory 7–9, 43, 53, 56, 64, 174
responsibilities 13, 25, 190
retrofitting 180

risk assessment 1, 13, 14, 24, 178, 180
room air velocities 84
room control 98, 145, 151, 152
room pressurization 51, 68, 69, 73, 151, 152, 167
runaround loop 138, 140

S

safety features 77, 178
sanitation 2, 4, 44, 58, 74, 75, 81, 168, 171, 175, 191–193
sash movement 68, 69
sash position 68, 69, 70, 71, 80, 89, 95, 146, 147, 148, 149, 153, 165
sash stops 68, 69, 80
scheduling 29, 30
Scientific Equipment and Furniture Association 4, 34, 45, 81, 168, 175
scrubbing 34, 117, 122
selection parameters 137, 142
sensor accuracy 152
separation of areas 191
series pumping 114, 115
slot hood 77, 79
smoke testing 165, 167
snorkel exhausts 145, 149
source reduction 33
space pressurization and airflow 191
special animal conditions 191
special purpose hood 77
special space considerations 53
stack design 2, 37
stack tip downwash 132, 133
stack velocity 125, 126, 131, 132
standard types of animals 186
state/local ambient concentration limit 118
storage 10, 12, 13, 25, 27, 29, 30, 33, 34, 37, 40, 54, 56, 58, 61, 63, 89, 113, 133, 151, 155, 158, 180, 193
supply air 26, 29, 31–34, 37, 40, 42, 49–52, 59, 60, 71, 72, 74, 75, 76, 78, 88, 96–100, 138, 139, 142, 143, 151, 155, 158, 159, 160, 162, 170, 175, 182, 186, 190, 191, 193
supply air quantity 96, 97
supply air/exhaust air treatment requirements 34
supply ducts 107, 108, 143
support areas 158, 193

support shop 62, 63

support space 7, 10, 11, 30, 53, 62

symptoms of inadequate performance 119

system complexity 42

system redundancy 42, 192

system response time 149

system verification 179

systems design procedure 115

T

TAB requirements 52

task lighting 27, 29

teaching laboratory 7, 8, 53, 55, 56

temperature 2, 8–10, 15, 16, 25, 26, 28, 29, 31, 32, 35, 48, 49, 52, 56, 58, 59, 60, 61, 63, 64, 72, 73, 79, 87, 88, 94, 97, 98, 100–104, 107, 108, 111–113, 115, 118, 119, 121–123, 131, 139, 140, 142, 143, 145, 151, 152, 154, 155, 160–163, 171, 172, 177, 182, 183, 186, 187, 190–193

temperature gradients 160, 161

temperature sensor location 152

thermosiphon 140, 142

tracking 33, 34, 65

trade magazines 4

training 2, 3, 24, 25, 52, 72, 133, 164, 168, 174, 180, 185

transfer air 26, 31, 32, 159, 160

turbulent airflow 68, 132

two-speed/variable speed pumping 115

U

unitary system 97, 99, 100

United States Department of Health and Human Services 4

utility corridors 37, 38

utility distribution 37–40

V

vacuum and diffusion pumps 111, 112

variable 2, 81, 145

variable air volume 51, 54, 67, 69, 71, 77, 81, 86, 89, 90, 96, 97, 98, 99, 104, 108, 120, 131, 146, 148, 182, 191

variable air volume fume hood 42, 67, 71, 81, 148, 149

variable volume fume hood 146, 147, 153, 170, 181, 183

ventilation rates 32, 40, 42, 152, 155, 158, 163, 190

ventilation systems 33, 35, 36, 64, 169, 171

vibration 28, 29, 59, 61, 62, 103, 104, 125, 131, 132, 167, 171, 172

W

walk-in hood 77, 78

waste-handling facility 63, 64

water treatment and quality requirements 111, 112

water-cooled loads 2, 111

water-to-air heat recovery 2, 141

weather control 125, 131

weighing station 77, 78

wind tunnel modeling 133, 134

workstation placement 40, 42

workstations and specific functions 40